Handbook of Geotechnical Investigation and Design Tables

Handbook of Geotechnical Investigation and Design Tables

Second Edition

Burt G. Look

Geotechnical Practice Leader, Sinclair Knight Merz, Australia

CRC Press is an imprint of the
Taylor & Francis Group, an **informa** business

A BALKEMA BOOK

CRC Press/Balkema is an imprint of the Taylor & Francis Group, an informa business

Typeset by MPS Limited, Chennai, India

Library of Congress Cataloging-in-Publication Data

Look, Burt.
Handbook of geotechnical investigation and design tables / Burt G. Look, geotechnical practice leader, Sinclair Knight Merz, Australia.—Second edition.
pages cm
Includes bibliographical references and index.
ISBN 978-1-138-00139-8 (hardback : alk. paper)
1. Engineering geology—Handbooks, manuals, etc. 2. Earthwork. I. Title.
TA705.L66 2014
624.1'51—dc23

2013039021

Published by: CRC Press/Balkema
P.O. Box 11320, 2301 EH Leiden, The Netherlands
e-mail: Pub.NL@taylorandfrancis.com
www.crcpress.com – www.taylorandfrancis.com

ISBN: 978-1-138-00139-8 (pbk)
ISBN: 978-1-315-81323-3 (ebk)

Table of Contents

Preface

The first edition of this book was published in 2007.

There has been very positive feedback from practicing engineers, who need a "go to" book, and who have suggested various additions. This encouragement now results in this 2nd edition which uses the similar format of summary data tables and expands the book with additional tables as well as some corrections and additional notes to the tables of the 1st edition.

This book is is principally a data book for the practicing Geotechnical Engineer and Engineering Geologist, which covers:

- The planning of the site investigation.
- The classification of soil and rock.
- Common testing and the associated variability.
- The strength and deformation properties associated with the test results.
- The engineering assessment of these geotechnical parameters for both soil and rock.
- Applications of data and theory in geotechnical design

This data is presented by a series of tables and correlations to be used by experienced geotechnical professionals. These tables are supplemented by dot points (notes style) explanations. The reader must consult the references provided for the full explanations of applicability and to derive a better understanding of the concepts. The complexities of the ground cannot be over-simplified, and while this data book is intended to be a reference to obtain and interpret essential geotechnical data and design, it should not be used without an understanding of the fundamental concepts. This book does not provide details on fundamental soil and rock mechanics theories as this information can be sourced from elsewhere.

The geotechnical engineer provides predictions, often based on limited data. By cross checking with different methods, the engineer can then bracket the results as often different prediction models produces different results. Typical values are provided for various situations and types of data to enable the engineer to proceed with the site investigation, its interpretation and related design implications. This bracketing of results by different methods provides a validity check as a geotechnical report or design can often have different interpretations simply because of the method used. Even in some sections of this book a different answer can be produced (for similar data) based on the various references, and illustrates the point on variations based on different methods. While an attempt has been made herein to rationalise some of these inconsistencies between various texts and papers, there are still many unresolved issues. This book does not attempt to avoid all such inconsistencies.

In many cases the preliminary assessments made in the field are used for the final design, without further investigation or sometimes, even laboratory testing. These results in a conservative and non-optimal design at best, but also can lead to under-design. Examples of these include:

- Preliminary boreholes used in the final design without added geotechnical investigation.
- Field SPT values being used directly without the necessary correction factors or site specific correlations, which can change the soil parameters adopted.
- Preliminary bearing capacities given in the geotechnical report. These allowable bearing capacities are usually based on the soil conditions only for a "typical" surface footing only, while the detailed design parameter requires a consideration of the depth of embedment, size and type of footing, location, etc.

Additionally there seems to be a significant chasm in the interfaces in geotechnical engineering. These are:

- The collection of geotechnical data and the application of such data. For example, Geologists can take an enormous time providing detailed rock descriptions on rock joints, spacing, infills, etc. Yet its relevance is often unknown by many (even those who provide that data), except to say that it is good practice to have detailed rock core logging. This book should assist to bridge that data – application interface, in showing the relevance of such data to design.
- Analysis and detailed design. The analysis is a framework to rationalise the intent of the design. However after that analysis and reporting, this intent must be transferred to a working drawing. There are many detailing design issues that the analysis does not cover, yet has to be included in design drawings for construction purposes. These are many rules of thumbs, and this book provides some of these design details, as this is seldom found in a standard soil mechanics text.

Geotechnical concepts are usually presented in a sequential fashion for learning. This book adopts a more random approach by assuming that the reader has a grasp of fundamentals of engineering geology, soil and rock mechanics. The cross – correlations can then occur with only a minor introduction to the terminology.

Some of the data tables have been extracted from spread sheets using known formulae, while some data tables are from existing graphs. This does mean that many users who have a preference for reading of the values in such graphs will find themselves in an uncomfortable non visual environment where that graph has been "tabulated" in keeping with the philosophy of the book title.

Many of the design inputs here have been derived from experience, and extrapolation from the literature. There would be many variations to these suggested values, and I look forward to comments to refine such inputs and provide the inevitable exceptions, that occur. Only common geotechnical issues are covered and more specialist areas have been excluded. Many of these guidelines evolved over many years, as notes to me. In so doing if any table inadvertently has an unacknowledged source then this is not intentional, but a blur between experience and extrapolation/application of an original reference.

Again it cannot be overstated, recommendations and data tables presented herein, including slope batters, material specifications, etc. are given as a guide only on the key issues to be considered, and must be factored for local conditions and specific projects for final design purposes. The range of applications and ground conditions are too varied to compress soil and rock mechanics into a cook – book approach. These tabulated correlations; investigation and design rules of thumbs should act as a guideline, and is not a substitute for a project specific assessment.

Acknowledgements

I acknowledge the many engineers and work colleagues who constantly challenge for an answer, as many of these notes evolved from such working discussions.

In this 2nd edition, Ashley Bullas provided the graphics support for the diagrams.

B.G.L.
July 2013

Chapter 1

Site investigation

1.1 Geotechnical engineer

- Geotechnical engineering (Geotechnics/Geomechanics) requires further specialisation from Civil engineering or Geology degrees either from experience and/or post graduate studies. Needs understanding of soil and rock mechanics (Figure 1.1). Traditionally from the Civil engineering stream.
- Significant overlap between Engineering geologist and Geotechnical engineer – but these professions are not the same with the strengths being Geological and Geotechnical Models, respectively.
- The geological model is historical/factual while the geotechnical model is prediction of the response of the ground to changes.

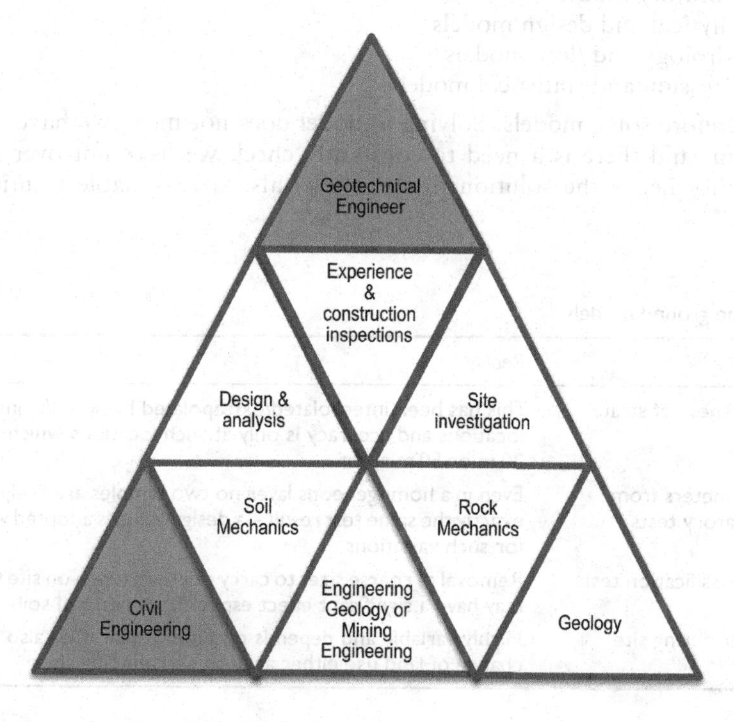

Figure 1.1 The formation of a geotechnical engineer.

Table 1.1 Ground model inputs by specialists.

Geological model	*Ground model*	*Geotechnical model*
Engineering geologist		*Geotechnical engineer*
Model by Geotechnical engineer in simple cases Obtain site data Develop geological model – relevant geological structures Provide key ground related issues; site history Identify regional issues affecting site Work with Geotechnical engineer to develop site specific solution		Model by Engineering geologist in simple cases Carry out testing to an acceptable level Develop geotechnical models (parameters for design) Recognise and provide site constraints – e.g. slopes or allowable bearing capacity; site future use Suggest ground solutions Work with Engineering geologist to implement ground issues into the design

1.2 Developing models

- Reality is too complicated. We therefore use models: – a simplification of reality. These can be
  - Financial and scheduling models
  - Geological and geotechnical models
  - Laboratory models
  - Analytical and design models
  - Hydrology and flow models
  - Regression and statistical models
- We therefore solve models. Solving a model does not mean we have solved the problem, and there is a need to constantly check we have not over simplified the reality i.e. is the solution to the model also a reasonable solution to the reality.

Table 1.2 Some ground models.

Models	*Reality*
Depth of thickness of strata	This has been interpolated/extrapolated between/from test locations and accuracy is only at such locations which could be 20 m or 50 m apart
Strength parameters from field or laboratory tests	Even in a homogeneous layer, no two samples are likely to produce exactly the same test result – a design value is adopted which allows for such variations
Laboratory classification tests	Removal of coarse sizes to carry out tests when on site this quantity may have a significant effect especially in residual soils
Ground water during site investigations	Highly variable and depends on time of year. Can also vary with change of land use either at or in surrounding sites

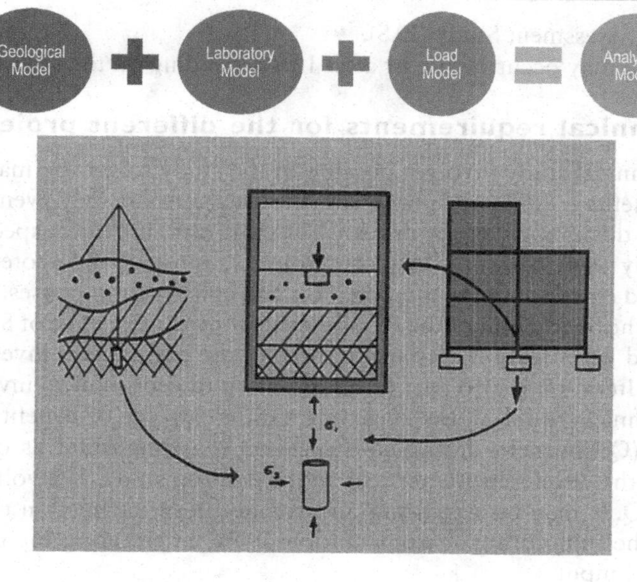

Figure 1.2 Example of a geotechnical model.

1.3 Geotechnical involvement

- There are two approaches for acquiring geotechnical data
  - Accept the ground conditions as a design element, i.e. based on the structure/development design location and configuration, and then obtain the relevant ground conditions to design for/against. This is the traditional approach.
  - Geotechnical input throughout the project by planning the structure/development with the ground as a considered input, i.e. the design, layout and configuration is influenced by the ground conditions. This is the recommended approach for minimisation of overall project costs.
- Geotechnical involvement should occur throughout the life of the project. The input varies depending on phase of project.
- The phasing of the investigation provides the benefit of improved quality and relevance of the geotechnical data to the project.

Table 1.3 Geotechnical involvement.

Project phase	*Geotechnical study for types of projects*		
	Small	*Medium*	*Large*
Feasibility/IAS Planning	Desktop study/ site investigation	Desktop study	Desktop study Definition of needs
Preliminary engineering Detailed design		Site Investigation (S.I.)	Preliminary site investigation Detailed site investigation
Construction Maintenance	Inspection	Monitoring/inspection Inspection	Monitoring/inspection

- Impact Assessment Study (IAS).
- Planning may occur before or after IAS depending on the type of project.

1.4 Geotechnical requirements for the different project phases

- The geotechnical study involves phasing of the study to get the maximum benefit. The benefits (~20% per phase) are shown approximately evenly distributed throughout the lifecycle of the project – but this ratio is project specific.
- Traditionally (and currently in most projects), most of the geotechnical effort (>90%) and costs are in the investigation and construction phases.
- Phasing of the SI determines the requirements for the next stage of SI (Figure 1.3).
- The detailed investigation may make some of the preliminary investigation data redundant. Iteration is also part of optimisation of geotechnical investigations.
- The geotechnical input at any stage has a different type of benefit. The Quality Assurance (QA) benefit during construction, is as important as optimising the location of the development correctly in the desktop study. The volume of testing as part of Q.A. may be significant and has not been included in the table. This considers the monitoring/instrumentation as the engineering input and not the testing (QA) input
- The optimal approach during construction may allow reduced factors of safety to be applied and so reduce the overall project costs. That approach may also be required near critical areas without any reduction in factors of safety.

Table 1.4 Geotechnical requirements.

Geotechnical study	*Key model*	*Relative (100% total)*		*Key data*	*Comments*
		Effort	*Benefit*		
Desktop study	Geological model	<5%	~20%	Geological setting, existing data, site history, aerial photographs and terrain assessment	Minor SI costs (site reconnaissance) with significant planning benefits
Definition of needs		<5%	~20%	Justify Investigation requirements and anticipated costs	Safety plans and services checks. Physical, environmental and allowable site access
Preliminary investigation	Geological & geotechnical model	10–20%	~20%	Depth, thickness and composition of soils and strata	Preliminary investigation of ≤20% of planned detailed site investigation
Detailed site investigation	Geotechnical model	60–80%	~20%	Quantitative and characterisation of critical or founding strata	Laboratory analysis of 20% of detailed soil profile
Monitoring/ inspection		<10%	~20%	Instrumentation as required. QA testing	Confirms models adopted or requirements to adjust assumptions. Increased effort for observational design approach

- Construction costs ~85% to 95% of total capital project costs.
- Design costs ~5% to 10% of total capital costs.
- Geotechnical costs ~ 0.4% to 4% of total capital costs.

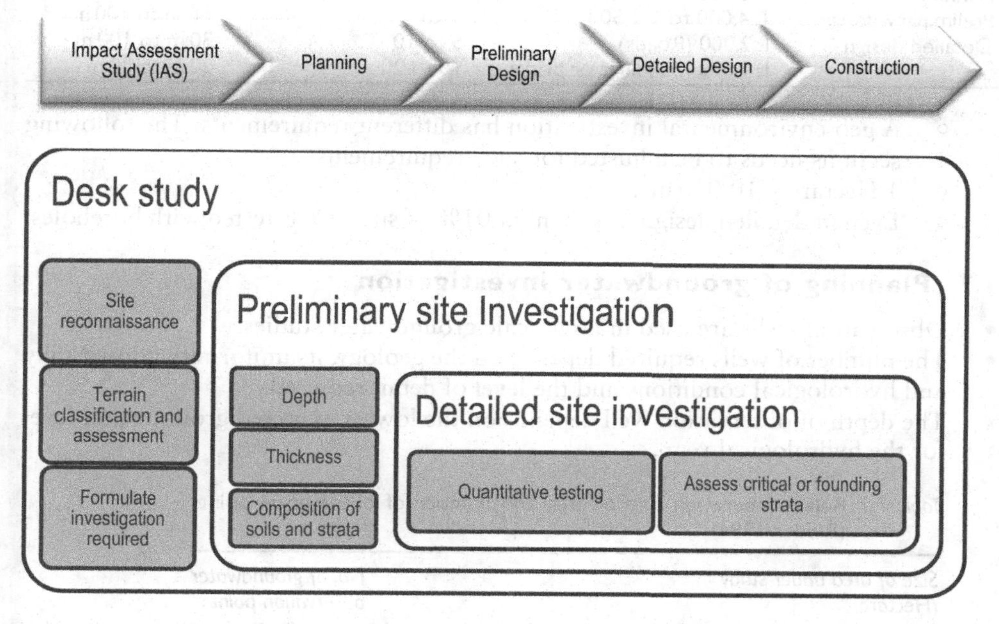

Figure 1.3 Steps in effective use of geotechnical input throughout all phases of the project.

1.5 Relevance of scale

- At each stage of the project, a different scale effect applies to the investigation.

Table 1.5 Relevance of scale.

Size study	*Typical scale*	*Typical phase of project*	*Relevance*
Regional	1:100,000	Regional studies	GIS analysis/hazard assessment
Medium	1:25,000	Feasibility studies	Land units/hazard analysis
Large	1:10,000	Planning/IAS	Terrain/risk assessment
Detailed	1:2,000	Detailed design	Detailed development. Risk analysis

1.6 Planning of site investigation

- The SI depends on the phase of the project.
- The testing intensity should reflect the map scale of the current phase of the study.

Table 1.6 Suggested test spacing.

Phase of project	Typical map scale	Boreholes per hectare	Approximate spacing
IAS	1:10,000	0.1 to 0.2	200 m to 400 m
Planning	1:5,000	0.5 to 1.0	100 m to 200 m
Preliminary design	1:4,000 to 1:2,500	1 to 5	50 m to 100 m
Detailed design	1:2,000 (Roads)	5 to 10	30 m to 100 m
	1:1,000 (Buildings or bridges)	10 to 20	20 m to 30 m

- A geo-environmental investigation has different requirements. The following sections needs to be adjusted for such requirements.
- 1 Hectare = 10,000 m^2.
- Even in detailed design less than 0.001% of site area is tested with boreholes.

1.7 Planning of groundwater investigation

- Observation wells are used in large scale groundwater studies.
- The number of wells required depends on the geology, its uniformity, topography and hydrological conditions and the level of detail required.
- The depth of observation well depends on the lowest expected groundwater level for the hydrological year.

Table 1.7 Relation between size of area and number of observation points (Ridder, 1994).

Size of area under study (Hectare)	No. of groundwater observation points
100	20
1,000	40
10,000	100
100,000	200

1.8 Level of investigation

- Define the geotechnical category in planning the investigation. This determines
  - The level of investigation required, then
  - Define the extent of investigation required, then
  - Hire/use appropriate drilling/testing equipment

Table 1.8 Geotechnical categories (GC) of investigation.

Description	Geotechnical category		
	GC1	GC2	GC3
Nature and size of construction	Small & relatively simple – conventional loadings	Conventional structures – no abnormal loadings	Large or unusual structures
Surroundings	No risk of damage to neighbouring buildings, utilities, etc.	Risk of damage to neighbouring structures	Extreme risk to neighbouring structures

(Continued)

Table 1.8 (Continued)

Description	Geotechnical category		
	GC1	GC2	GC3
Ground conditions	Straightforward. Does not apply to refuse, uncompacted fill, loose or highly compressible soils	Routine procedures for field and laboratory testing	Specialist testing
Ground water conditions	No excavation below water table required	Below water table. Lasting damage cannot be caused without prior warning	Extremely permeable layers
Seismicity	Non-Seismic	Low seismicity	High seismic areas
Capital cost	<$1.0 Million (Aust. – 2013)		>$50 Milion (Aust. – 2013)
SI Cost as % of capital cost	0.1%–0.5% or $1,000 (Aust. min)	0.25%–1%	0.5%–2% or $100,000 (Aust. min)
Type of study	Qualitative investigation may be adequate	Quantitative geotechnical studies	Two stage investigation required
Minimum level of expertise	Staff under supervision by an experienced Geotechnical specialist	Experienced Geotechnical Engineer/Engineering Geologist	Geotechnical Engineer with relevant experience to work with Engineering Geologist and specialist Geotechnical/Tunnel/ Geo – Environmental engineer/Hydrogeologist/ etc.
Examples	o Sign supports o Walls <2 m o Single or 2-storey Buildings o Domestic buildings; Light structures with column loads up to 250 kN or walls loaded to 100 kN/m o Minor roads	o Light – medium Industrial & Commercial buildings o Major roads o Roads > 1 km o Small/medium Bridges o Rail lines o Alluvial areas o Water nearby	o Dams o Hydraulic structures o Tunnels o Ports o Large bridges & buildings o Heavy machinery Foundations o Offshore platforms o Deep basements o Silos

- Geotechnical design or geo-environmental investigation costs not included in above.
- $1 Aust. ~ $1 U.S ~ £0.6 ~ €0.7 (2013) but this varies significantly.

1.9 Planning prior to ground truthing

- Prepare preliminary site investigation and test location plans prior to any ground truthing. This may need to be adjusted on site.
- Services searches (dial before you dig) are mandatory prior to ground truthing.

- Further services location tests and/or isolations may be required on site. Typically mandatory for any service within 3 m of the test location.
- Obtain utility services plans both above and below the ground are required. For example, an above ground electrical line may dictate either the proximity of the borehole, or a drilling rig/with a certain mast height and permission from the electrical safety authority before proceeding.
- The planning should allow for any physical obstructions such as coring of a concrete slab, and its subsequent repair after coring.

Table 1.9 Planning checklist.

Type	*Items*
Informative	Timing. Authority to proceed. Inform all relevant stakeholders. Environmental approvals. Access. Site history. Physical obstructions. Positional accuracy requirements (taped, GPS, survey)
Site specific safety plans	Traffic Controls. Services checks. Possible shut down of nearby operational plant. Isolations required. Site specific inductions
SI Management	Checklists. Coordination and contacts. GPS or map locator. Aims of investigation understood by all. Logging templates. Depth and sampling requirements. Sampling bags, core trays, permanent markers. Budget limits where client needs to be advised if additional SI required. Repair or sealing of test locations.

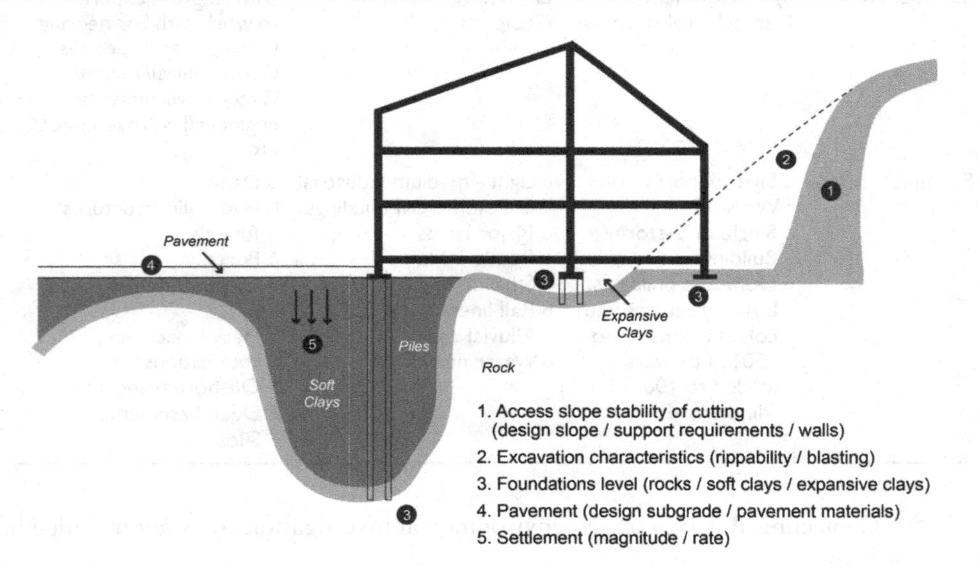

Figure 1.4 Site ground considerations.

1.10 Extent of investigation

- The extent of the investigation should be based on the relationship between the competent strata and the type of loading/sensitivity of structure. Usually this information is limited at the start of the project. Hence the argument for a 2 phased investigation approach for all but small (GC1) projects. For example in a piled foundation design.

- The preliminary investigation or existing nearby data (if available) determines the likely founding level
- And the detailed investigation provides quantitative assessment, targeting testing at that founding level

• The load considerations should determine the depth of the investigation
  - >1.5 × width (B) of loaded area for square footings (pressure bulb ~0.2q where q = applied load)
  - >3.0 × width (B) of loaded area for strip footings (pressure bulb ~0.2q)
  - However the loading/width is often not known prior to the investigation, and one should assess to a "competent" strata
• The ground considerations intersected should also determine the depth of the investigation as the ground truthing must provide
  - Information of the competent strata, and probe below any compressible layer
  - Spacing dependent on uniformity of sub-surface conditions and type of structure

Table 1.10a Guideline to extent of investigation (structures).

Development	*Test spacing*	*Approximate depth of investigation*
Building	20 m to 50 m	o 2B–4B for shallow footings (Pads and Strip, respectively) o 3 m or 3 pile diameters below the expected founding level for piles. If rock intersected target N* > 100 and RQD > 25% o 1.5B (building width) for rafts or closely spaced shallow footings o 1.5B below 2/3D (pile depth) for pile rafts
Bridges	At each pier location	o 4B–5B for shallow footings o 10 pile diameters in competent strata, or o Consideration of the following if bedrock intersected – 3 m minimum rock coring – 3 Pile diameters below target founding level based on o N* > 150 o RQD > 50% o Moderately weathered or better o Medium strength or better
Monopoles and transmission towers	At each location	– 0 m to 20 m high: D = 4.5 m – 20 m to 30 m high: D = 6.0 m – 30 m to 40 m high: D = 7.5 m – 40 m to 50 m high: D = 9.0 m – 60 m to 70 m high: D = 10.5 m – 70 m to 80 m high: D = 15.0 m Applies to medium dense to dense sands and stiff to very stiff clays. Based on assumption on very lightly loaded structure and lateral loads are the main considerations Reduce D by 20% to 50% if hard clays, very dense sands or competent rock Increase D by >30% for loose sands and soft clays

- N* Inferred SPT value
- RQD – Rock Quality Designation
- H – Height of slope
- D – Depth of investigation
- Ensure boulders or layers of cemented soils are not mistaken for bedrock by penetrating approximately 3 m into bedrock.
- Where water bearing sand strata, there is a need to seal exploratory boreholes especially in dams, tunnels and environmental studies.
- Any destructive tests on operational surfaces (travelled lane of roadways) needs repair
- In soft/compressible layers and fills, the SI may need to extend BHs in all cases to the full depth of that layer.
- Samples/testing every 1.5 m spacing or changes in strata.
- Obtain undisturbed samples in clays and carry out SPT tests in granular material.
- Use of the structure also determines whether a GC2 or GC3 investigation applies. For example, a building for a nuclear facility (GC3) requires a closer spacing than for an industrial (GC2) building.

Table 1.10b Guideline to extent of investigation (earthworks).

Development	*Test spacing*	*Approximate depth of Investigation*
Embankments	25 m to 50 m (critical areas) 100 m to 500 m as in roads	Beyond base of compressible alluvium at critical loaded/suspect areas: otherwise as in roads
Cut slopes	25 m to 50 m for H > 5 m 50 m to 100 m for H < 5 m	5 m below toe of slope or 3 m into bedrock below toe whichever is shallower
Landslip	3 BHs or test pits minimum along critical section	Below slide zone. As a guide (as the slide zone may not be known) use 2 × height of slope or width of zone of movement. 5 m below toe of slope or 3m into bedrock below toe whichever is shallower
Pavements/roads Local roads < 150 m Local roads > 150 m	250 m to 500 m 2 to 3 locations 50 m to 100 m	2 m below formation level for subgrade assessment. Deeper if compressible material intersected
Runways and rails	250 m to 500 m	3 m below formation level for subgrade assessment.
Pipelines	250 m to 500 m	1 m below invert level for subgrade assessment.
Tunnels	25 m to 50 m	3 m below invert level or 1 tunnel diameter, whichever is deeper: greater depths where contiguous piles for retentions
	Deep tunnels need special consideration	Target 0.5–1.5 linear metres drilling per route metre of alignment. Lower figure over water or difficult to access urban areas
Dams	25 m to 50 m	2 × height of dam, 5 m below toe of slope or 3m into bedrock below toe whichever is greater. Extend to zone of low permeability

(*Continued*)

Table 1.10b (Continued)

Development	Test spacing	Approximate depth of Investigation
Canals	100 m to 200 m	3 m minimum below invert level or to a zone of low permeability
Culverts <20 m width 20 m–40 m >40 m width	1 Borehole One at each end One at each end and 1 in the middle with maximum spacing of 20 m between boreholes	2B–4B but below base of compressible layer
Car parks	2 Bhs for <50 parks 3 Bhs for 50–100 4 Bhs for 100–200 5 Bhs for 200–400 6 Bhs for >400 parks	2 m below formation level

1.11 Site investigation for driven piles to rock

- Below is just one of many foundation specifics that may apply as a variant to the previous table.
- The table below allows for investigating to the zone of influence depth below the pile refusal level.
- Often judgement is required to determine the depth to a "competent" level as neither loads nor type of foundation may be known at the time of investigation. For piles driven to refusal the competent material may be moderately weathered rock or as per table below.

Table 1.11 Depth of investigation for piles driven to refusal (Adams et al., 2010).

Material type	Minimum drilling depth
Sandstone	SPT (N*) $\geq 90 + 2$ m
Shale	SPT (N*) $\geq 160 + 2$ m

- Applies to bridge piles (in Queensland, Australia) which are typically 450 mm to 550 mm prestressed concrete piles. Depth of extremely weathered rock is typically 14 m but ranging from 9 m to 25 m.
- Steel piles can be driven further than indicated in table.

1.12 Volume sampled

- The volume sampled varies with the size of load and the project.
- Overall the volume sampled/volume loaded ratio varied from 10^4 to 10^6, i.e. less than 10^{-4}% is sampled.
- Earthen systems have a greater sampling intensity.

Table 1.12 Relative volume sampled (simplified from graph in Kulhawy, 1992)

Type of development	*Typical volume sampled*	*Typical volume loaded*	*Relative volume sampled/ volume loaded*
Buildings	$0.4\,m^3$	$2 \times 10^4\,m^3$	2×10^{-5}
Concrete dam	$10\,m^3$	$5 \times 10^5\,m^3$	2×10^{-5}
Earth dam	$100\,m^3$	$5 \times 10^6\,m^3$	2×10^{-5}

1.13 Relative risk ranking of developments

- The risk is very project and site specific, i.e. varies from project to project, location and its size.
- The investigation should therefore theoretically reflect overall risk.
- Geotechnical Category (GC) rating as per Table 1.8 can also be assessed by the development risk.
- The variability or unknown factors has the highest risk rank (F), while certainty has the least risk rank (A)
    - Projects with significant environmental and water considerations should be treated as a higher risk development
    - Developments with uncertainty of loading are also considered higher risk, although higher loading partial factors of safety usually apply
- The table is a guide in assessing the likely risk factor for the extent and emphasis of the geotechnical data requirements.

Table 1.13 Risk categories.

Development	*Risk factor considerations*						
	Loading	*Environment*	*Water*	*Ground*	*Economic*	*Life*	*Overall*
Offshore platforms	F	F	F	F	F	E	
Earth dam > 15 m	E	E	E	E	E	F	
Tunnels	E	E	E	E	E	F	
Power stations	E	E	D	D	F	E	High
Ports & coastal developments	F	E	F	F	E	E	GC3
Nuclear, chemical, & biological complexes	D	F	D	D	D	F	
Concrete dams	D	D	E	E	E	E	
Contaminated land	B	F	D	E	C	F	
Tailing dams	D	E	E	E	D	D	
Mining	E	D	D	D	D	D	
Hydraulic structures	D	D	E	E	D	D	
Buildings storing hazardous goods	D	E	C	C	C	E	
Landfills	B	D	D	D	D	E	Serious
Sub-stations	D	D	C	C	D	E	GC3
Rail embankments	D	C	D	D	D	E	
Earth dams 5 m–15 m	D	D	D	D	D	D	
Cofferdams	E	D	E	E	C	D	
Cuttings/walls >7 m	D	C	D	D	D	D	
Railway bridges	D	C	C	C	D	D	
Petrol stations	C	D	C	C	C	D	

(Continued)

Table 1.13 *(Continued)*

Development	*Risk factor considerations*						
	Loading	*Environment*	*Water*	*Ground*	*Economic*	*Life*	*Overall*
Road embankments	C	C	D	D	C	D	
Mining waste	C	D	D	D	C	D	
Highway bridges	C	C	C	C	D	D	
Transmission lines	C	D	A	D	D	C	
Deep basements	D	C	E	C	C	C	Moderate
Office buildings > 15 levels	C	C	B	A	E	D	GC2
Earth dams < 5 m	C	C	D	C	C	C	
Apartment buildings > 15 levels	C	C	B	C	D	D	
Roads/Pavements	C	B	D	D	C	C	
Public buildings	C	B	B	B	D	D	
Furnaces	D	C	B	C	B	C	
Culverts	C	C	D	C	C	B	
Towers	C	C	B	D	C	B	
Silos	E	C	C	D	C	A	
Heavy machinery	E	C	C	D	B	B	
Office buildings 5–15 levels	B	B	B	A	D	C	Usual
Warehouses, buildings storing non-hazardous goods	C	C	C	C	B	B	GC2
Apartment buildings 5–15 levels	B	B	B	B	D	C	
Apartment buildings < 5 levels	A	B	B	C	C	C	
Office Buildings < 5 levels	B	B	C	A	C	C	
Light industrial buildings	B	C	C	B	B	B	
Sign supports	D	A	A	C	A	A	Low
Cuttings/walls < 2 m	A	A	B	C	A	A	GC1
Domestic buildings	B	A	C	B	B	A	

- The table has attempted to sub-divide into approximate equal risk categories. It is therefore relative risk rather than absolute, i.e. there will always be unknowns even in the low risk category.

1.14 Sample amount

- The samples and testing should occur every 1.5 m spacing or changes in strata.
- Obtain undisturbed samples in clays and carry out penetration tests in granular material.
- Do not reuse samples e.g. do not carry out another re-compaction of a sample after completing a compaction test as degradation may have occurred.

Table 1.14 Disturbed sample quantity.

Test	*Minimum quantity*
Soil stabilisation	100 kg
CBR	40 kg
Compaction (Moisture density curves)	20 kg
Particle sizes above 20 mm (Coarse gravel and above)	10 kg
Particle sizes less than 20 mm (Medium gravel and below)	2 kg
Particle sizes less than 6 mm (Fine gravel and below)	0.5 kg
Hydrometer test – particle size less than 2 mm (Coarse sand and below)	0.25 kg
Atterberg Tests	0.5 kg

1.15 Sample disturbance

- Due to stress relief during sampling, some changes in strength may occur in laboratory tests.

Table 1.15 Sample disturbances (Vaughan et al., 1993).

Material type	*Plasticity*	*Effect on undrained shear strength*
Soft clay	Low	Very large decrease
	High	Large decrease
Stiff clay	Low	Negligible
	High	Large increase

1.16 Sample size

- The sample size should reflect the intent of the test and the sample structure.
- Because the soil structure can be unknown (local experience guides these decisions), then prudent to phase the investigations as suggested in Table 1.3.

Table 1.16 Specimen size (Rowe, 1972).

Clay type	*Macro-fabric*	*Mass permeability, km/s*	*Parameter*	*Specimen size (mm)*
Non-fissured sensitivity < 5	None	10^{-10}	$c_u, c', Ø'$	37
			m_v, c_v	76
	Pedal, silt, sand layers, inclusions, organic veins	10^{-9} to 10^{-6}	c_u	100–250
			$c', Ø'$	37
			m_v	75
			c_v	250
	Sand layers >2 mm at <0.2 m spacing	10^{-6} to 10^{-5}	$c', Ø'$	37
			m_v, c_v	75
Sensitivity > 5	Cemented with any above		$c_u, c', Ø', m_v, c_v$	50–250
Fissured	Plain fissures	10^{-10}	c_u	250
			$c', Ø'$	100
			m_v, c_v	75
	Silt or sand fissures	10^{-9} to 10^{-6}	c_u	250
			$c', Ø'$	100
			m_v, c_v	75
Jointed	Open joints		$Ø'$	100
Pre-existing slip			$c_r, Ø_r$	150 or remoulded

- This table highlights the conundrum between an economic investigation (smallest size sample) vs. quality of testing (large size sample) when neither the soil type nor fabric may be known prior to the investigation.

1.17 Quality of site investigation

- The quality of an investigation is primarily dependent on the experience and ability of the drilling personnel, supervising geotechnical engineer, and adequacy of the plant being used. This is not necessarily evident in a cost only consideration.

- The Table below therefore represents only the secondary factors upon which to judge the quality of an investigation.
- A good investigation would have at least 50% of the influencing factors shown, i.e. does not necessarily contain all the factors as this is project and site dependent.

Table 1.17 Quality of a detailed investigation.

Influencing factors	*Quality of site investigation*			*Comments*
	Good	*Fair/Normal*	*Poor*	
Quantity of factors	>70%	40% to 70%	<40%	10 factors provided herein
Phasing of investigation		Yes	No	Refer Table 1.4
Safety and environmental plan		Yes	No	Refer Table 1.9
Test/hectare • Buildings/bridges • Roads	 >20 >10	 ≥10 ≥5	 <10 <5	Refer Table 1.6 for detailed design. Tests can be boreholes, test pits, cone penetration tests, etc. Relevant tests from previous phasing included
Extent of investigation reflects type of development		Yes	No	Refer Table 1.10
Depth of investigation adequate to ground		Yes	No	Refer Table 1.10 and Table 1.11
Sample amount sufficient for lab testing		Yes	No	Refer Table 1.14
Specimen size accounting for soil structure		Yes	No	Refer Table 1.16
% of samples testing in the laboratory	≥20%	≥10%	<10%	Assuming quality samples obtained in every TP and every 1.5 m in BHs
Sample tested at relevant stress range		Yes	No	This involves knowing the depth of sample (for current overburden pressure), and expected loading
Budget as % of capital works		≥0.2%	<0.2%	Value should be significantly higher for dams, and critical projects (Table 1.14)

- An equal ranking has been provided although some factors are of greater importance than others in the table. This is however project specific.
- The table can be expanded to include other factors such a local experience, prior knowledge of project/site, experience with such projects, etc.

1.18 Costing of investigation

- The cost of an investigation depends on the site access, local rates, experience of driller and equipment available.
- There would be additional cost requirements for safety inductions, traffic control, creating site access, distance between test locations, and remoteness of the site.

Table 1.18 Typical productivity for costing (Queensland Australia).

Drilling		*Soil*	*Soft rock*	*Hard rock*
Land based drilling		20 m/day	15 m/day	10 m coring/day
Cone penetration testing (excludes dissipation testing)		100 m/day	Not applicable	Not applicable
		(Highly dependent on weather/tides/location)		
Floating barge		Non-cyclonic months	Cyclonic month	
	Open water	Land based × 50%	Land based × 30 %	
	Sheltered water	Land based × 70%	Land based × 50 %	
		(Dependent on weather/Location)		
Jack up barge		Non-cyclonic months	Cyclonic month	
	Open water	Land based × 70%	Land based × 50%	
	Sheltered water	Land based × 90%	Land based ×70%	

- Over water drilling costed on daily rates as cost is barge dependent rather than metres drilled.
- Jack up barge has significant mobilisation cost associated – depends on location from source.
- Smaller or lighter equipment are often required for limited access or ground support sites. Production rates would be reduced.

1.19 Site investigation costs

- Often an owner needs to budget items (to obtain at least preliminary funding). The cost of the SI can be initially estimated depending on the type of project.
- The actual SI costs will then be refined during the definition of needs phase depending on the type of work, terrain and existing data.
- A geo-environmental investigation is costed separately.

Table 1.19 Site investigation costs (Rowe, 1972).

Type of work	*% of capital cost of works*	*% of earthworks and foundation costs*
Earth dams	0.89–3.30	1.14–5.20
Railways	0.60–2.00	3.5
Roads	0.20–1.55	1.60–5.67
Docks	0.23–0.50	0.42–1.67
Bridges	0.12–0.50	0.26–1.30
Embankments	0.12–0.19	0.16–0.20
Buildings	0.05–0.22	0.50–2.00
Overall mean	0.7	1.5

- Overall the % value for buildings seems low and assumes some prior knowledge of the site.
- A value of 0.2% of capital works should be the minimum budgeted for sufficient information.
- The laboratory testing for a site investigation is typically 10% to 20% of the testing costs, while the field investigation is the remaining 80% to 90%, but

this varies depending on site access. This excludes the professional services of supervision and reporting.

- There is an unfortunate trend to reduce the laboratory testing, with inferred properties from the visual classification and/or field testing only. This is a commercial/competitive bidding decision rather than the best for project/ optimal geotechnical data. It also takes away the field/ laboratory check essential for calibration of the field assessment and for the development and training of geotechnical engineers.

1.20 The business of site investigation

- The geotechnical business can be divided into 3 parts (professional, field and laboratory).
- Each business can be combined, i.e. consultancy with laboratory, or exploratory with laboratory testing.

Table 1.20 The three "businesses" of site investigation (adapted from Marsh, 1999).

The services	*Provision of professional services*	*Exploratory holes*	*Laboratory testing*
Employ	Engineers and scientists	Drillers and fitters	Lab technicians
Use	Brain power and computers	Rigs, plant and equipment	Equipment
Live in	Offices	Plant yards and workshops	Laboratories and stores
QA with	CPEng	Licensed driller, ADIA	NATA
Invest in	CPD and software	Plant and equipment	Lab equipment
Worry about achieving	<1600 chargeable hours a year per member of staff	<1600 m drilled a year per drill rig	<1600 Plasticity Index tested per year per technician

- CPEng – Chartered Professional Engineer
- CPD – Continuous Professional Development
- NATA – National Association of Testing Authorities
- ADIA – Australian Drilling Industry Association

Chapter 2

Soil classification and description

2.1 Important information

- Each project is likely to have a different emphasis on the required soil information required.
- This table offers indicators on the relative importance of the data to be obtained.

Table 2.1 Relative importance of ground information.

Relative importance	*Information*	*Refer*
1	Referencing: site, location, test number, sheet number of total	Table 2.2
2	Drilling information: depth, drilling and test method (test pit, borehole), water level	Tables 2.2 to 2.5
3	Soil or rock type	Table 2.6: Chapter 3 for rock
4	Soil classification: Unified Soil Classification (USC) symbol	Table 2.8
5	Soil strength: Consistency	Table 2.16 & 2.17
6	Soil description: Plasticity and particle description, moisture condition, colour	Table 2.12 to 2.14
7	Origin and structure	Table 2.18 and Table 2.20

- o Soil contamination is not discussed in this text.
- o Relative importance should not be misinterpreted as not required as all information listed should be provided. However describing the soil type without its location highlights the relative order of precedence.

2.2 Soil borehole record

- The borelog term is liberally used here for all logs, but can be a test pit or borehole log.

Table 2.2 Borelog.

Drilling information				*Soil description*					*Field testing*					*Strata information*			
Depth	Drilling method	Water level	Sample type	USC symbol/soil type	Colour	Plasticity/particle description	Structure	Consistency	Moisture	Standard penetration type	Shear vane test	Pocket penetrometer	Dynamic cone penetrometer	Origin	Graphic log	Elevation	Depth

- Identification of the Test log is also required with the following data
  - Client
  - Project description
  - Project location
  - Project number
  - Sheet No. __ of __
  - Reference: Easting, Northing, Elevation, Inclination
  - Reference map or drawing
  - Date started and completed
  - Geomechanical details only. Environmental details not covered
  - Comment on access, services or weather also useful
- The strata information (graphic symbol and elevation) is added in final log. The additional depth column is often useful for ease of referencing (and not duplication).

2.3 Borehole record in the field

- The above is an example of a template of a final log to be used by designer. The sequence of entering field, data its level of detail and relevance can be different.
- Advantage of dissimilar borehole template in the field is
  - A specific field log allows greater space to capture field information relevant to a quality log but also administrative details not relevant to the designer (final version).
  - The design engineer prefers both a different sequence of information and different details from the field log, i.e. the field log may include some administrative details for payment purpose that is not relevant to the designer
  - A designer often uses the borelog information right to left, i.e. assessing key issues on the right of the page when thumbing through logs, then looking at details to the left, while the field supervisor logs left to right, i.e. progressively more details are added left to right.

- In this regard a landscape layout is better for writing the field logs while a portrait layout is better for the final report.

- However many prefer the field log to look the same as the final produced borehole record.

Table 2.3 Borehole record in the field.

Drilling information				*Sampling and testing*			*Soil description*						*Comments and origin*
Depth	Drilling method	Time of drilling	Water level	Sample type	Amount of recovery	Field test – type (PP< SPT, SV, PP, DCP)	USC symbol/soil type	Colour	Plasticity/particle description	Structure	Consistency	Moisture	

- o Pocket and Palm PCs are increasingly being used. Many practitioners prefer not to rely only on an electronic version. These devices are usually not suitable for logging simultaneously with fast production rates of drilling, even with coded entries. These devices are useful in mapping cuttings and for relatively slow rock coring on site, or for cores already drilled.
- o Tablet Technology should increase this input task.

2.4 Drilling information

- The table shows typical symbols only. Many consultants or authorities may have their own variation.

Table 2.4 Typical drilling data symbols.

Symbol	*Equipment*
BH	Backhoe bucket (rubber tyred machine)
EX	Excavator bucket (tracked machine)
HA	Hand auger
AV	Auger drilling with steel "V" bit
AT	Auger drilling with Tungsten Carbide (TC) bit
HOA	Hollow auger
R	Rotary drilling with flushing of cuttings using
RA	– air circulation
RM	– bentonite or polymer mud circulation
RC	– water circulation
	Support using
C	– Casing
M	– Mud
W	– Water

2.5 Water level

- The importance of this measurement on all sites cannot be over-emphasised.
- Weather and rainfall conditions at the time of the investigation are also relevant.

Table 2.5 Water level.

Symbol	*Water measurement*
$\underline{\nabla}$	Measurement: standing water level and date
$\overline{\nabla}$	Water noted
▷	Water inflow
◁	Water/drilling fluid loss

2.6 Soil type

- The soil type is the main input in describing the ground profile.
- Individual particle sizes <0.075 mm (silts and clays), are indistinguishable by the eye alone.
- Some codes use the 60 μm instead of the 75 μm, which is consistent with the other particle sizes.
- Numbers shown are rounded for simplicity, i.e. codes may show 63 mm instead of 60 mm.

Table 2.6 Soil type and particle size (simplified).

Major divisions		*Symbols*	*Subdivision*	*Particle size*
	Boulders			>200 mm
	Cobbles			60 mm–200 mm
Coarse grained soils (more than half of material is larger than 0.075 mm)	Gravels (more than half of coarse fraction is larger than 2 mm)	G	Coarse Medium Fine	20 mm–60 mm 6 mm–20 mm 2 mm–6 mm
	Sands (more than half of coarse fraction is smaller than 2 mm)	S	Coarse Medium Fine	0.6 mm–2 mm 0.2 mm–0.6 mm 75 μm–0.2 mm
Fine grained soils (more than half of material is smaller than 0.075 mm)	Silts Clays Organic	M C O	High/low Plasticity	<75 μm

- Sieve sizes do not necessarily match the boundaries shown.
- U.S. sieve sizes use the No. 4 (4.75 mm) for coarse sand to fine gravel and No. 200 (75 μm) for fine sand to fine grained soils.

2.7 Major and minor components of soil descriptions

- Soil types are usually identified with the 2 main components. The last term describes the primary material while the first term is the secondary material.
- Minor components are described separately and not with 3 descriptors.

Table 2.7 Soil description.

Components	Material description	Comment
Primary	Clay	One material
With secondary	Sandy clay	Sand is the secondary component while clay is the primary material
Mixed	Gravelly silty sandy clay	Too many descriptors
Mixed – with some	Sandy clay with some gravel	Coarse grained 5–12% Fine grained 12–30%
Mixed – with trace	Sandy clay with a trace of gravel	Coarse grained <5% Fine grained <12%

2.8 Field guide identification

- Soil types are identified in different ways in the field.

Table 2.8 Initial field guide identification of soil types.

Material type	Field Identification
Gravels or large sizes	Easily visible to the naked eye
Coarse to fine sand	Coarse sand may be visually identifiable, but below this size requires tactile evaluation on coarseness or other identification
Silts and clays	Tactile evaluation on its plasticity. Not visible by the naked eye
Organics	Smell or visual presence of roots or other fibrous matter

2.9 Sedimentation test

- The proportion of sizes >2 mm (gravel sizes) can be easily distinguished within the bulk samples.
- Sizes <2 mm (sands, silts and clays) are not easily distinguished in a bulk sample.
- Soil grittiness distinguishes between a coarse and fine grained.
- A sedimentation test is useful in this regard for an initial assessment.
- For a full classification, a hydrometer and sieve test is required.

Table 2.9 Sedimentation tests for initial assessment of particle sizes.

Material type	Approximate time for particles to settle in 100 mm of water
Coarse sand	1 second
Fine sand	10 seconds
Silt	1–10 minutes
Clay	1 hour

  - Shaking the jar with soil sample + 100 mm of water should show the coarse particles settling after 30 seconds. Clear water after this period indicates little to no fine sizes.

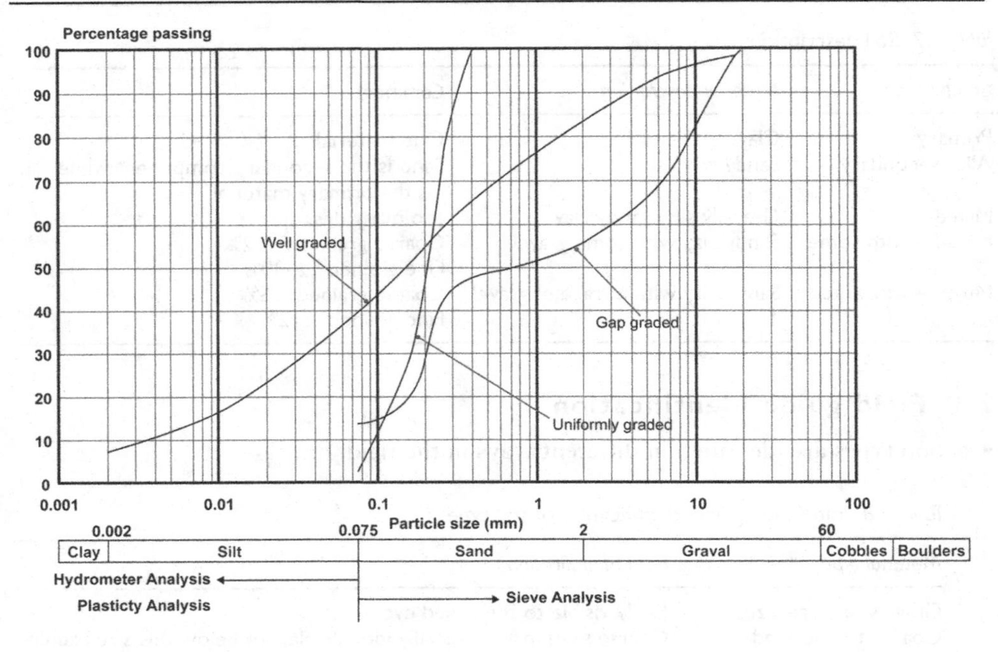

Figure 2.1 Grading curve.

2.10 Unified soil classification

- The soil is classified in the field initially, but must be validated by some laboratory testing.
- Without any laboratory validation test, then any classification is an "opinion". Even with confirmatory laboratory testing, then the log is still an interpolation on validity.
- Laboratory testing is essential in borderline cases, e.g. silty sand vs. sandy silt.

Table 2.10 Unified Soil Classification (USC) group symbols.

Soil type	*Description*	*USC symbol*
Gravels	Well graded	GW
	Poorly graded	GP
	Silty	GM
	Clayey	GC
Sands	Well graded	SW
	Poorly graded	SP
	Silty	SM
	Clayey	SC
Inorganic silts	Low plasticity	ML
	High plasticity	MH
Inorganic clays	Low plasticity	CL
	High plasticity	CH
Organic	With silts/clays of low plasticity	OL
	With silts/clays of high plasticity	OH
Peat	Highly organic soils	Pt

- Once classified many inferences on the behaviour and use of the soil is made.
- Medium plasticity uses symbols mixed or intermediate symbols, e.g. CL/CH or CI (Intermediate).

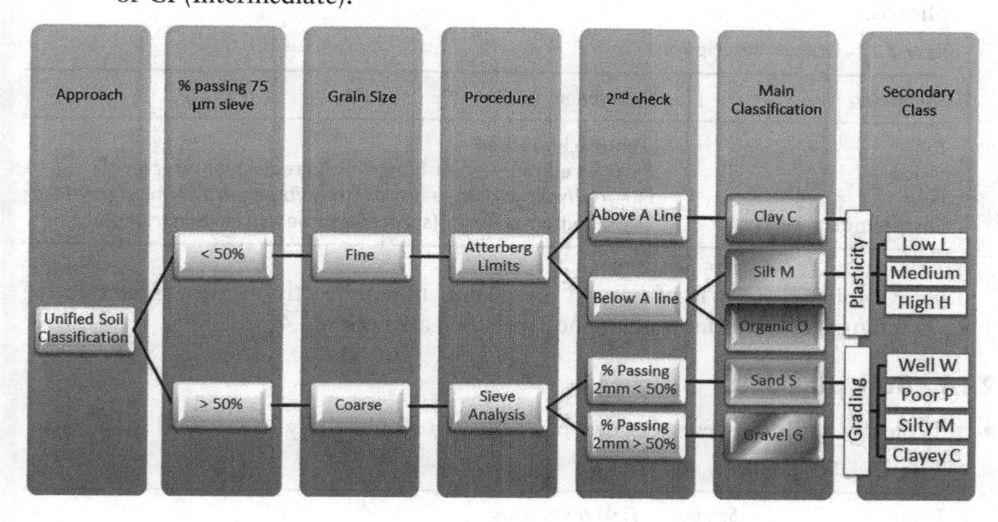

Figure 2.2 Soil classification.

2.11 Particle description

- The particle description is usually carried out in the field.

Table 2.11 Particle distribution.

Particle description	*Subdivision*
Large size (Boulders, cobbles, gravels, sands)	Coarse/medium/fine
Fine size (Silts, clays)	Plasticity
Spread (Gradation)	Well/poorly/gap/uniform
Shape	Rounded/sub-rounded/sub-angular/angular

- These simple descriptions can influence the design considerably. For example an angular grain has a larger frictional value than a rounded grain.

2.12 Gradings

- While some field descriptions can be made on the spread of the particle distribution, the laboratory testing provides a quantitative assessment for design.
- Visual observation of the grading curve is often more useful. With a near vertical or spread curve suggesting uniform or well graded, respectively.

Table 2.12 Gradings.

Symbol	*Description*	*Comments*
D_{10} (mm)	Effective size – 10% passing sieve	
D_{60} (mm)	Median size – 60% passing sieve	
U	Uniformity coefficient $= D_{60}/D_{10}$	Uniformly graded $U < 5$
C	Coefficient of curvature $= D_{30}^2/(D_{60}\,D_{10})$	Well graded $U > 5$ and $C = 1$ to 3

2.13 Colour

- Colour charts may be useful to standardise descriptions and adjacent to core photos.

Table 2.13 Colour description.

Parameter	*Description*
Tone	Light/dark/mottled
Shade	Pinkish/reddish/yellowish/brownish/greenish/bluish/greyish
Hue	Pink/red/yellow/orange/brown/green/blue/purple/white/grey/black
Distribution	Uniform/non-uniform (spotted/mottled/streaked/striped)

  - Colour by itself is of limited direct use, but indirectly used to infer relative change of material type or moisture. See also table 7.4.

2.14 Soil plasticity

- Typically a good assessment can be made of soil plasticity in the field.

Table 2.14 Soil plasticity.

Term	*Symbol*	*Field assessment*	
Non-plastic	–	Falls apart in hand	
Low plasticity	L	Cannot be rolled into (3 mm) threads when moist	
Medium plasticity	L/H	Can be rolled into threads when moist.	Shows some shrinkage on drying
High plasticity	H		Considerable shrinkage on drying. Greasy to touch. Cracks in dry material

  - Some classification systems uses the Intermediate (I) symbol instead of the L/H. The latter is an economy of symbols.

2.15 Atterberg limits

- Laboratory testing for the Atterberg limits confirms the soil plasticity descriptors provided in the field.

Table 2.15 Atterberg limits.

Symbol	*Description*	*Comments*
LL	Liquid limit – minimum moisture content at which a soil will flow under its own weight	Cone penetrometer test or Casagrande apparatus
PL	Plastic limit – minimum moisture content at which a 3 mm thread of soil can be rolled with the hand without breaking up	Test
SL	Shrinkage limit – maximum moisture content at which a further decrease of moisture content does not cause a decrease in volume of the soils	Test
PI	Plasticity Index = LL – PL	Derived from other tests
LS	Linear shrinkage is the minimum moisture content for soil to be mouldable	Test. Used where difficult to establish PL and LL. PI = 2.13 LS

- These tests are performed on the % passing the 425 micron sieve. This % should be reported, especially in residual soils. There are examples of "rock" sites having a high PI, when 90% of the sample has been discarded in the test.

2.16 Consistency of cohesive soils

- Field assessments are typically used with a tactile criterion. The pocket penetrometer can also be used to quantify the values, but it has limitations due to scale effects, conversions, sample used on and the soil type. Refer section 5.

Table 2.16 Consistency of cohesive soil.

Term	*Symbol*	*Field assessment*	*Thumb pressure penetration*	*Undrained shear strength (kPa)*
Very soft	VS	Exudes between fingers when squeezed. Likely to have free water	>25 mm	<12
Soft	S	Can be moulded by light finger pressure. Typically wet	>10 mm	12–25
Firm	F	Can be moulded by strong finger pressure.	<10 mm	25–50
Stiff	St	Cannot be moulded by fingers. Can be indented by thumb pressure.	<5 mm	50–100
Very stiff	VSt	Can be indented by thumb nail.	<1 mm	100–200
Hard	H	Difficult to be indented by thumb nail.	~0 mm	>200

- Hard clays can have values over 500 kPa. However above that value the material may be referred to as a claystone or mudstone, i.e. an extremely low strength rock.
- British Standards classify soft up to 40 kPa and at the other end of the spectrum very stiff is 150 kPa to 300 kPa.

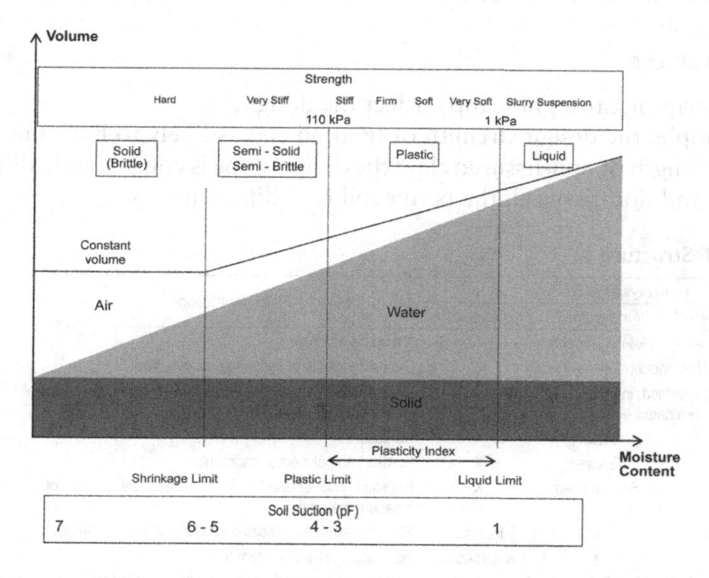

Figure 2.3 Consistency limits.

2.17 Consistency of non-cohesive soils

- The SPT value in this table is a first approximation only using the uncorrected SPT value.
- The SPT values are an upper bound for coarse granular materials for field assessment only. Correction factors are required for detailed design.
- The SPT needs to be corrected for overburden, energy ratio and particle size. This correction is provided in later chapters.

Table 2.17 Consistency of non-cohesive soil.

Term	*Symbol*	*Field assessment*		*SPT N-value*	*Density index (%)*
Very loose	VL	50 mm peg easily driven.	Foot imprints easily	<4	<15
Loose	L	12 mm reinforcing bar easily pushed by hand	Shovels easily	4–10	15–35
Medium dense	MD	12 mm bar needs hammer to drive >200 mm	Shovelling difficult	10–30	35–65
Dense	D	50 mm peg hard to drive 12 mm bar needs hammer to drive <200 mm	Needs pick for excavation	30–50	65–85
Very dense	VD	12mm bar needs hammer to drive <60 mm	Picking difficult	>50	>85
Cemented	C	12 mm bar needs hammer to drive <20 mm	Cemented, indurated or large size particles	>50	N/A

- Cemented is shown in the table, as an extension to what is shown in most references.
- N-values >50 often considered as rock.
- Table applies to medium grain size sand. Material finer or coarser may have a different value. Correction factors also need to be applied. Refer Chapter 5.

2.18 Structure

- This descriptor can significantly affect the design.
- For example, the design strength of fissured clay is likely to have only 2/3 of the design strength of non-fissured clay; the design slope is considerably different from fissured and non-fissured; the permeability is different.

Table 2.18 Structure.

Term applies to soil type			*Field identification*
Coarse grained	*Fine grained*	*Organic*	
←------------------ Heterogeneous ----------------→			A mixture of types
←-------- Homogenous -------→			Deposit consists of essentially of one type
←---- Interstratified, interbedded, interlaminated ---→		X	Alternating layers of varying types or with bands or lenses of other materials
X	Intact	X	No fissures
X	Fissured	X	Breaks into polyhedral fragments
X	Slickensided	X	Polished and striated defects caused by motion of adjacent material
X	X	Fibrous	Plant remains recognisable and retains some strength
X	X	Amorphous	No recognisable plant remains
Saprolytic / residual Soils		X	Totally decomposed rock with no identifiable parent rock structure

2.19 Moisture content

- This is separate from the water level observations. There are cases of a soil described as wet above the water table and dry below the water table.
- The assessor must distinguish between natural moisture content and moisture content due to drilling fluids used.

Table 2.19 Moisture content.

Term	*Symbol*	*Field assessment*	
		Cohesive soils	*Granular soils*
Dry	D	Hard and friable or powdery	Runs freely through hands
Moist	M	Feels cool, darkened in colour Can be moulded	Tend to cohere.
Wet	W	Feels cool, darkened in colour Free water forms on hands when handling	Tend to cohere

- Some reports provide the moisture content in terms of the plastic limit. This however introduces the possibility of 2 errors in the one assessment, Refer chapter 10 for inherent variability in soil measurement for the moisture content and plastic limit.

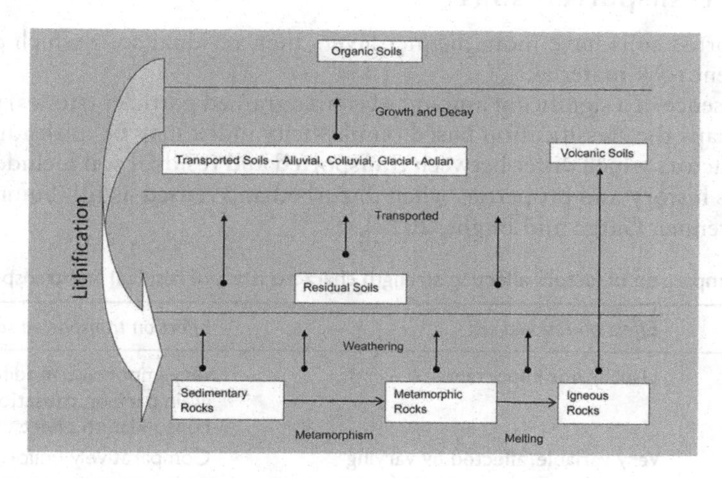

Figure 2.4 Soil and rock origins.

2.20 Origin

- This can be obtained from geology maps as well as from site and material observations.
- Soils are usually classified broadly as transported and residual soils.
- The transporting mechanism determines its further classification
  - Alluvial – deposited by water
  - Glacial – deposited by ice
  - Aeolian – deposited by wind

- Colluvial – deposited by gravity
- Fill – deposited by man.

Table 2.20 Classification according to origin.

Classification	*Process of formation and nature of deposit*
Residual	Chemical weathering of parent rock. More stony and less weathering with increasing depth
Alluvial	Materials transported and deposited by water. Usually pronounced stratification. Gravels are rounded
Colluvial	Material transported by gravity. Heterogeneous with a large range of particle sizes
Glacial	Material transported by glacial ice. Broad gradings. Gravels are typically angular
Aeolian	Material transported by wind. Highly uniform gradings. Typically silts or fine sands
Organic	Formed in place by growth and decay of plants. Peats are dark coloured
Volcanic	Ash and pumice deposited in volcanic eruptions. Highly angular. Weathering produces highly plastic, sometimes expansive clay
Evaporites	Materials precipitated or evaporated from solutions of high salt contents. Evaporites form as a hard crust just below the surface in arid regions

2.21 Comparison of characteristics between residual and transported soils

- Transported soils have more distinct layers than residual soils which grade into the parent rock material.
- The presence of a significant amount of coarse grained particles (stones) in residual soils means the classification based on plasticity index may be misleading.
- Other factors which differ between transported and residual soil include the effect of stress history and properties when disturbed and reused as fill (Summary table from Brenner, Garga and Blight, 2012).

Table 2.21 Comparison of factors affecting strength characteristics of residual and transported soils.

Factor	*Effect on residual soil*	*Effect on transported soil*
Stress history	Usually not important.	Very important, modifies initial grain packing, causation of over consolidation effect.
Grain/particle strength	Very variable, affected by varying mineralogy and presence of many weak grains is possible.	Comparatively uniform; few weak grains generally present as weak particles eliminated during transport.
Bonding	Important component of strength mostly due to residual bonds or cementation; causes cohesion intercept and yield stress; can be easily destroyed by soil disturbance.	Occurs with geologically aged deposits, produces cohesion intercept and yield stress, can be destroyed by disturbance.
Relict structure & discontinuities	Develop from pre-existing structure or structure features in parent rock, include bedding, flow structures, joints, slickensides.	Develop from deposition cycles and from stress history, formation of slickensided surfaces possible.
Anisotropy	Usually derived from relict rock fabric, e.g. bedding.	Derived from deposition and stress history of soil.
Void ratio/density	Depends on state reached in weathering process and independent of stress history.	Depends directly on stress history.

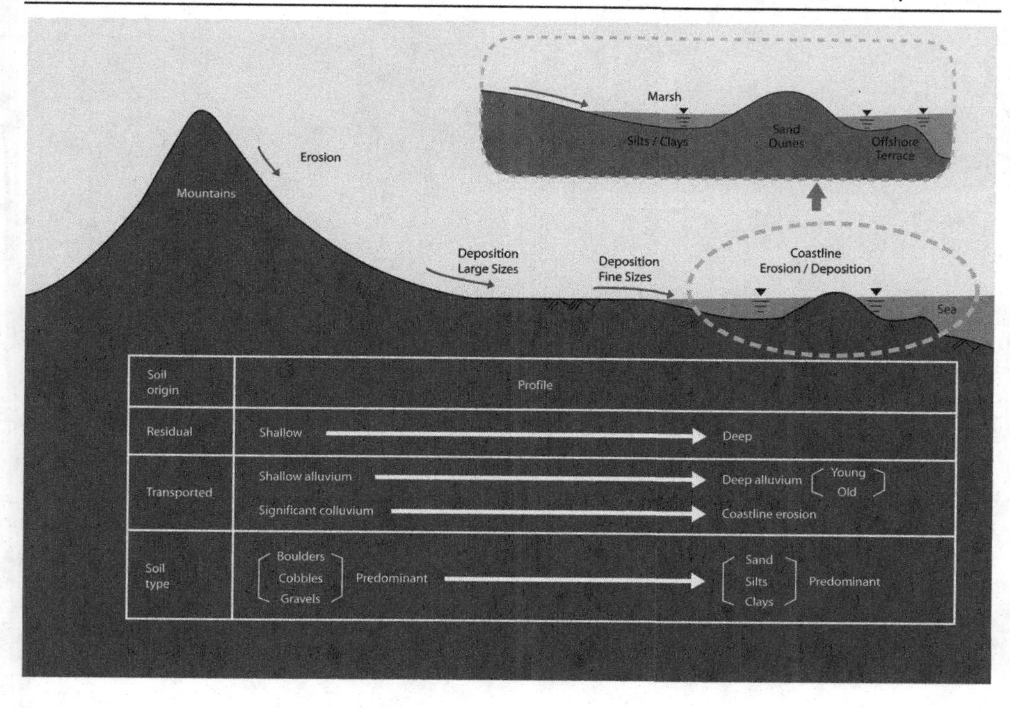

Figure 2.5 Predominance of soil type.

2.22 Classification of residual soils by its primary mode of occurrence

- Residual soils are formed in situ.
- The primary rock type affects its behaviour as a soil.

Table 2.22 Classification of residual soils by its primary origin (Hunt, 2005).

Primary occurrence	*Secondary occurrence*	*Typical residual soils*
Granite Diorite	Saprolite	Low activity clays and granular soils
Gabbro Basalt Dolerite	Saprolite	High activity clays
Gneiss Schist	Saprolite	Low activity clays and granular soils
Phyllite		Very soft rock
Sandstone		Thin cover depends on impurities. Older sandstones would have thicker cover
Shales	Red	Thin clayey cover
	Black, marine	Friable and weak mass high activity clays
Carbonates	Pure	No soil, rock dissolves
	Impure	Low to high activity clays

Chapter 3

Rock classification

3.1 Important rock information

- This table offers indicators on the relative importance of the data to be obtained. All information is required.

Table 3.1 Relative importance of ground information.

Relative importance	*Information*	*Refer*
1	Referencing: site, location, test number, sheet number of total	Table 3.2
2	Drilling information: depth, drilling and test method (test pit, borehole), water level	Tables 3.2 to 2.3
3/4	Rock weathering	Table 3.5
3/4	Rock strength and rock quality designation (RQD)	Table 3.8 and 3.9
5	Rock description: colour, defects, spacing	Table 3.11 to 3.14
6	Rock type/Origin	Table 3.15 and 3.16

3.2 Rock description

- Rocks are generally described in the borelog using the following sequence of terms shown, and with delineation between the soil log and rock log.
  - Identification of the test log is also required with the following data
    - Client
    - Project description
    - Project location
    - Project number
    - Sheet No. __ of __
    - Reference: Easting, northing, elevation, inclination
    - Date started and completed

Table 3.2 Borelog.

Drilling information				*Rock description*					*Intact strength*				*Rock mass defects*		*Strata information*			
Depth	Drilling method	Water level	Core recovery	Weathering grade	Colour	Structure	Rock quality designation (RQD)	Moisture	Estimated strength	Point load index (axial)	Point load index (diametral)	Unconfined compressive strength	Defect spacing	Defect description (depth, type, angle, roughness, infill, thickness)	Origin	Graphic log	Elevation	Depth

3.3 Field rock core log

- The field core log may be different from the final report log. Refer previous notes (Section 2.2) on field log versus final log.
- The field log variation is based on the strength tests not being completed at the time of boxing the cores.

Table 3.3 Field borelog.

Drilling information					*Rock description*					*Testing*		*Rock mass defects*		*Comments and origin*
Depth	Drilling method	Time of drilling	Water level	Core recovery	Weathering grade	Colour	Structure	Estimated strength	Rock quality designation (RQD)	Point load index (axial/diametral)	Other	Defect spacing	Defect description (depth, type angle, roughness, infill, thickness)	

- Due to the relatively slow rate of obtaining samples (as compared to soil) then there would be time to make some assessments. However, some supervisors prefer to log all samples in the laboratory, as there is a benefit in observing the full core length at one session.
  - For example, the rock quality designation (RQD). If individual box cores are used, the assessment is on the core run length. If all boxes for a particular borehole are logged simultaneously, the assessment RQD is on the domain length (preferable).
- Typical bore box sizes are 1.0 m to 1.5 m in length with 5 to 4 slots. This controls the weight to 25 kg maximum for lifting purposes.
  - Rock origins are in 3 groups
    - Sedimentary rocks
    - Igneous rocks
    - Metamorphic rocks

3.4 Drilling information

- The typical symbols only are shown. Each consultant or authority has their own variation.

Table 3.4 Typical symbols used for rock drilling equipment.

Symbol	*Equipment*
PQ	Coring using 85 mm (inside diameter) core barrel
HQ	Coring using 63 mm core barrel
NMLC	Coring using 52 mm core barrel
NQ	Coring using 47 mm core barrel
RR	Tricone (Rock Roller) Bit
DB	Drag Bit

3.5 Rock weathering

- The rock weathering is the most likely parameter to be assessed.
- Weathering is often used to assess strength as a quick and easily identifiable approach – but should not be used as a standalone. This approach must be first suitably calibrated with the assessment of other rock properties such as intact strength, and defects.

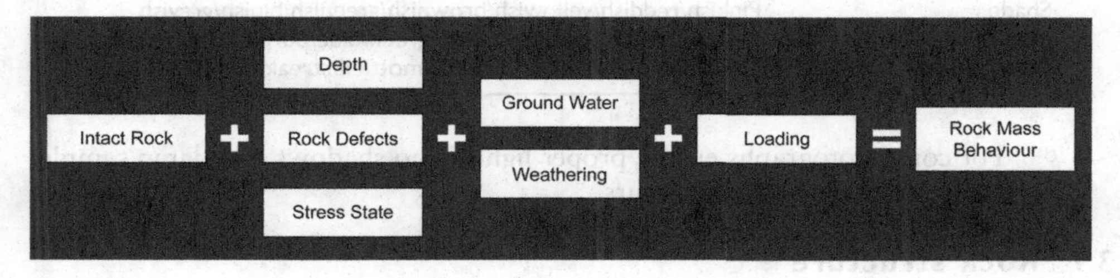

Figure 3.1 Rock mass behaviour.

Table 3.5 Rock weathering classification.

Term	*Symbol*	*Field assessment*
Residual soil	RS	Soil developed on extremely weathered rock; the mass structure and substance fabric are no longer evident; there is a large change in volume but the soil has not been significantly transported. Described with soil properties on the log
Extremely Weathered	XW	Soil is weathered to such an extent that it has 'soil' properties i.e. it either disintegrates or can be remoulded, in water. May be described with soil properties
Distinctly Weathered	DW (MW/HW)	Rock strength usually changed by weathering. The rock may be highly discoloured, usually by iron staining. Porosity may be increased by leaching, or may be decreased due to deposition of weathering products in pores.
Slightly Weathered	SW	Rock is slightly discoloured but shows little or no change of strength from fresh rock.
Fresh	FR	Rock shows no sign of decomposition or staining.

- RS is not a rock type and represents the completely weathered product in situ
- Sometimes aspect is important with deeper weathering in the warmth of northern sunlight (for countries in the southern hemisphere)
- Distinctly weathered may be further classified into Highly (HW) and moderately weathered (MW). The former represents greater than 50% soil, while the latter represents less than 50% soil.
- This table is appropriate for field assessment. Detailed testing on rock strength (Table 6.9) show that strength can vary between intact samples of SW and FR weathered rock.

3.6 Colour

- Colour charts are useful for core photography.

Table 3.6 Colour description.

Parameter	*Description*
Tone	Light/dark/mottled
Shade	Pinkish/reddish/yellowish/brownish/greenish/bluish/greyish
Hue	Pink/red/yellow/orange/brown/green/blue/purple/white/grey/black
Distribution	Uniform/non-uniform (spotted/mottled/streaked/striped)

- For core photographs ensure proper lighting/no shadows and damp samples to highlight defects and colours.

3.7 Rock structure

- The rock structure describes the frequency of discontinuity spacing and thickness of bedding.

- The use of defects descriptors typically used in place of below individual descriptors.
- Persistence reflects the joint continuity.

Table 3.7 Rock structure.

Rock structure	*Description*	*Dimensions*
Thickness of bedding	Massive	>2.0 m
	Thick-bedded	0.6 to 2.0 m
	Mid-bedded	0.2 to 0.6 m
	Thin-bedded	0.06 m to 0.2 m
	Very thinly bedded/laminated	<0.06 m
Degree of fracturing/jointing	Unfractured	>2.0 m
	Slightly fractured	0.6 to 2.0 m
	Moderately fractured	0.2 to 0.6 m
	Highly fractured	0.06–0.2 m
	Intensely fractured	<0.06 m
Dip of bed or fracture	Flat	0 to 15 degrees
	Gently dipping	15 to 45 degrees
	Steeply dipping	45 to 90 degrees
Persistence	Very high	>20 m
	High	10–20 m
	Medium	3–10 m
	Low	1–3 m
	Very low	>1 m

3.8 Rock quality designation

- RQD (%) is a measure of the degree of fracturing. This is influenced also by quality of drilling, and handling of the rock cores.

Table 3.8 Rock quality designation.

RQD (%)	*Rock description*	*Definition*
0–25	Very poor	
25–50	Poor	
50–75	Fair	$\text{RQD (\%)} = \frac{\text{Sound core pieces} > 100\,\text{mm}}{\text{Total core run length}} * 100$
75–90	Good	
>90	Excellent	

  - Many variations for measurement of this supposedly simple measurement.
  - Drilling induced fractures should not be included in the RQD measurement.
  - The domain rather than the core length should be used to assess the RQD. Different values result if the RQD is measured in a per-metre length or a domain area. The latter represents the true RQD values while the former would have an averaging effect.
  - Figure 3.2 shows the RQD measurement procedure.
  - RQD is dependent on the borehole orientation. An inclined borehole adjacent to a vertical borehole is expected to give a different RQD value.

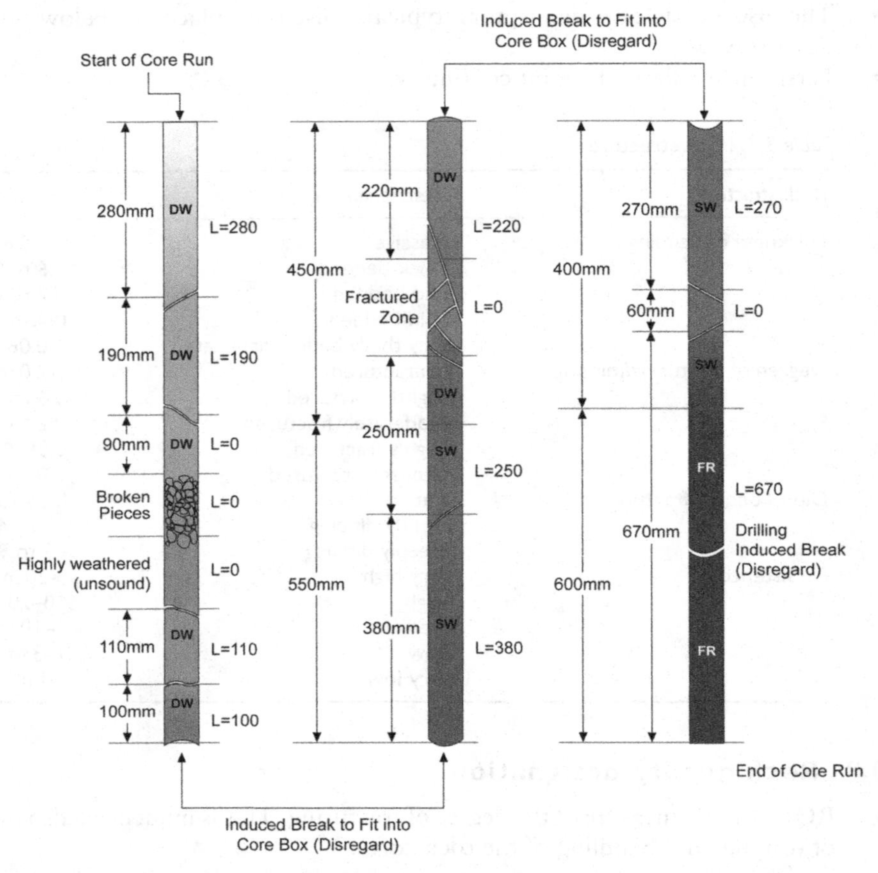

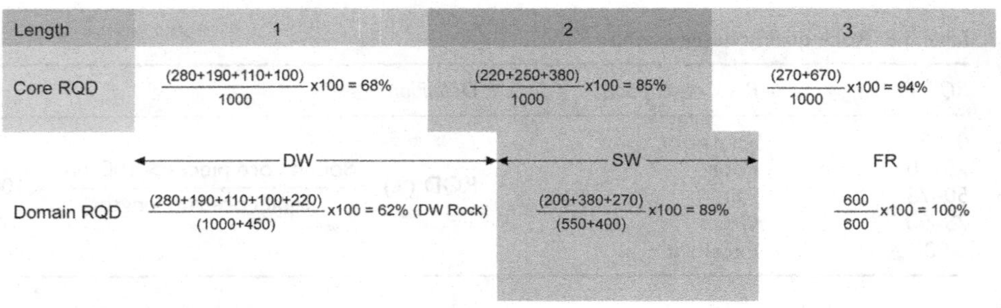

Length	1	2	3
Core RQD	$\frac{(280+190+110+100)}{1000}$ x100 = 68%	$\frac{(220+250+380)}{1000}$ x100 = 85%	$\frac{(270+670)}{1000}$ x100 = 94%
	DW	SW	FR
Domain RQD	$\frac{(280+190+110+100+220)}{(1000+450)}$ x100 = 62% (DW Rock)	$\frac{(200+380+270)}{(550+400)}$ x100 = 89%	$\frac{600}{600}$ x100 = 100%

NOTE: MINOR DIFFERENCES IN LOGGING CORE LENGTH (1000mm IN EXAMPLE) AND LOGGING DOMAIN

Figure 3.2 RQD measurement.

3.9 Rock strength

- This table refers to the strength of the intact rock material and not to the strength of the rock mass, which may be considerably weaker due to the effect of rock defects.

Table 3.9 Rock strength.

Strength	*Symbol*	*Field assessment*	
		By hand	*Hammer with hand held specimen*
Extremely low	EL	Easily remoulded to a material with soil properties.	
Very low	VL	Easily crumbled in 1 hand.	
Low	L	Broken into pieces in 1 hand.	
Medium	M	Broken with difficulty in 2 hands.	Easily broken with light blow (thud).
High	H		1 firm blow to break (rings).
Very high	VH		>1 blow to break (rings)
Extremely high	EH		Many blows to break (rings).

3.10 Rock hardness

- The rock hardness is not the same as the rock strength.

Table 3.10 Field assessment of hardness.

Description of hardness	*Moh's hardness*	*Characteristic using pocket knife*		
		Rock dust	*Scratch marks*	*Knife damage*
Friable	1–2	Little powder	None. Easily crumbled. Too soft to cut. Crumbled by hand	
Low	2–4	Heavy trace	Deeply gouged	No damage
Moderately Hard	4–6	Significant trace of powder	Readily visible (after powder blown away)	
Hard	6–8	Little powder	Faintly visible	Slight damaged; trace of steel on rock
Very hard	8–10	None	None	Damaged; steel left on rock

3.11 Discontinuity scale effects

- The scale effects are an order of magnitude only, with significant overlap.

Table 3.11 Discontinuity scale effects.

Discontinuity group	*Typical range*	*Typical scale*
Defect thickness	2 mm to 60 cm	20 mm
Bedding, foliation, jointing	0.2 m to 60 m	2 m
Major shear zones, seams	20 m to 6 km	200 m
Regional fault zones	2 km to 600 km	20 km

3.12 Rock defects spacing

- The rock defects are generally described using the following sequence of terms:
  - [Defect spacing]; [Depth (metres from surface), defect type, defect angle (degrees from horizontal), surface roughness, infill, defect thickness (mm)].

Table 3.12 Defect spacing.

Description	*Spacing*
Extremely closely spaced (Crushed)	<20 mm
Very closely spaced	20 mm to 60 mm
Closely spaced (Fractured)	60 mm to 200 mm
Medium spaced	0.2 m to 0.6 m
Widely spaced (Blocky)	0.6 m to 2.0 m
Very widely spaced	2.0 m to 6.0 m
Extremely widely spaced (Solid)	>6.0 m

3.13 Rock defects description

- The defects are also called discontinuities.
- The continuity of discontinuities is difficult to judge in rock cores. An open exposure is required to evaluate (trench, existing cutting).
- Even in an existing cutting, the defects in the vertical and lateral direction can be measured, but the continuity into the face is not readily evident.

Table 3.13 Rock defect descriptors.

Rock defects	*Descriptors*	*Typical details*
Joints	Type	Bedding, cleavage, foliation, schistiosity
	Joint wall separation	Open (size of open) or Closed (zero size) Filled or clean
	Roughness	Macro surface (stepped, curved, undulating. Irregular, planar); Micro surface (rough, smooth, slicken-sided)
	Infilling	Clays (low friction); crushed rock (medium to high friction); Calcite/Gypsum (may dissolve)
Faults and shear zones	Extent	Thickness
	Character	Coating, infill, crushed rock, clay infilling

- Continuity may be relative to the type of structure, loading or cutting.
- Discontinuities considered continuous under structures if it is equal to the base width, when sliding can be possible.

3.14 Rock defect symbols

- Typical symbols only. Each consultant or authority has their own variation.

Table 3.14 Defect description.

Defect type	*Surface roughness*		*Coating or infill*
	Macro-surface geometry	*Micro-surface geometry*	
Bp – Bedding parting	St – Stepped	Ro – Rough	cn – clean
Fp – Foliation parting	Cu – Curved	Sm – Smooth	sn – stained
Jo – Joint	Un – Undulating	Sl – Slickensided	vn – veneer
Sh – Sheared zone	Ir – Irregular		cg – coating
Cs – Crushed seam	Pl – Planar		
Ds – Decomposed seam			
Is – Infilled seam			

- The application of this data is considered in later chapters.
- For example, friction angle of an infill fracture < for a smooth fracture < rough fracture. But the orientation and continuity of the defects would determine whether it is a valid release mechanism.
- The opening size and number of the joints would determine its permeability.

3.15 Sedimentary and pyroclastic rock types

- The grain size and shape as used to describe soils can be also used for rocks.
- Sedimentary rocks are the most common rock type at the earth's surface and sea floor. They are formed from soil sediments or organic remains of plants and animals which have been lithified under significant heat and pressure of the overburden, or by chemical reactions.
- This rock type tends to be bedded.
- Pyroclastic Rocks are a type of igneous rock. Pyroclasts have been formed by an explosive volcanic origin, falling back to the earth, and becoming indurated. The particle sizes thrown into the air can vary from 1000 tonne block sizes to a very fine ash (Tuff).

Table 3.15 Rock type descriptor (adapted from AS 1726 – 1993, Mayne, 2001 and Geoguide 3, 1988).

Description		*Sedimentary*					*Pyroclastic*
Superficial deposits	Grain size (mm)		Clastic (Sediment)		Chemically formed	Organic remains	
Boulders		Rudaceous	Conglomerate (Rounded fragments) Breccia (Angular fragments)		Halite Gypsum		Agglomerate (Round grains) Volcanic Breccia (Angular grains)
	200						
Cobbles							
	60						
Coarse gravel							
	20						
Medium gravel							
	6						
Fine gravel							
	2						
Coarse sand		Arenaceous	Sandstone Quartzite Arkose Greywacke		Limestone Dolomite		Coarse grained Tuff
	0.6						
Medium sand							
	0.2						
Fine sand							
	.06						
Silt		Argillaceous	Mudstone	Siltstone		Chalk, Lignite, Coal	Fine grained Tuff
	.002						
Clay			Shale	Claystone			Very fine grained Tuff

- Even for rocks in a similar descriptor other factors may determine its overall strength properties.
- For example, Sandstone, Arkose and Greywacke are similarly classed, but sandstone would usually have rounded grains, which are one size, Arkose would be Sub-angular and well graded while Greywacke would be angular and well graded. This results in an intact Greywacke being stronger than sandstone.

3.16 Metamorphic and igneous rock types

- The grain sizes are more appropriate (measurable) for the assessment of the sedimentary rocks. However the size is shown in the table below for comparison purposes.
- Igneous rocks are formed when hot molten rock solidifies. Igneous rocks are classified mainly on its mineral content and texture.
- Metamorphic rocks are formed from other rock types, when they undergo pressure and/or temperature changes. Metamorphic rocks are classed as foliated and non-foliated.

Table 3.16 Rock type descriptor (adapted from AS 1726 – 1993, Mayne, 2001 and Geoguide 3, 1988).

Description		*Igneous (quartz content)* Pale → Dark			*Metamorphic*	
Superficial deposits	Grain Size (mm)	Acid (Much)	Intermediate (Some)	Basic (Little to none)	Foliated	Non- Foliated
Boulders						
	200					
Cobbles						
	60					
Coarse gravel		Granite Aplite	Granodiorite Diorite	Gabbro Peridotite	Gneiss Migmatite	Marble Quartzite Granulite Hornfels
	20					
Medium gravel						
	6					
Fine gravel						
	2					
Coarse sand						
	0.6					
Medium sand		Microgranite	Microdiorite	Dolerite	Schist	Serpentine
	0.2					
Fine sand						
	.06					
Silt						
	.002	Rhyolite Dacite	Andesite Quartz Trachyte	Basalt	Phyllite Slate	
Clay						

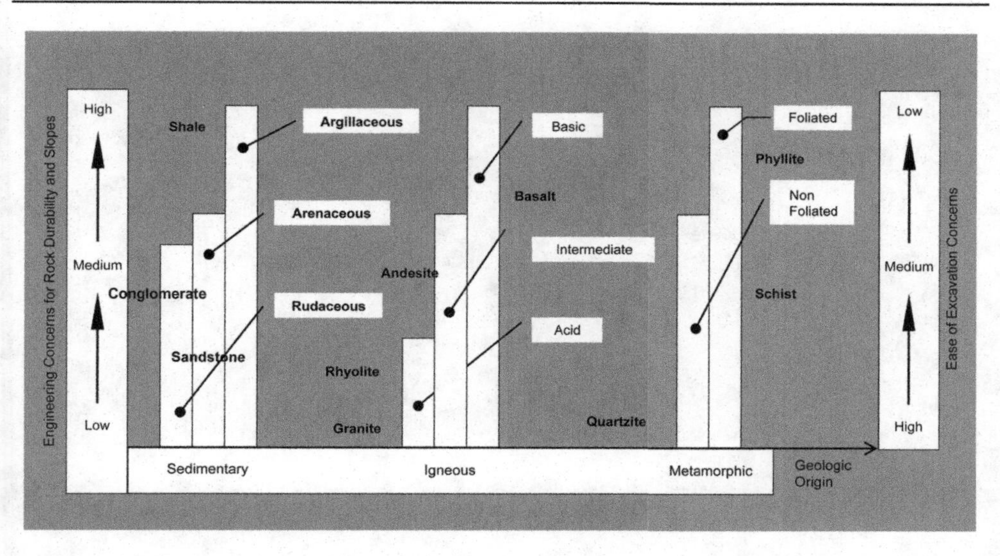

Figure 3.3 Engineering concerns of various rock types for durability slope stability and excavatability. Aggregate and stones are seldom selected on basis of rock type alone. Preliminary consideration only.

Chapter 4

Field sampling and testing

4.1 Types of sampling

- The samples are recovered to classify the material and for further laboratory testing.
- Refer Chapter 1 for the effect of size of sampling and disturbance.

Table 4.1 Types of sampling.

Sample type	*Quality*	*Uses*
Disturbed	Low	Samples from the auger and wash boring, which may produce mixing of material. Complete destruction of the fabric and structure. Identify strata changes
Representative	Medium	Partially deformed such as in split barrel sampler. Fabric/ structure, strength compressibility and permeability expected to be changed. Classification tests
Continuous	Medium/high	Hole is advanced using continuous piston or thin walled tube sampling. Obtains a full strata profile
Undisturbed	High	Tube or block samples for strength and deformation testing. Tube samples are obtained from boreholes and block samples from test pits

- ○ Disturbed samples obtained from augers, wash boring returns on chippings from percussion drilling.
- ○ Split barrel sampler used in the standard penetration test (SPT).
- ○ Tube samplers are usually thin walled with a cutting edge, but with piston samplers in soft to firm material.
- ○ Undisturbed tube samples are not possible in sands, and split barrel sampling is used.
- ○ In situ testing (e.g. CPTu) has been found to be more cost effective and provides better information in some soil profiles rather than sampling.

4.2 Boring types

- Various operations are used to advance the borehole, before obtaining samples.
- Hole clean outs are required before sampling.
- Common drilling methods are presented in the Table.

Table 4.2 Boring types.

Boring type	*Uses*
Hand auger	Shallow depths. Hammering required for "undisturbed" samples means lower quality obtained. Unlikely to be able to hand auger below DCP ~ 8 (very stiff clays or dense sands)
Solid stem auger	Used in dry holes in competent materials. May need to use casing for collapsing material
Hollow stem auger	Similar to solid stem (continuous flight) auger drilling, except hollow stem is screwed into to ground and acts as casing. Sampling and testing from inside of auger. Penetration in strong soils/gravel layers difficult
Wash boring	Used to advance the borehole and keep the hole open below the water table. Fluid may be mud (polymer) or water depending on the soil conditions. Maintains hydrostatic head
Rock coring	Hardened cutting bit with a core barrel used to obtain intact rock samples
Air track probes	Provides a rapid determination of rock quality/depth to rock based on the time to advance the hole. Rock assessment is difficult as rock chippings only obtained

- Maintaining a hydrostatic head below the water prevents blow out of the base of the hole, with a resulting inconsistency in the SPT result.
- Similarly if the base of the hole is loosened by over washing in sands.

4.3 Field sampling

- Typical symbols only. Each consultant has their own variation.
- The symbols are used to speed up on site documentation.
- This requires an explanatory note on symbols to accompany any test record.

Table 4.3 Type of sampling.

Symbol	*Sample or test*
TP	Test pit sample
W	Water sample
D	Disturbed sample
B	Bulk disturbed sample
SPT	Standard Penetration Test sample
C	Core sample
U (50)	Undisturbed sample (50 mm diameter tube)
U (75)	Undisturbed sample (75 mm diameter tube)
U (100)	Undisturbed sample (100 mm diameter tube)

- The use of electronic hand held devices for logging is becoming more popular. These devices are useful for static situations such as existing rock cuttings and exposures, or laboratory core logging.
- In dynamic situations such as field logging with a high production rate of say 20 metres/day, these electronic devices are not as efficient and flexible as the conventional handwritten methods – although the use of tablets shows

promise. The preference of having a hard copy and not relying on electronic logging only in these situations is another argument not in its favour in such cases. The use of coded symbols aids in faster input of the data.

4.4 Field testing

- The common field testing is shown in the table.

Table 4.4 Type of field testing.

Symbol	*Test*	*Measurement*
DCP	Dynamic cone penetrometer	Blows/100 mm
SPT	Standard penetration test	Blows/300 mm
CPT	Cone penetration test	Cone resistance q_c (MPa); Friction ratio F.R. (%) q_c;
CPTu	Cone penetration test with pore pressure measurement (Piezocone)	F.R.; Pore pressure (kPa). Time for pore pressure dissipation t (sec)
PT	Pressuremeter test	Lift-off and limit pressures (kPa), Volume change (cm^3)
PLT	Plate loading test	Load (kN), Deflection (mm)
DMT	Dilatometer test	Lift-off and expansion pressures (kPa)
PP	Pocket penetrometer test	kPa
VST	Vane shear test	Nm, kPa
BB	Benkelman Beam	Deflection mm converted to CBR value
LFWD	Light falling weight deflectometer	MPa (Modulus)
CIST	Clegg impact soil tester	Impact value
WPT	Water pressure (Packer) test	Lugeons

- There are many variations of tests in different countries. For examples the DCP, has differences in weight, drop and rods used. The CPT has mechanical and electric types with differences in interpretation.
- Vane shear test may have a direct read out for near surface samples, but with rods with a torque measurement for samples at depth.
- Benkelman beams may have automated variations such as delectographs which are based on a wheel load but has differences in the deflection bowl. Falling weight deflectometers (FWD) uses a falling weight rather than a wheel load. The Benkelman Beam uses a ½ axle load of the standard 80 kN vehicle loading on a dual tyre single axle with a tyre pressure of 550 kPa. The Benkelman Beam is recommended only for 450 mm maximum thickness flexible granular pavements (and not other pavement types).
- Deflectometers has a test production rate of 3 times that of Benkelman beams
- The use of readings from the LFWD and CIST are described in this chapter as these instruments provide direct reading values as compared to most others where some post processing or interpretation may be required.

4.5 Comparison of in situ tests

- The appropriateness and variability of each test should be considered. An appropriate test for ground profiling may not be appropriate for determining the soil modulus.
- Variability in testing is discussed in section 10.

Table 4.5 In situ test methods and general application (Bowles, 1996).

Test	*Area of ground interest*												
	Soil identification	*Establish vertical profile*	*Relative density D_r*	*Angle of friction ϕ*	*Undrained shear strength S_u*	*Pore pressure u*	*Stress history OCR and Ko*	*Modulus: Es, G'*	*Compressibility mv and Cc*	*Consolidation c_h and c_v*	*Permeability k*	*Stress-strain curve*	*Liquefaction resistance*
Acoustic probe	C	B	B	C	C		C	C					C
Borehole permeability	C					A				B	A		
Cone													
Dynamic	C	A	B	C	C		C						C
Electrical friction	B	A	B	C	B		C	B	C				B
Electrical piezocone	A	A	B	B	B	A	A	B	B	A	B	B	A
Mechanical	B	A	B	C	B		C	B	C				B
Seismic down hole	C	C	C					A				B	B
Dilatometer (DMT)	B	A	B	C	B		B	B	C			C	B
Hydraulic fracture						B	B			C	C		
Nuclear density tests			A	B				C					
Plate load tests	C	C	B	B	C		B	A	B	C	C	B	B
Pressure meter Menard	B	B	C	B	B		C	B	B			C	C
Self-boring pressure	B	B	A	A	A	A	A	A	A	A	B	A	A
Screw plate	C	C	B	C	B		B	A	B	C	C	B	B
Seismic down-hole	C	C	C					A				B	B
Seismic refraction	C	C						B					B
Shear vane	B	C			A		B						
Standard penetration test (SPT)	B	B	B	C	C				C				A

c_h = Vertical consolidation with horizontal drainage; c_v = Vertical consolidation with vertical drainage.
Code: A = most applicable; B = may be used; C = least applicable.

- The SPT was considered the "reference" insitu test, but as seen above more reliable and more informative testing is provided with "modern" equipment such cone penetration tests although no sample is obtained.

4.6 Standard penetration test in soils

- The SPT evolved as a sampler first, but then later quantified with the blow counts to obtain the sample. As noted in the table above improved quantification of site can be obtained with CPTs
- In soils, the SPT is usually terminated with 30 blows / 100 mm in the seating drive as a refusal level for the Australian Standard AS 1289 – 6.3.1 – 1993. AS 1289 specifically states standard penetration tests for soils. In rock this refusal level is insufficient data. British Standards BS 1377:1990 and ASTM Standard D1586–84 allow further blows before discontinuing the test, and thus are more appropriate for improved extrapolating its SPT value in rock.

- The first 150 mm is the seating drive, which allows for possible material fall in at the base of the hole and /or loosening of base material. Comparison between each 150 mm increment should be made to assess any inconsistencies. For example N values 1, 7, 23 suggests:
  - An interface (examine sample recovery if possible); or
  - Loose material falling into the base of the borehole, and the initial seating and first increment drive representing blow counts in a non in situ material.

Table 4.6 Standard penetration test in soils.

Symbol	*Test*
7, 11, 12 (e.g.)	Example of blows per 150 mm penetration.
N = 23 (e.g.) or N_{SPT}	Penetration resistance (blows for 300 mm penetration following 150 mm seating drive, Example of 11 + 12 = 23 = N_{SPT} (actual field value with no correction factors).
N > 60	Total hammer blows exceed 60.
7, 11, 25/20 mm (e.g.)	Partial Penetration, example of blows for the measured penetration (examine sample as either change in material here or fall in at top of test).
N′	Corrected N-value for silty sands below the water table.
N*	Inferred SPT value.
RW	Rod Weight only causing penetration (N < 1).
HW	Hammer and rod weight only causing full penetration (N < 1).
HB	Hammer Bouncing (Typically N* > 50).
$(N_o)_{60}$	Penetration resistance normalized to an effective overburden of 100 kPa, and an energy of 60% of theoretical free fall energy. $(N_o)_{60} = C_N\ C_{ER}\ N_{SPT}$.
$C_N\ C_{ER}$	Correction factor for overburden (C_N) and energy ratio (C_{ER}).

  - Correction factors need to be applied for overburden in granular soils and type of hammers.
  - Typically $(N_o)_{60} < 60$ for soils. Above this value, the material is likely cemented sand, coarse gravels, cobbles, boulders or rock. However these materials may still be present for N-values less than 60.
  - While the SPT N – value is the summation of the 300 mm test drive, the incremental change should also be noted, as this may signify loose fall in of material (i.e. incorrect values) or change in strength (or layer) profile over that 450 mm.
  - Some codes use 75 mm increments to be measured. Other than soft soils ($c_u < 25$ kPa), this is unlikely to provide added benefit and with the dis-benefit of additional effort

4.7 Standard penetration test in rock

- The SPT procedure in rock is similar to that in soils but extending the refusal blows to refusal. This is at least 30 blows in less than 100 mm for both a seating and a test drive before discontinuing the test.
- Tabulate both the seating and the test drive. The driller may complain about damage to the equipment.
- A solid cone (apex angle of 60°) is used for tests in gravelly soils, boulders and soft weathered rock.

- Values of N > 50 while useful in classifying a very dense soil has limited value in residual soils or extremely weathered rock. N* should be extrapolated to a value of 120 or above to be of quantitative value to the designer for assessing rock strength.

Table 4.7 Standard penetration test in rock.

Symbol	*Test*
N = 23 (e.g.)	Penetration resistance (blows for 300 mm penetration following 150 mm seating drive, Example of 11 + 12 = 23).
−30/50 mm, 30/20 mm (e.g.)	Partial penetration, example of blows for the measured penetration, but allowing for measuring both seating and test drive.
N*	Inferred SPT Value.

- ○ There is a continuous debate on whether inferred values should be placed on a factual log. However, the debate then extends to how much on the log is factual. For example, is the colour description (person dependent) more factual than N*.

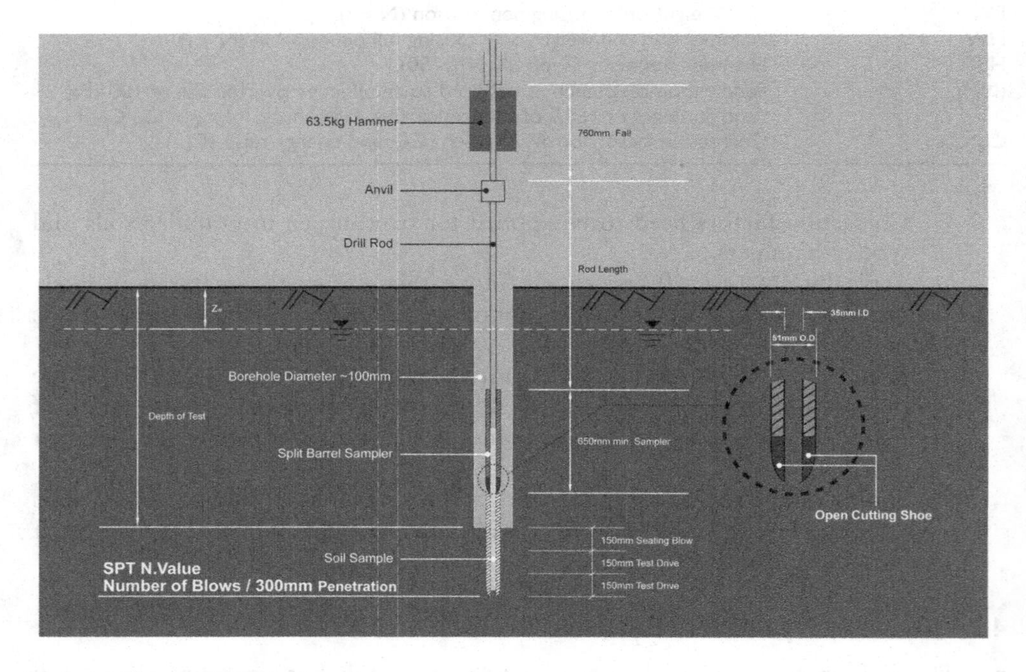

Figure 4.1 Standard penetration test.

4.8 Overburden correction factors to SPT result

- An overburden correction factor applies for granular materials.
- $N_o = C_N N$.
- Depth of water below surface $= z_w$.

Table 4.8 SPT correction factors to account for overburden pressure (adapted from Skempton, 1986).

Effective overburden (kPa)	*Correction factor, C_N*		*Approximate depth of soil (metres) to achieve nominated effective overburden pressure for various ground water level (z_w)*			
	Fine sands	*Coarse sands*	*At surface $z_w = 0$ m*	*$z_w = 2$ m*	*$z_w = 5$ m*	*$z_w = 10$ m*
0	2.0	1.5	0.0 m	0.0 m	0.0 m	0.0 m
25	1.6	1.3	3.1 m	1.4 m	1.4 m	1.4 m
50	1.3	1.2	6.2 m	3.7 m	2.8 m	2.8 m
100	1.0	1.0	12.5 m	10.0 m	6.2 m	5.6 m
200	0.7	0.8	25.0 m	22.5 m	18.8 m	12.5 m
300	0.5	0.6	37.5 m	35.0 m	31.2 m	25.0 m
400	0.5	0.5	50.0 m	47.5 m	43.7 m	37.5 m

- Average saturated unit weight of 18 kN/m^3 used in table. Unit weight can vary.
- Borehole water balance is required for tests below the water table to avoid blow out at the base of the hole with loosening of the soil, and a resulting non-representative low N-value.
- In very fine or silty sands below the water table, a pore pressure may develop and an additional correction factor applies for $N' > 15$. $N = 15 + ½(N' - 15)$.

4.9 Equipment and borehole correction factors for SPT result

- An equipment correction and borehole size correction factors apply.
- The effect of borehole diameter is negligible for cohesive soils, and no correction factor is required.
- The energy ratio is normalized to 60% of total energy.
- $(N_o)_{60} = C_N C_{ER} N$.
- $C_{ER} = C_H C_R C_s C_B$.

Table 4.9 Energy ratio correction factors to be applied to SPT value to account for equipment and borehole size (adapted from Skempton, 1986 and Takimatsu and Seed, 1987).

To account for	*Parameter* *Hammer – Release – Country*	*Correction factor*
Hammer (C_H)	Donut – free fall (Tombi) – Japan	1.3
	Donut – rope and pulley – Japan	1.1
	Safety – rope and pulley – USA	1.0
	Donut – free fall (Trip) – Europe, China, Australia	1.0
	Donut – rope and pulley – China	0.8
	Donut – rope and pulley – USA	0.75
Rod Length (C_R)	10 m	1.0
	10 m to 6 m	0.95
	6 m to 4 m	0.85
	4 m to 3 m	0.75
Sampler (C_s)	Standard	1.0
	US sampler without liners	1.2
Borehole Diameter (C_B)	65 mm – 115 mm	1.0
	150 mm	1.05
	200 mm	1.15

4.10 Cone penetration test

- There are several variations of the cone penetration test (CPT). Electric and mechanical cones should be interpreted differently.
- The CPTu data is tabled below. The CPT would not have any of the pore pressure measurements.
- The CPT has a high production rate (typically 100 m / day but varies depending on number, soil type, distance between tests, accessibility, etc.) compared to other profile testing.

Table 4.10 Cone penetration tests.

Symbol	*Test*
q_c	Measured cone resistance (MPa)
q_T	Corrected cone tip resistance (MPa): $q_T = q_c + (1 - a_N)u_b$
a_N	Net area ratio provided by manufacturer $0.75 < a_N < 0.82$ for most $10\,cm^2$ penetrometers $0.65 < a_N < 0.8$ for most $15\,cm^2$ penetrometers
F_s	Sleeve frictional resistance
FR	Friction ratio $= F_s/q_c$
u_0	In-situ pore pressure
B_q	Pore pressure parameter – excess pore pressure ratio $B_q = (u_d - u_0)/(q_T - P'_o)$
P'_o	Effective overburden pressure
u_d	Measured pore pressure (kPa)
Δu	$\Delta u = u_d - u_0$
T	Time for pore pressure dissipation (sec)
t_{50}	Time for 50% dissipation (minutes)

- The dissipation tests which can take a few minutes to a few hours has proven more reliable in determining the coefficient of consolidation, than obtaining that parameter from a consolidation test.

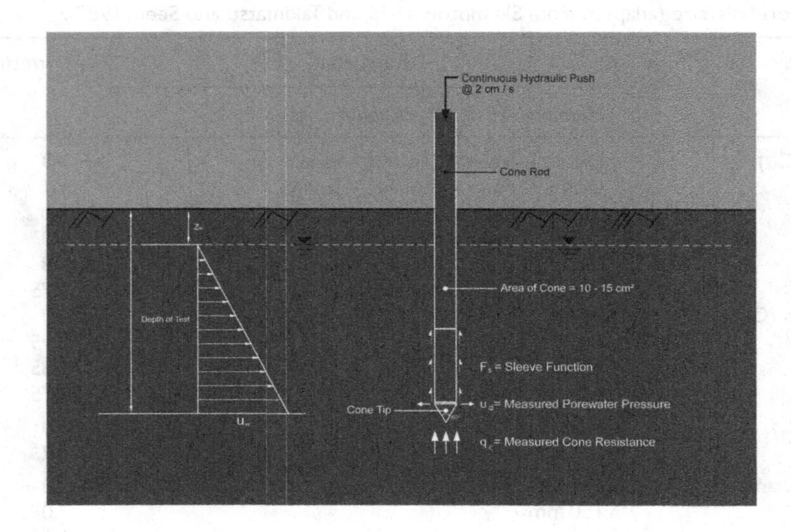

Figure 4.2 Cone penetration test.

4.11 Dilatometer

- A Dilatometer test is most useful when used with a CPT.
- It has a very high production rate, but below that of the CPT.

Table 4.11 Dilatometer testing.

Symbol	*Test*
p_o (MPa)	Lift-off pressure (Corrected A – Reading)
p_l (MPa)	Expansion pressure (Corrected B – Reading)
I_D	Material index $(I_D) = (p_l - p_o)/(p_o - u_0)$
u_0	Hydrostatic pore water pressure
E_D	Dilatometer modulus $(E_D) = 34.7\ (p_l - p_o)$
K_D	Horizontal stress index $(K_D) = (p_o - u_0)/\sigma'_{vo}$
σ'_{vo}	Effective vertical overburden stress

4.12 Pressuremeter test

- The Pressuremeter test should be carried out with the appropriate stress range.
- It is useful for in situ measurement of deformation.

Table 4.12 Pressuremeter testing.

Symbol	*Test*
P_o (MPa)	Lift-off pressure
P_L (MPa)	Limit pressure
P_0	Total horizontal stress $\sigma_{ho} = P_0$
E_{PMT}	Young's modulus $(E_{PMT}) = 2(1 + \nu)(V/\Delta V)\Delta P$
ν	Poisson's ratio
V	Current volume of probe $= V_0 + \Delta V$
V_0	Initial probe volume $= V_0$
ΔV	Measured change in volume
ΔP	Change in pressure in elastic region

4.13 Vane shear

- Some shear vanes have a direct read-out (kPa). These are usually limited to shallow depth testing.
- Values change depending on shape of vane.

Table 4.13 Vane shear testing.

Symbol	*Test*
s_{uv} (kPa)	Vane strength $(s_{uv} = 6T_{max}/(7\pi D^3)$ for $H/D = 2$
D	Blade diameter
H	Blade height
T_{max}	Maximum recorded torque
s_{uv} (peak)	Maximum strength
s_{uv} (remoulded)	Remoulded strength (residual value) – vane is rotated through 10 revolutions
μ	Vane shear correction factor
s_{uv} (corr)	$s_{uv}\ (corr) = \mu s_{uv}$

4.14 Vane shear correction factor

- A correction factor should be applied to the vane shear test result for the value to be meaningful.
  - Rate of shear can influence the result.
  - Embankments on soft ground using large equipment are usually associated with 1 week construction time (loading) – 10,000 minutes, Chandler (1988).

Table 4.14 Vane shear correction factor (based on Bjerrum, 1972).

Plasticity index (%)	*Vane correction factor (μ)*
<20%	1.0
30%	0.9
40%	0.85
50%	0.75
60%–70%	0.70
80%–100%	0.65

4.15 Dynamic cone penetrometer tests

- This DCP test is measured in two ways as shown in the table.
- There are variations of the DCP in terms of its hammer weight and drop height. The two variations have similar energy characteristics as shown in Figure 4.3. However the 8 kg hammer has a 30 degree cone.

Table 4.15 Dynamic cone penetrometer tests.

Measurement	*Example*	*Comments*
Blows/100 mm	10 Blows/100 mm	Equivalent reading
Penetration (mm)/blow	10 mm/blow	

  - The DCP is most useful as profiling tool, although it is used to determine the strength properties and with correlations to the CBR. The blows/ 100 mm is the profiling approach, while the penetration/blow is the strength approach.

4.16 Light weight falling deflectometer

- A falling weight deflectometer (FWD) is used on roadway applications to measure deflection over large distances with significant data, while LFWD (Light falling weight deflectometer) is a point data. However, the relatively small frame of the LFWD means it is much easier to move, and more applicable to small projects with site constraints.
- LFWD measures the deflection of the material and estimates the insitu modulus of that material; subgrade and unbound bases. This provides near instantaneous

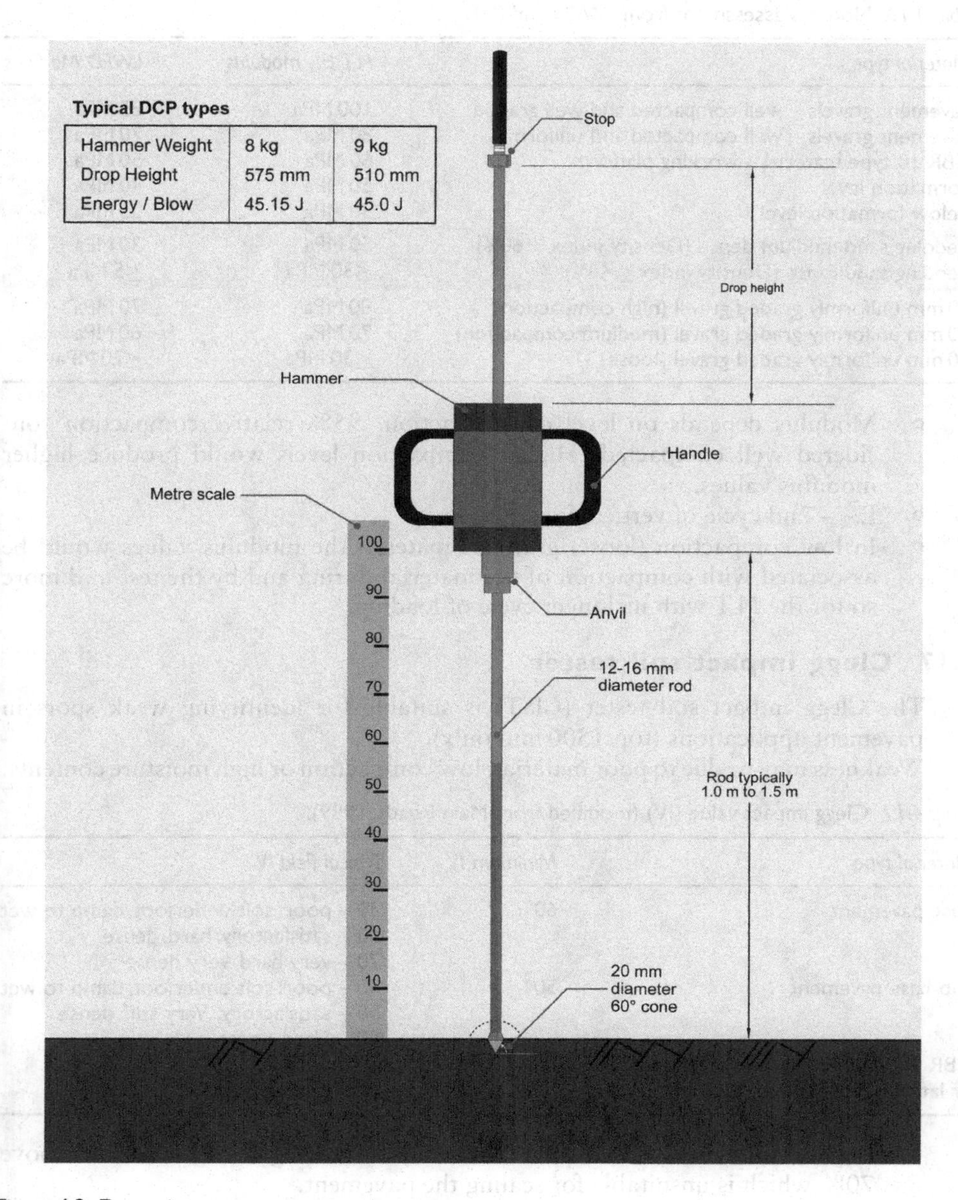

Figure 4.3 Dynamic cone penetrometer test.

advice regarding the comparative stiffness of road formations without the need for any surface disturbance. This may be used for quality control checking purposes in place of CBR tests.

- The PLT modulus is typically 1.15 times the LWFD modulus for gravels – using a Prima 100 equipment (but this can vary for type of material, level of compaction and equipment). The value for a SC fill was 1.7 (Lacey et al., 2013).

Table 4.16 Modulus assessment from LWFD and PLT.

Material type	*PLT E_{v2} modulus*	*LWFD Modulus*
Pavement gravels – well compacted and well graded	100 MPa	85 MPa
Pavement gravels – well compacted and uniform	80 MPa	70 MPa
CBR 10 type material – working platform	60 MPa	50 MPa
Formation level	50 MPa	40 MPa
Below formation level	30 MPa	25 MPa
Bedding sand medium dense (Density index = 60%)	50 MPa	30 MPa
Bedding sand loose (Density index ≤ 40%)	≤30 MPa	≤5 MPa
20 mm uniformly graded gravel (high compaction)	80 MPa	70 MPa
20 mm uniformly graded gravel (medium compaction)	70 MPa	60 MPa
20 mm uniformly graded gravel (loose)	≤30 MPa	≤20 MPa

- Modulus depends on level of compaction. 95% relative compaction considered well compacted. Higher compaction levels would produce higher modulus values.
- E_{v2} – 2nd cycle of vertical loading
- In low compaction (loose) granular material the modulus values would be associated with compaction of the material during and by the test and more so for the PLT with its longer cycle of loading.

4.17 Clegg impact soil tester

- The Clegg impact soil tester (CIST) is suitable for identifying weak spots in pavement applications (top 1500 mm only).
- Weakness may be due to poor material, low compaction or high moisture contents.

Table 4.17 Clegg impact value (IV) (modified from Main Roads, 1989).

Material type	*Minimum IV*	*Typical field IV*
Base pavement	60	30 – poor; soft underfoot, damp to wet 60 – satisfactory; hard, dense 70 – very hard, very dense
Sub-base pavement	50	30 – poor; soft underfoot, damp to wet 50 – satisfactory; Very stiff, dense 60 – Hard, dense
CBR 10 material – working platform or laterite gravels	25	20 – poor 25 – satisfactory

- Impact values below 30 in pavements may be due degree of saturations above 70% which is unsuitable for sealing the pavement.
- Table is for a standard compaction. Lower values apply for modified compaction.
- Clegg hammers come in different sizes (weights) – and different IV then applies. Table is for a standard 4.5 kg hammer but 0.5 kg (light), to 10 kg and 20 kg (heavy) versions are available.

4.18 Surface strength from site walk over

- The pressure exerted by a person walking on the ground is based on their mass and foot size.

Table 4.18 Surface strength from site walk over.

Pressure from person	*Typical undrained shear strength (kPa) support*			*Factor of safety (Bearing)*
	Light	*Medium*	*Heavy*	
Typical pressure	20 kPa	30–40 kPa	50 kPa	
No visible depressions	15 kPa	20–25 kPa	30 kPa	2.0
Some and visible depressions	10 kPa	15–20 kPa	25 kPa	1.5
Large depressions	5 kPa	10–15 kPa	15 kPa	1.0

- For the table:
  - a heavy person is used as above 80 kg with small shoe size.
  - a light is person is below 60 kg with a large shoe size.
- All others are medium pressure
- Very soft clays (<12 kPa) will have some to large depressions even with a light person pressure.
- Soft clays will have visible depressions except for a light person. Depressions for all other persons.
- Firm to stiff clay typically required for most (medium) pressure persons so as not to leave visible depressions.
- A heavy person pressure requires stiff clay, so as not to leave visible depressions.

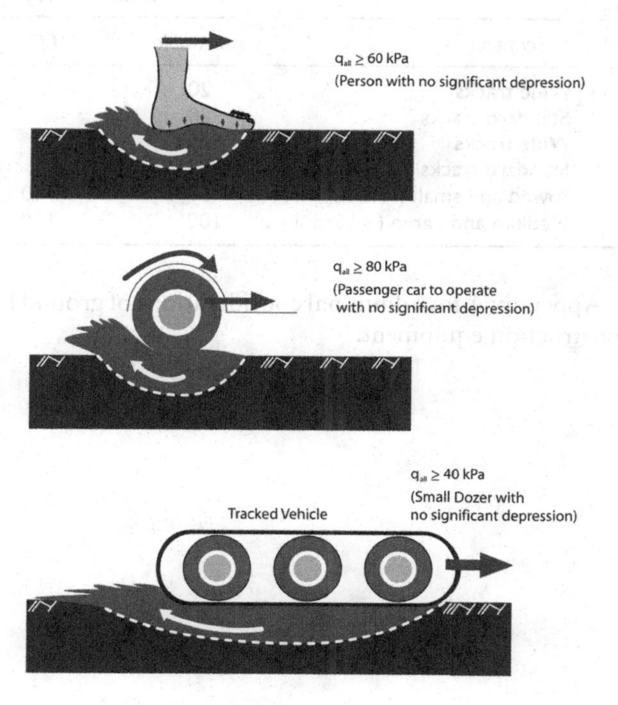

Figure 4.4 Surface depression from human and traffic movement.

4.19 Surface strength from vehicle drive over

- The likely minimum strength of the ground may also be assessed from the type of vehicle used.

Table 4.19 Trafficability of common vehicles.

Vehicle type	*Minimum strength for vehicle to operate*
Passenger car	40 kPa – firm
10 tonne (6 * 4) truck	30 kPa – firm
3 tonne (4 * 4) truck	25 kPa – firm to soft
1 tonne 4 wheel drive vehicle	20 kPa – soft

4.20 Operation of earth moving plant

- Many earth moving equipment use large tyres or tracks to reduce the ground pressure. The table provides the shear strength requirement for such equipment to operate:
  - Feasible – deepest rut of 200 mm after a single pass of machine.
  - Efficient – rut $\leq$50 mm after a single pass.

Table 4.20 Typical strength required for vehicle drive over (from Farrar and Daley, 1975).

Plant		*Minimum shear strength (kPa)*	
Type	*Description*	*Feasible*	*Efficient*
Small Dozer	Wide tracks	20	
	Standard tracks	30	
Large Dozer	Wide tracks	30	
	Standard tracks	35	
Scrapers	Towed and small (<15 m^3)	60	140
	Medium and Large (>15 m^3)	100	170

  - Refer to Appendix A for additional considerations of ground bearing pressures from construction equipment.

Chapter 5

Soil strength parameters from classification and testing

5.1 Errors in measurement

- The industry trend is to minimise laboratory testing in favour of correlations from borelogs. This is driven by commercial incentives to reduce the investigation costs and win the project.
- This approach can often lead to conservative, but sometimes incorrect designs.

Table 5.1 Errors in measurement.

Type of error	*Comment*
Inherent soil variability	Sufficient number of tests can minimise this error.
Sampling error	Correct size sample/type of sampler to account for soil structure and sensitivity. In situ testing for granular material.
Measurement error	Not all test results from even accredited laboratories should be used directly. Sufficient number of laboratory tests to show up "outliers". Understand limitation of the tests. Validate with correlation tests. However appreciate significant variation in correlations.
Statistical variation	Use results knowing that result does vary (Chapter 10). Use of values appropriate to the risk and confidence of test results.

- Clay strength is typically 50% to 100% of value obtained from a 38 mm sample. Larger samples capture the soil structure effect (refer Table 1.16).

5.2 Clay strength from pocket penetrometer

- The pocket penetrometer (PP) is the simplest quantitative test used as an alternative to the tactile classification of strength (Table 2.16).
- The approximation of PP value $= 2\,C_u$ is commonly used. C_u (kPa) $= q_u/2$. However this varies for the type of soil as shown in the table.
- Some considerations in using this tool are:

- It does not consider scale effects.
- Caution on use of results when used in gravelly clays. This is not an appropriate test in granular materials.

Table 5.2 Evaluating strength from PP values (Look, 2004).

Material	*Unconfined compressive strength q_u*
In general	0.8 PP
Fills	1.15 PP
Fissured clays	0.6 PP

- For soils: three pocket penetrometer (PP) readings on undisturbed tube sample (base of tube): Report the PP value – do not convert to a C_u on the borelog.
  - Some field supervisors are known to use the PP on SPT samples – this practice is to be avoided as the PP value is meaningless on a disturbed sample from the effects of SPT driving (Table 4.1). Soft to firm samples are compressed and often provide stiff to very stiff results and hard samples are shattered and also provide stiff to very stiff results.

5.3 Clay strength from SPT data

- As a first approximation $C_u = 5$ SPT is commonly used. This correlation is known to vary from 2 to 8.
- The overburden correction is not required for SPT values in clays.
- Sensitivity of clay affects the results.

Table 5.3 Clay strength from SPT data.

Material	*Description*	*SPT – N (blows/300 mm)*	*Strength*
Clay	Very Soft	≤2	0–12 kPa
	Soft	2–5	12–25 kPa
	Firm	5–10	25–50 kPa
	Stiff	10–20	50–100 kPa
	Very Stiff	20–40	100–200 kPa
	Hard	>40	>200 kPa

- An indication of the variability of the correlation in the literature is as follows:
  - Sower's graphs uses $C_u = 4N$ for high plasticity clays and increasing to about 15N for low plasticity clays.
  - Contrast with Stroud and Butler's (1975) graph which shows Cu = 4.5N for PI >30%, and increasing to Cu = 8N for low plasticity clays (PI = 15%).
- Therefore use with caution, and with some local correlations.

5.4 Residual soils strength from SPT data

- Residual soils are more heterogeneous than transported soils with more variability in properties (Chapter 10).
- The relationship of the previous table is compared for specific residual soils in south east Queensland (Priddle et al., 2013).

Table 5.4 Clay strength from SPT data.

SPT – N (Blows/300 mm)	Strength (kPa) for XW rock type					
	Residual soils	Sandstone	Mudstone	Greywacke	Phyllite	Tuff
<10	<50	<65	<75	–	–	–
10–20	50–125	65–130	75–150	110	125	100
20–40	125–225	130–250	150–275	110	125	100
>40	>225	>250	>275	110	125	100

- Residual soils derived from Phyllite and Tuff has a low relationship with the SPT – N value.
- The Greywacke corrected N-value seemed to be overburden dependent rather than indicative of N-value to strength relationship.

5.5 Clean sand strength from SPT data

- The values vary from corrected to uncorrected N values and type of sand.
- The SPT – value can be used to determine the degree of compactness of a cohesionless soil. However, it is the soil friction angle that is used as the strength parameter.

Table 5.5 Strength from SPT on clean medium size sands only.

Description	Relative density D_r	SPT – N (blows/300 mm)		Strength
		Uncorrected field value	Corrected value	Friction angle
Very loose	<15%	$N \leq 4$	$(N_o)_{60} \leq 3$	$\phi < 28°$
Loose	15–35%	$N = 4–10$	$(N_o)_{60} = 3–8$	$\phi = 28–30°$
Med dense	35–65%	$N = 10–30$	$(N_o)_{60} = 8–25$	$\phi = 30–40°$
Dense	65–85%	$N = 30–50$	$(N_o)_{60} = 25–43$	$\phi = 40–45°$
Very dense	>85%	$N > 50$	$(N_o)_{60} > 43$	$\phi = 45°$

- Reduce ϕ by ~ 5° for clayey sand.
- Increase ϕ by ~ 5° for gravelly sand.

5.6 Fine and coarse sand strength from SPT data

- Fine sands have reduced values from the table above while coarse sand has an increased strength value.
- The corrected N value is used in the table below.

Table 5.6 Strength from corrected SPT value on clean fine and coarse size sands.

Description	*Relative density D_r*	*Corrected SPT – N (blows/300 mm)*			*Strength*
		Fine sand	*Medium*	*Coarse sand*	
V. loose	<15%	$(N_o)_{60} \le 3$	$(N_o)_{60} \le 3$	$(N_o)_{60} \le 3$	$\phi < 28°$
Loose	15–35%	$(N_o)_{60} = 3–7$	$(N_o)_{60} = 3–8$	$(N_o)_{60} = 3–8$	$\phi = 28–30°$
Med dense	35–65%	$(N_o)_{60} = 7–23$	$(N_o)_{60} = 8–25$	$(N_o)_{60} = 8–27$	$\phi = 30–40°$
Dense	65–85%	$(N_o)_{60} = 23–40$	$(N_o)_{60} = 25–43$	$(N_o)_{60} = 27–47$	$\phi = 40–45°$
V. dense	>85%	$(N_o)_{60} > 40$	$(N_o)_{60} > 43$	$(N_o)_{60} > 47$	$\phi = 45–50°$
	100%	$(N_o)_{60} = 55$	$(N_o)_{60} = 60$	$(N_o)_{60} = 65$	$\phi = 50°$

- Above is based on Skempton (1988):
  - $(N_o)_{60}/D_r^2 = 55$ for Fine Sands.
  - $(N_o)_{60}/D_r^2 = 60$ for Medium Sands.
  - $(N_o)_{60}/D_r^2 = 65$ for Coarse Sands.

5.7 Effect of aging

- The SPT in recent fills and natural deposits should be interpreted differently.
- Typically the usual correlations and interpretations are for natural materials. Fills and remoulded samples should be assessed different.

Table 5.7 Effect of aging (Skempton, 1988).

Description	*Age (years)*	*$(N_o)_{60}/D_r^2$*
Laboratory tests	10^{-2}	35
Recent fills	10	40
Natural deposits	$>10^2$	55

- Fills with a corrected N value of 5 can therefore be considered medium dense, while in a natural deposit, this value would be interpreted as a loose sand.

5.8 Effect of angularity and grading on strength

- Inclusion of gradations and particle description on borelogs can influence strength interpretation.
- These two factors combined affect the friction angle almost as much as the density itself as measured by the SPT N-value.

Table 5.8 Effect of angularity and grading on siliceous sand and gravel strength BS 8002 (1994).

Particle description	*Sub division*	*Angle increase*
Angularity	Rounded	A = 0
	Sub-angular	A = 2
	Angular	A = 4
Grading	Uniform soil ($D_{60}/D_{10} < 2$)	B = 0
	Moderate grading ($2 \le D_{60}/D_{10} \le 6$)	B = 2
	Well graded ($D_{60}/D_{10} > 6$)	B = 4

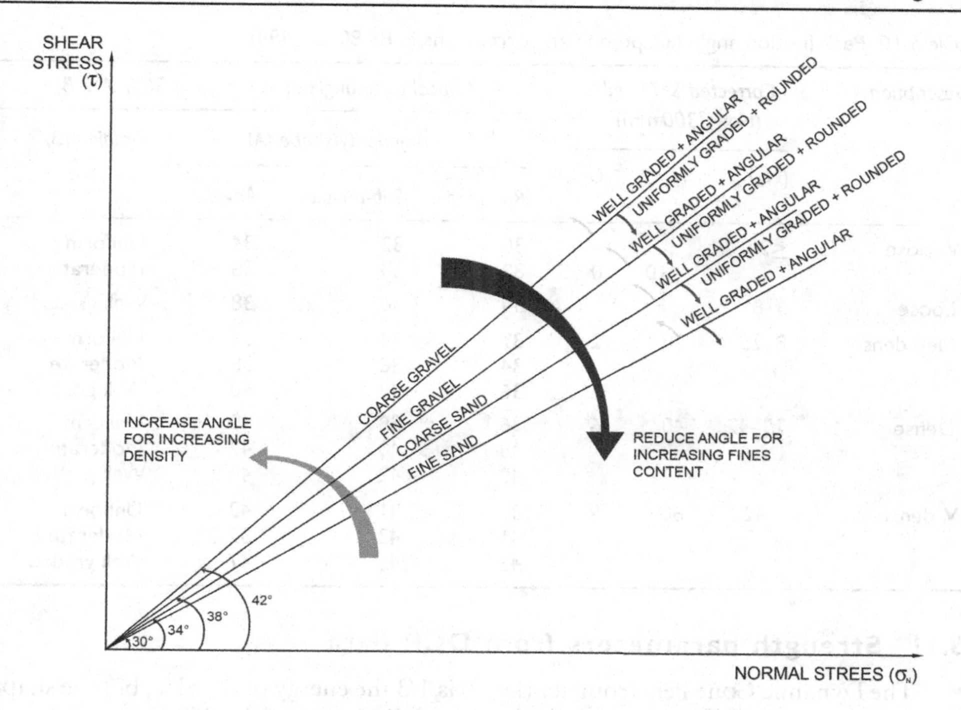

Figure 5.1 Indicative variation of sand friction angle with gradation, size and density.

5.9 Critical state angles in sands

- The critical state angle of soil (ϕ_{crit}) = 30 + A + B.
- This is the constant volume friction angle. The density of the soil provides an additional frictional value but may change depending on its strain level.

Table 5.9 Critical state angle.

Particle distribution Grading		*Critical state angle of soil (ϕ_{crit}) = 30 + A + B*		
		Angularity		
	B	*Rounded* A = 0	*Sub-Angular* A = 2	*Angular* A = 4
Uniform soil ($D_{60}/D_{10} < 2$)	B = 0	30	32	34
Moderate grading ($2 \leq D_{60}/D_{10} \leq 6$)	B = 2	32	34	36
Well graded ($D_{60}/D_{10} > 6$)	B = 4	34	36	38

5.10 Peak and critical state angles in sands

- The table applies for siliceous sands and gravels.
- Using above Table for A and B, the peak friction angle (ϕ_{peak}) = 30 + A + B + C.

Table 5.10 Peak friction angle (adapted from correlations in BS 8002, 1994).

Description	*Corrected SPT – N' (blows/300 mm)*			*Critical state angle of soil* (ϕ_{crit}) = 30 + A + B			
				Angularity/shape (A)			*Grading (B)*
	$(N_o)_{60}$	*N'*	*C*	*Rounded*	*Sub-angular*	*Angular*	
V. loose	≤3			30	32	34	Uniform
		≤10	0	32	34	36	Moderate
Loose	3–8			34	36	38	Well graded
Med dense	8–25	20	2	32	34	36	Uniform
				34	36	38	Moderate
				36	38	40	Well graded
Dense	25–42	40	6	36	38	40	Uniform
				38	40	42	Moderate
				40	42	44	Well graded
V. dense	>42	60	9	39	41	43	Uniform
				41	43	52	Moderate
				43	45	47	Well graded

5.11 Strength parameters from DCP data

- The Dynamic Cone Penetrometer (DCP) is 1/3 the energy of the SPT, but the shape of the cone results in less friction than the Split Spoon of the SPT.
- $n \sim 1/3(N_o)_{60}$ used in the Table below. Due to easier penetration of cone, then nominally less than converted n-values shown for clay strength classification
- The top 0.5 m to 1.0 m of most clay profiles can have a lower DCP value and is indicative of the depth of the desiccation cracks.

Table 5.11 Soil and rock parameters from DCP data.

Material	*Description*	*DCP – n (blows/100 mm)*	*Strength*
Clays	V. soft	0–1	C_u = 0–12 kPa
	Soft	0–1	C_u = 12–25 kPa
	Firm	1–2	C_u = 25–50 kPa
	Stiff	2–5	C_u = 50–100 kPa
	V. stiff	6–9	C_u = 100–200 kPa
	Hard	>10	C_u > 200 kPa
Sands	V. loose	0–1	ϕ < 30°
	Loose	1–3	ϕ = 30–35°
	Med dense	3–8	ϕ = 35–40°
	Dense	8–15	ϕ = 40–45°
	V. dense	>15	ϕ > 45°
Gravels, cobbles, boulders*		>10	ϕ = 35°
		>20	ϕ > 40°
Rock		>10	c' = 25 kPa, ϕ > 30°
		>20	c' > 50 kPa, ϕ > 30°

*Lowest value applies, erratic and high values are common in this material.

- Table should be interpreted left to right for clays. For example, firm clay has n-value of 1 to 2. A value of 1 to 2 is not necessarily firm clay e.g. could also be a stiff to very stiff clay in fills and residual profiles.

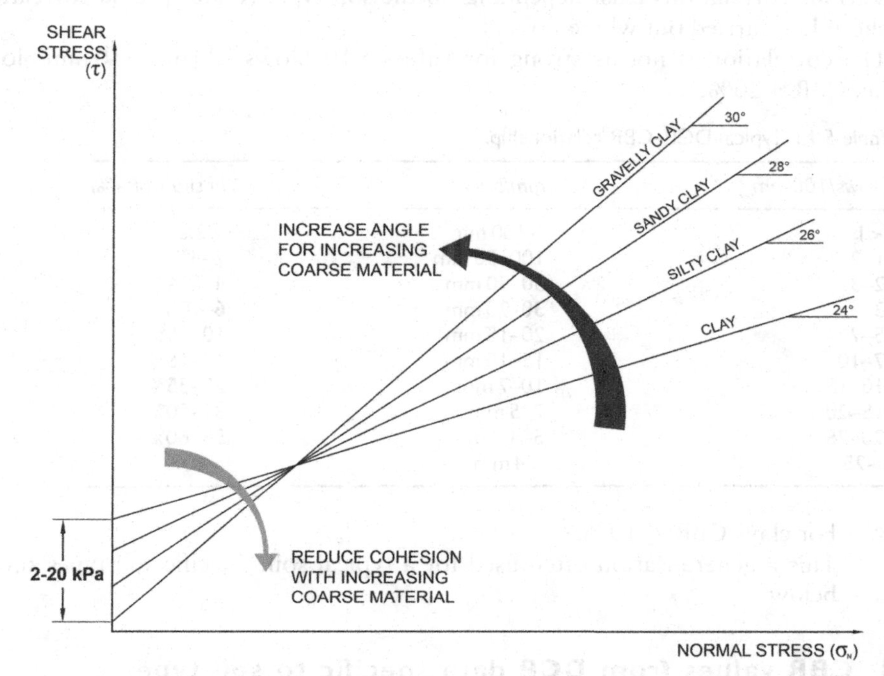

Figure 5.2 Indicative variation of clay strength with changing granular content.

5.12 CBR value from soil classification test

- California Bearing Ratio (CBR) is an index of strength used to assess subgrades and pavement materials.
- Refer to Table 13.6 for further description.

Table 5.12 CBR value from soil classification description.

Description	*Typical CBR (%)*	*Comment*
Very soft	≤1	Remove and replace
Soft	1–2	Needs working platform or geotextile
Firm	2–5	Rutting may occur from construction traffic
Stiff	5–10	
Very stiff	10–20	Working platform quality
Hard	≥20	Sub-base quality

- "Soft spots" are often removed and replaced when identified from proof rolling. This is a non-technical term used in industry that should not be translated as equivalent to a "soft clay" (undrained strength less than 25 kPa). In some case stiff to firm clayey materials are also covered by this term which means "unsuitable" or below the design value.

5.13 CBR value from DCP data

- The DCP is often used for the determination of the in situ CBR.
- Various correlations exist depending on the soil type. A site specific correlation should be carried out where possible.
- The correlation is not as strong for values $\geq$ 10 blows/100 mm (10 mm/blow), i.e. CBR > 20%.

Table 5.13 Typical DCP–CBR relationship.

Blows/100 mm	*mm/blow*	*In situ CBR (%)*
<1	>100 mm	<2%
1–2	100–50 mm	2–4%
2–3	50–30 mm	4–6 %
3–5	30–20 mm	6–10%
5–7	20–15 mm	10–15%
7–10	15–10 mm	15–25%
10–15	10–7 mm	25–35%
15–20	7–5 mm	35–50%
20–25	5–4 mm	50–60%
>25	<4 mm	>60%

- For clays CBR $\leq$ 10%.
- This a generalisation often used for a typical soil. Specific soil types shown below.

5.14 CBR values from DCP data specific to soil type

- General soils – Log(CBR) = 2.465 – 1.12(Log(DCP)) – Webster et al. (1992)
- For various soils types Webster et al. (1994) suggests
  - Gravel, sand, and silt – CBR = $292/(DPI)^{1.12}$. This is similar to Austroads.
  - High plasticity clays CBR = 1/0.002871 DCP. CBR values above 25% not shown in table.
  - Low plasticity clays – CBR = $1/(0.017\ DCP)^2$.

Table 5.14 DCP–CBR relationships for varying soil types (Webster et al. (1992).

Blows/100 mm	*mm/blow*	*In situ CBR (%)*				
		Austroads	*General*	*Gravel, sand, and silt*	*Low plasticity clay*	*High plasticity clay*
<1	>100 mm	<2%	<1%	<2%	<1%	<3%
1–2	100–50 mm	2–4%	1–3%	2–4%	~1%	3–7%
2–3	50–30 mm	4–6%	3–4%	4–6%	1–3%	7–10%
3–5	30–20 mm	6–10%	4–8%	6–10%	3–9%	10–15%
5–7	20–15 mm	10–15%	8–11%	10–15%	9–17%	15%
7–10	15–10 mm	15–25%	11–20%	15–25%	17–25%	15%
10–15	10–7 mm	25–35%	20–30%	25–35%	25%	15%
15–20	7–5 mm	35–50%	30–40%	35–50%	25%	15%
>20	<5 mm	>50%	>40%	>50%	25%	15%

- For subgrades CBR ≤ 25% and 15% for low and high plasticity clays, respectively. Hence upper limit despite above equations.
- Webster's work is based on the 8 kg cone as per Figure 4.3 while Austroads is the 9 kg cone. The energy is similar but cone tip is different.

5.15 Allowable bearing capacity from DCP tests

- The DCP may be used as in the field assessment of allowable bearing capacity for shallow footings.
- Indicative values will vary with soil type.
- Factor of safety of 3 used in Table.

Table 5.15 Allowable bearing capacity from DCP.

Blows/100 mm	*Allowable bearing capacity (kPa)*	*Typical material*
≤1	≤50 kPa	Very soft to soft clays, very loose sands
1–2	50–100 kPa	Firm clays, loose sands
2–5	100–200 kPa	Stiff clays, medium dense sands
6–9	200–400 kPa	Very stiff clays, medium dense to dense sands
≥10	>400 kPa	Hard clays, dense to very dense sands

- For high and low plasticity clays the allowable bearing capacity may be lower and higher, respectively.

5.16 Soil classification from cone penetration tests

- This is an ideal tool for profiling to identify lensing and thin layers.
- It is most useful in alluvial areas.
- The table shows simplified interpretative approach. The actual classification and strength is based on the combination of both the friction ratio and the measured cone resistance, and cross checked with pore pressure parameters.

Table 5.16 Soil classification (adapted from Meigh, 1987 and Robertson et al., 1986).

Parameter	*Value*	*Non-cohesive soil type*	*Cohesive soil type*
Measured cone resistance, q_c	< 1.2 MPa	–	Normally to lightly over consolidated
	> 1.2 MPa	Sands	Over consolidated
Friction Ratio (FR)	< 1.5%	Non-cohesive	–
	> 3.0%	–	Cohesive
Pore pressure parameter, B_q	0.0 to 0.2	Dense sand (q_T > 5 MPa)	Hard/stiff soil (O.C) (q_T > 10 MPa)
	0.0 to 0.4	Medium/loose sand (2 MPa < q_T < 5 MPa)	Stiff clay/silt (1 MPa < q_T < 2 MPa)
	0.2 to 0.8		Firm clay/fine silt (q_T < 1 MPa)
	0.8 to 1.0		Soft clay (q_T < 0.5 MPa)
	>0.8		Very soft clay (q_T < 0.2 MPa)

(Continued)

Table 5.16 (Continued)

Parameter	*Value*	*Non-cohesive soil type*	*Cohesive soil type*
Measured pore pressure (u_d – kPa)	~0	Dense sand ($q_T - P'_o > 12$ MPa) Medium sand ($q_T - P'_o > 5$ MPa) Loose sand ($q_T - P'_o > 2$ MPa)	
	50 to 200 kPa		Silt/stiff clay ($q_T - P'_o > 1$ MPa)
	>100 kPa		Soft to firm clay ($q_T - P'_o < 1$ MPa)

- ○ Applies to electric cone and different values apply for mechanical cones. Refer to Figures 5.3 and 5.4 for different interpretations of the CPT results.

5.17 Soil type from friction ratios

- The likely soil types based on friction ratios only are presented in the table below.
- This is a preliminary assessment only and the relative values with the cone resistance, needs to be also considered in the final analysis.

Table 5.17 Soil type based on friction ratios.

Friction ratio (%)	*Soil type*
<1	Coarse to medium sand
1–2	Fine sand, silty to clayey sands
2–5	Sandy clays, silty clays, clays, organic clays
>5	Peat

5.18 Clay parameters from cone penetration tests

- The cone factor conversion can have significant influence on the interpretation of results.
- For critical conditions and realistic designs, there is a need to calibrate this testing with a laboratory strength testing.

Table 5.18 Clay parameters from cone penetration test.

Parameter	*Relationship*	*Comments*
Undrained strength (c_u – kPa)	$c_u = q_c/N_k$ $c_u = \Delta u/N_u$	Cone factor $(N_k) = 17$ to 20 17–18 for normally consolidated clays 20 for over-consolidated clays Cone factor $(N_u) = 2$ to 8
Undrained strength (c_u – kPa), corrected for overburden	$C_u = (q_c - P'_o)/N'_k$	Cone factor $(N'_k) = 15$ to 19 15–16 for normally consolidated clays 18–19 for over-consolidated clays
Coefficient of horizontal consolidation (c_h – m^2/year)	$c_h = 300/t_{50}$	t_{50} – minutes (time for 50% dissipation)
Coefficient of vertical consolidation (c_v – m^2/year)	$c_h = 2\,c_v$	Value may vary from 1 to 10

5.19 Clay strength from cone penetration tests

- The table below uses the above relationships to establish the clay likely strength.

Table 5.19 Soil strength from cone penetration test.

Soil classification		*Approximate q_c (MPa)*	*Assumptions. Not corrected for overburden*
Very soft	$C_u = 0–12$ kPa	<0.2	$N_k = 17$ (Normally consolidated)
Soft	$C_u = 12–25$ kPa	0.2–0.4	$N_k = 17$ (Normally consolidated)
Firm	$C_u = 25–50$ kPa	0.4–0.9	$N_k = 18$ (Lightly over consolidated)
Stiff	$C_u = 50–100$ kPa	0.9–2.0	$N_k = 18$ (Lightly over consolidated)
Very stiff	$C_u = 100–200$ kPa	2.0–4.2	$N_k = 19$ (Over consolidated)
Hard	$C_u \geq 200$ kPa	>4.0	$N_k = 20$ (Over consolidated)

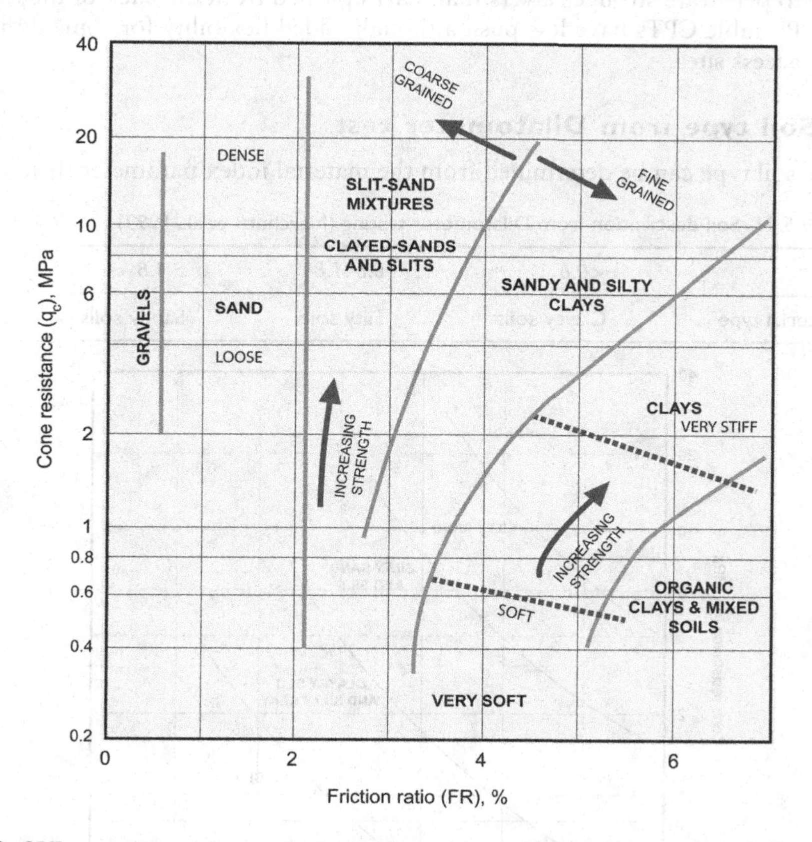

Figure 5.3 CPT properties, and strength changes for mechanical cones (Schertmann, 1978).

5.20 Simplified sand strength assessment from cone penetration tests

- A simplified version is presented below for a preliminary assessment of soil strength in coarse grained material.
- This may vary depending on the depth of the effective overburden and type of coarse grained material.

Table 5.20 Preliminary sand strength from cone penetration tests.

Relative density Dr (%)		*Cone resistance, q_c (MPa)*	*Typical ϕ°*
Very loose	$D_r < 15$	<2.5	<30°
Loose	$D_r = 15–35$	2.5–5.0	30–35°
Med dense	$D_r = 35–65$	5.0–10	35–40°
Dense	$D_r = 65–85$	10–20	40–45°
Very dense	$D_r > 85$	>20	>45°

- The cone may reach refusal in very dense/cemented sands depending on the thrust (weight) of the rigs.
- Rigs with the CPT pushed through its centre of gravity are usually expected to penetrate stronger layers than CPTs pushed from the back of the rigs.
- Portable CPTs have less push although added flexibility for some difficult to access sites.

5.21 Soil type from Dilatometer test

- The soil type can be determined from the material index parameter (I_D).

Table 5.21 Soil description from Dilatometer testing (Marchetti et al., 1993).

I_D	*<0.6*	*0.6–1.8*	*>1.8*
Material type	Clayey soils	Silty soils	Sandy soils

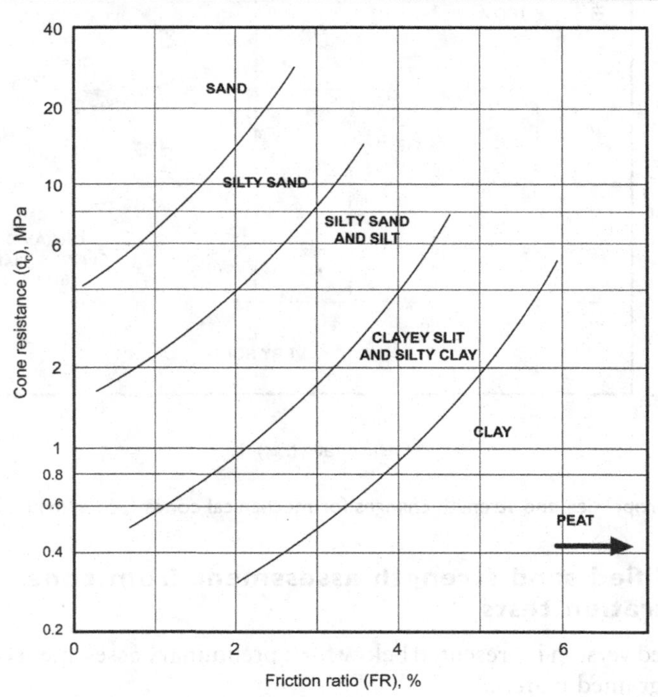

Figure 5.4 CPT properties, and strength changes for electrical cones (Robertson and Campanella, 1983).

5.22 Lateral soil pressure from Dilatometer test

- The DMT can be used to determine the lateral stress.
- Lateral stress coefficient K_o = effective lateral stress/effective overburden stress.
- Lateral Stress index $K_D = (p_o - u_0)/\sigma_{vo}$.

Table 5.22 Lateral soil pressure from Dilatometer test (Kulhawy and Mayne, 1990).

Type of clay	*Empirical parameter β_k*	*Lateral stress coefficient K_o*				
		Formulae	*2*	*5*	*10*	*15*
Insensitive clays	1.5	$(K_D/1.5)^{0.47} - 0.6$	0.5	1.2	1.8	2.4
Sensitive clays	2.0	$(K_D/2.0)^{0.47} - 0.6$	0.4	0.9	1.5	N/A
Glacial till	3.0	$(K_D/3.0)^{0.47} - 0.6$	N/A	0.7	1.2	1.5
Fissured clays	0.9	$(K_D/0.9)^{0.47} - 0.6$	N/A	1.6	2.5	3.2

- $K_D < 2$ indicates a possible slip surface in slope stability investigations (Marchetti et al., 1993).

5.23 Soil strength of sand from Dilatometer test

- Local relationships should always be developed to use with greater confidence.

Table 5.23 Soil strength of sand from Dilatometer testing.

Description	*Strength – relative density D_r (%) and friction angle*		*K_D*
Very loose	$D_r < 15\%$	$\phi < 30°$	<1.5
Loose	$D_r = 15–35\%$	$\phi = 30–35°$	1.5–2.5
Med dense	$D_r = 35–65\%$	$\phi = 35–40°$	2.5–4.5
Dense	$D_r = 65–85\%$	$\phi = 40–45°$	4.5–9.0
Very dense	$D_r > 85\%$	$\phi > 45°$	>9.0

5.24 Clay strength from effective overburden

- This relationship is also useful to determine degree of over consolidation based on measured strength.

Table 5.24 Estimate of a normally consolidated clay shear strength from effective overburden (adapted from Skempton, 1957).

Effective overburden (kN/m³)		*Undrained shear strength of a normally consolidated clay $C_u = (0.11 + 0.0037PI)\ \sigma'_v$*					
	C_u/σ'_v =	*0.18*	*0.26*	*0.30*	*0.33*	*0.41*	*0.48*
	Likely OCR	*<2*		*2–4*		*3–8*	
	PI =	*20%*	*40%*	*50%*	*60%*	*80%*	*100%*
10–50	Very soft to soft	2–9	3–13	3–15	3–17	4–20	5–24
50–100	Very soft to firm	9–18	13–26	15–30	17–33	20–41	24–48
150–200	Firm to Stiff	28–37	39–52	44–59	50–66	61–81	72–96
300	Stiff to very stiff	55	77	89	100	122	144

- For values of $C_u/\sigma'_v > 0.5$, the soil is usually considered heavily over consolidated.
- Lightly over consolidated has OCR 2–4.
- OCR – Overconsolidation ratio.
- Typically $C_u/\sigma'_v = 0.23$ used for near normally consolidated clays (OCR < 2).
- C_u/σ'_v is also dependent on the soil type and the friction angle (refer Chapter 7).

5.25 Variation of undrained strength ratio

- The undrained strength ratio (C_u/σ'_v) provided in the previous table varies based on the test mode and applies for stability problems where simple shear or extension occurs.
- Variation of C_u/σ'_v with test methods/modes has been summarised by Mayne et al., 2009, for normally consolidated Boston Blue Clays.

Table 5.25 Variation of undrained strength ratio with test method (here from Mayne et al., 2009).

Test method	*Mode*	C_u/σ'_v
Plane strain	Compression	0.34
Triaxial	Compression	0.33
Iso-consolidated triaxial	Compression	0.32
Iso-consolidated triaxial	Extension	0.24
Direct simple shear		0.20
Plane strain	Extension	0.19
Unconsolidated undrained		0.185
Triaxial	Extension	0.16
Unconfined	Compression	0.14

Chapter 6

Rock strength parameters from classification and testing

6.1 Rock strength

- There are many definitions of strengths.
- The value depends on the extent of confinement and mode of failure.

Table 6.1 Rock strength descriptors

Rock strength	*Description*
Unconfined compressive strengths	A compression test strength under uniaxial load in an unconfined state UCS or q_u
Intact strength	Intact specimen without any defects
Rock mass strength	Depends on intact strength factored for its defects
Tensile strength	~5% to 25% UCS – use 10% UCS
Flexural strength	~2 × tensile strength
Point load index strengths	~UCS/20 but varies considerably. A tensile test
Brazilian strengths	A tensile test
Schmidt hammer strengths	Rebound value. A hardness test
Soft rock	UCS <10 MPa
Medium rock	UCS $=10$ to 20 MPa
Hard rock, typical concrete strength	UCS ≥ 20 MPa

6.2 Typical refusal levels of drilling rig

- The penetration rate, the type of drilling bit used and the type and size of drilling rig are useful indicators into the strength of material.
- Typical materials and strengths in south east Queensland is shown in the table.

Table 6.2 Typical refusal levels of drilling rigs in south east Queensland.

Property	*Typical material*			
Drill rig	*Weight of rig*	*V – Bit refusal*	*TC – Bit refusal*	*RR – Bit refusal*
Jacro 105	3.15 t	Very stiff to hard clays DCP = 8–10	XW Sandstone DCP = Refusal (~20)	N/A
Gemco HP7/ Jacro 200	6 t	XW Sandstone/Phyllite N* = 40–100	XW Sandstone/DW Phyllite: N* = 50–200	
Jacro 500	12 t	DW Phyllite N* = 50–200	N* = 70–300 I_s (50) ~ 2 MPa	DW Metasiltstone N* = 100–500

- N^* = Inferred SPT N-value.
- Drilling bits:
  - V – bit is hardened steel.
  - TC bit is a tungsten carbide.
  - RR – rock roller.
- DCP = 8–10 is also the limit of hand augering.

6.3 Parameters from drilling rig used

- This table uses the material strength implications from the refusal levels to provide an on-site indicator of the likely bearing capacity – a first assessment only.
- This must be used with other tests and observations.
- The intent throughout this text is to bracket the likely values in different ways, as any one method on its own may be misleading.

Table 6.3 Rock parameters from drilling rig.

Property		*Allowable bearing capacity (kPa)*		
Drill rig	*Weight of rig*	*V – Bit refusal*	*TC – Bit refusal*	*RR – bit refusal*
Jacro 105	3.15 t	300	500	N/A
Gemco HP7/Jacro 200	6 t	450	750	1500
Jacro 500	12 t	600	1000	2000
Typical material		Hard clay: C_u = 250 kPa XW Phyllite	DW Mudstone XW Greywacke	DW Sandstone DW Tuff

- Weight and size of drilling rig has different strength implications.
- Drilling Supervisor should ensure the driller uses different drill bits (T.C./V – Bit) as this is useful information.

6.4 Field evaluation of rock strength

- During the site investigation, various methods are used to assess the intact rock strength.
- Often SPT refusal is one of the first indicators of likely rock. However, the same SPT value in a different rock type or weathering grade may have different strength implications.

Table 6.4 Field evaluation of rock strength.

Strength	*Description*			*Approx. SPT*	*Is (50)*
	By hand	*Point of pick*	*Hammer with hand held specimen*	*N-Value*	*(MPa)*
Extremely Low	Easily crumbled in 1 hand	Crumbles		<100	Generally N/A
Very low				60–150	<0.1
Low	Broken into pieces in 1 hand	Deep indentations to 5 mm		100–350	0.1–0.3
Medium	Broken with difficulty in 2 hands	1 mm to 3 mm indentations	Easily broken with light blow (thud)	250–600	0.3–1

(Continued)

Table 6.4 (*Continued*)

Strength	*Description*			*Approx. SPT N-Value*	*Is (50) (MPa)*
	By hand	*Point of pick*	*Hammer with hand held specimen*		
High			1 firm blow to break (rings)	500–1000	1–3
Very high			> 1 blow to break (rings)		3–10
Extremely high			Many hammer blows to break (rings) – sparks	>600	>10

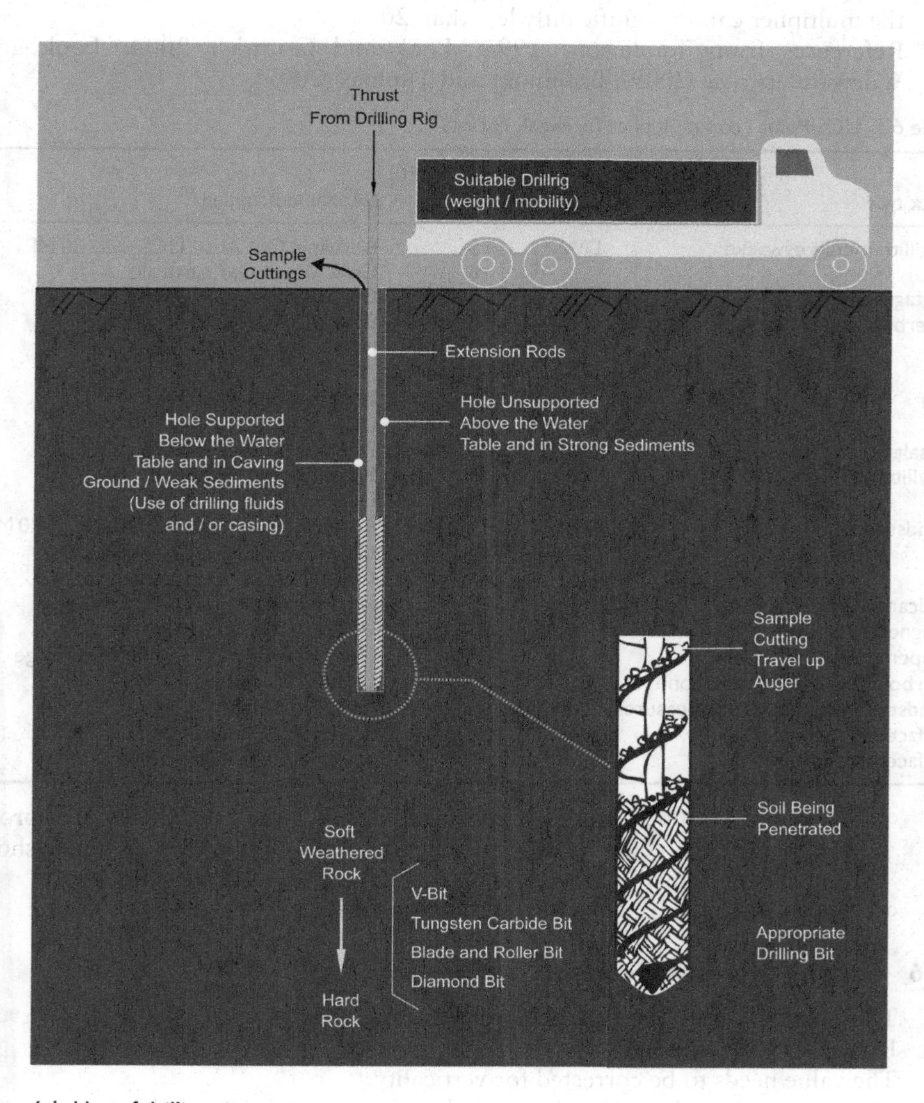

Figure 6.1 Use of drilling rigs.

- Anisotropy of rock material samples may affect the field assessment of strength.
- I_s (50) – Point load index value for a core diameter of 50 mm.
- The unconfined compressive strength is typically about $20 \times I_s$ (50), but the multiplier may vary widely for different rock types.

6.5 Rock strength from point load index values

- Point load index value is an index of strength. It is not a strength value.
- Multiplier typically taken as 23, but 20 as a simple first conversion. This is for high strength (Hard) rock. For lower strength rocks (UCS < 20 MPa, I_s (50) < 1 MPa) the multiplier can be significantly less than 20.
- References from Tomlinson, 1995; Look and Griffiths, 2004; Look and Wijeyakulasuriya (2009); Beaumont and Thomas (2007).

Table 6.5 UCS/Point Load multiplier for weak rocks.

Rock type	*Weathering*	*UCS/I_s (50) ratio*	*Location/description*
Argillite/metagreywacke	DW	5	Brisbane, Qld, Aust. UCS = 2–30 MPa
		8	Gold Coast, Qld, Australia
Metagreywacke	DW	15	Gold Coast, Qld, Australia
Interbedded sandstone – siltstone	DW	28 (axial)	Brisbane River, Qld, Australia
		42 (diametral)	UCS = 10–40 MPa
Tuff	DW	24	Brisbane, Qld, Australia
	SW/FR	18	UCS = 10–80 MPa
Basalt	DW	25	Brisbane, Qld, Australia
Phyllite/arenite	DW	9	Brisbane, Qld, Australia
	SW/FR	4	UCS = 1–20 MPa
Sandstones	DW	12	Brisbane, Q'ld, Australia UCS = 2–20 MPa
		10	Gold Coast, Queensland, Australia
		11	Central Queensland, Australia
Calcarenite limestone		7	Pilbara, W.A., Aust. UCS = 1–3 MPa
Magnesian limestone		25	UCS = 37 MPa average
Upper chalk		18	Humberside/UCS = 3–8 MPa average
Carbonate siltstone/mudstone		12	UAE/UCS = 2 MPa
Mudstone/siltstone (coal measures)		23	UCS = 23 MPa
Tuffaceous rhyolite		10	Korea/UCS = 20–70 MPa
Tuffaceous andesite		10	Korea/UCS = 40–140 MPa

- A ratio of 10 and 20 would be recommended as a non-calibrated first approximation for "soft" and "hard" rocks, respectively. But the values above show that the multiplier is dependent on rock type and is site specific.
- Queensland has a tropical weathered profile.

6.6 Strength from Schmidt hammer

- There are "N" and "L" Type Schmidt Hammers.
- $R_L = 0.605 + 0.677 R_N$.
- The value needs to be corrected for verticality.
- Minimum of 10 values at each sample location. Use 5 highest values.

Table 6.6 Rock strength using Schmidt "N" type hammer.

Strength	*Low*	*Medium*	*High*	*Very high*	*Extremely high*
UCS value (MPa)	<6	6–20	20–60	60–200	>200
Schmidt hammer rebound value	<10	10–25	25–40	40–60	>60
Typical weathering	XW	HW	MW	SW	FR

6.7 Strength assessment from RQD

- The rock quality designation (RQD) was described in section 3.7.
- The RQD can be used to assess the allowable bearing capacity as a first approximation – refer Table 22.1 and is also an indicator of field rock strength albeit biased toward the defects property.

Table 6.7 Strength assessment from RQD.

RQD (%)	*Rock Description*	*Strength*
0–25	Very poor	Very Low. Rock defects governs
25–50	Poor	Low
50–75	Fair	Medium. Strength of concrete
75–90	Good	High. High strength concrete
>90	Excellent	Typically greater than concrete. Intact strength governs

6.8 Relative change in strength between rock weathering grades

- The rock strengths change due to weathering and vary significantly depending on the type of rock.
- Rock weathering by itself, is not sufficient to define a bearing capacity. Phyllites do not show significant change in intact rock strength but often have a significant change in defects between weathering grades.

Table 6.8 Relative change in rock strengths between rock weathering grades (Look and Griffiths, 2004).

Rock		
Type	*Weathering*	*Relative change in intact strength*
Argillite/Greywacke	DW	1
	SW	2
	FR	6
Sandstone/Siltstone	DW	1
	SW	2
	FR	4
Phyllites	DW	1
	SW	1.5
	FR	2
Conglomerate/Agglomerate	DW	1
	SW	2
	FR	4
Tuff	DW	1
	SW	4
	FR	8

- The table shows a definite difference between intact rock strength for SW and FR rock despite that weathering description by definition, suggests that there is little difference in strength in the field (refer Table 3.4).

6.9 Parameters from rock weathering

- A geotechnical engineer is often called in the field to evaluate the likely bearing capacity of a foundation when excavated. Weathering grade is simple to identify, and can be used in conjunction with having assessed the site by other means (intact strength and structural defects).
- The field evaluation of rock weathering in the table presents generalised strengths.
- Different rock types have different strengths e.g. MW sandstone may have similar strength to HW granite. The table is therefore relative for a similar rock type.

Table 6.9 Field evaluation of rock weathering.

Properties	*Weathering*			
	XW	*DW*	*SW*	*FR*
Field description	Total discolouration. Readily disintegrates when gently shaken in water	Discolouration & strength loss, but not enough to allow small dry pieces to be broken across the fabric – MW Broken & crumbled by hand – HW	Strength seems similar to fresh rock, but more discoloured	No evidence of chemical weathering
Struck by hammer		Dull thud	Rings	Rings
q_{all}, other than rocks below	≤1 MPa	HW: 1–2 MPa MW: 2–4 MPa	5–6 MPa	8 MPa
q_{all} of argillaceous, organic & chemically formed sedimentary & foliated metamorphic rocks	≤0.75 MPa	HW: 0.75–1.0 MPa MW: 1.0–1.5 MPa	2–3 MPa	4 MPa

- Including rock type can make a more accurate assessment.
- Use of presumed bearing pressure (q_{all}) from weathering only is simple – but not very accurate – use only for preliminary estimate of foundation size.
- Weathered shales, sandstones and siltstones can deteriorate rapidly upon exposure or slake and soften when in contact with water. Final excavation in such materials should be deferred until just before construction of the retaining wall/foundation is ready to commence.
- Alternatively the exposed surface should be protected with a blinding layer immediately after excavation, provided water build up behind a wall is not a concern.
- A weathered rock can have higher intact rock strength than the less weathered grade of the same rock type, as a result of secondary cementation.

6.10 Rock classification

- The likely bearing capacity can be made based on the rock classification.
- There is approximately a ten-fold increase in allowable bearing capacity from an extremely weathered to a fresh rock.
- The table is for shallow footings.

Table 6.10 Rock classification.

Rock type	*Descriptor*	*Examples*	*Allowable bearing capacity (kPa)*
Igneous	Acid	Granite, Microgranite	800–8000
	Basic	Basalt, Dolerite	600–6000
	Pyroclastic	Tuff, Breccia	400–4000
Metamorphic	Non-Foliated	Quartzite, Gneiss	1,000–10,000
	Foliated	Phyllite, Slate, Schist	400–4000
Sedimentary	Hard	Limestone, Dolomite, Sandstone	500–5000
	Soft	Siltstone, Coal, Chalk, Shale	300–3000

- Intrusive igneous rocks are formed within the earth's crust and are coarse grained (e.g. granite).
- Extrusve igneous rocks form on the earth's crust and are fine grained (e.g. basalt).

6.11 Rock strength from slope stability

- The intact strength between different rock types is shown.
- For this book, the tables that follow are used to illustrate the relative strength. However this varies depending on the reference used.

Table 6.11 Variation of rock strength (Hoek and Bray, 1981).

Uniaxial compressive strength (MPa)	*Strength*	*Rock classification*		
		Sedimentary	*Metamorphic*	*Igneous*
40	Lowest		Phyllites	
50	↑	Clay – Shale		
60		Dolomites		
70		Siltstones	Micaschists	
80				Serpentinites
100			Quartzites	
110		Sandstones	Marbles	
120				Pegmatites
140				Granadiorites
150	↓			Granites
170	Highest			Rhyolites

6.12 Typical field geologist's rock strength

- Another example of rock strength variation, but with some variations to the previous table.

Table 6.12 Variation of rock strength (Berkman, 2001).

Uniaxial compressive strength (MPa)	*Strength*	*Rock classification*		
		Sedimentary	*Metamorphic*	*Igneous*
15	Lowest			Welded Tuff
20	↑	Sandstone		Porphyry
25		Shale		Granadiorite
30		Sandstone		
45		Limestone	Schist	
60		Dolomite		Granadiorite
70			Quartzite	Granite
80				Rhyolite
90		Limestone		Granite
100		Dolomite, Siltstone, Sandstone	Schist	
150				Granite
200	↓		Quartzite	
220	Highest			Diorite

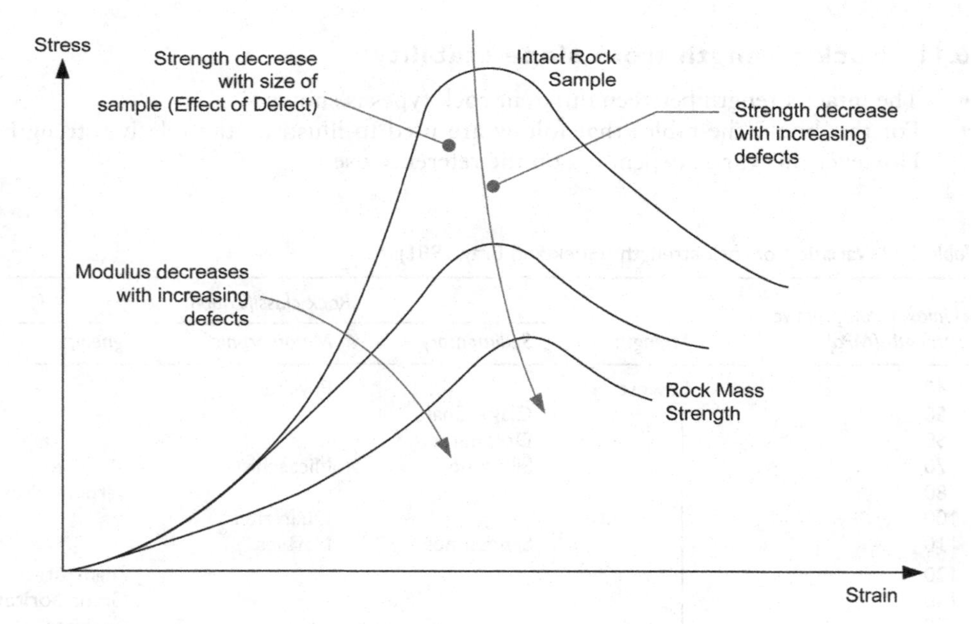

Figure 6.2 Rock type properties.

6.13 Typical engineering geology rock strengths

- Another example of rock strength variation, but with some variations to the previous table.

Table 6.13 Variation of rock strength (Walthman, 1994).

Uniaxial compressive strength (MPa)	*Strength*	*Rock classification*		
		Sedimentary	*Metamorphic*	*Igneous*
10	Lowest	Salt, Chalk		
20	↑	Shale, Coal, Gypsum, Triassic Sandstone, Jurassic Limestone		
40		Mudstone		
60		Carboniferous Sandstone	Schist	
80			Slate	
100		Carboniferous Limestone	Marble	
150		Greywackes	Gneiss	
200	↓			Granite
250	Highest		Hornfels	Basalt

6.14 Relative strength – combined considerations

- The above acknowledges that the description of rock strength from various sources does vary.
- Combining and averaging the rock strengths from various sources is included in this table.
- No doubt the variation is due to region specific results.

Table 6.14 Relative rock strength combining above variations.

Uniaxial compressive strength (MPa)	*Strength*	*Rock classification*		
		Sedimentary	*Metamorphic*	*Igneous*
10	Lowest	Salt, Chalk		Welded Tuff
20	↑	Shale, Coal, Gypsum, (2) Triassic Sandstone, Jurassic Limestone		Porphyry, Granadiorite
40		Mudstone, Sandstone, Clay – Shale	Phyllites	
60		Carboniferous Sandstone, Limestone, (2) Dolomite, Siltstones	(2) Schist, Micaschists	Granadiorite
80			Slate, Quartzite	Granite, Rhyolite Serpentinite
100		(2) Carboniferous Limestone, Dolomite, Siltstone, (2) Sandstone	(2) Marble, Schist Quartzites	Granite, Pegmatites
150		Greywackes	Gneiss	(2) Granite, Granadiorite, Rhyolite
200	↓		Quartzite	Granite, Diorite
250	Highest		Hornfels	Basalt

6.15 Parameters from rock type

- The table below uses the above considerations, by combining intact rock strengths with, rock type, structure and weathering.
- The rock weathering affects the rock strength. This table uses this consideration to provide the likely bearing capacity base on the weathering description, and rock type.
- The design values are a combination of both rock strength and defects.

Table 6.15 Estimate of allowable bearing capacity in rock.

Rock type	*Presumed allowable bearing capacity (kPa)*			
	XW	*DW*	*SW*	*FR*
Igneous				
Tuff	500	1,000	4,000	8,000
Rhyolite, Andesite, Basalt	800	2,000	5,000	9,000
Granite, Diorite	1,000	3,000	7,000	10,000
Metamorphic				
Schist, Phyllite, Slate	400	1,000	2,500	4,000
Gneiss, Migmatite	800	2,500	5,000	8,000
Marble, Hornfels, Quartzite	1,200	4,000	8,000	12,000
Sedimentary				
Shale, Mudstone, Siltstone	400	800	1,500	3,000
Limestone, Coral	600	1,000	2,000	4,000
Sandstone, Greywacke, Argillite	800	1,500	3,000	7,000
Conglomerate, Breccia	1,000	2,000	4,000	8,000

- The Igneous rocks which cooled rapidly with deep shrinkage cracks, such as the Basalts, tend to have a deep weathering profile.
- The foliated metamorphic rocks such as Phyllites can degrade when exposed with a resulting "softening" and loss of strength.

6.16 Rock durability

- Rock durability is important when the rock is exposed for a considerable time (in a cutting) or when to be used in earthworks (breakwater, or compaction).
- Sedimentary rocks are the main types of rocks which may degrade to a soil when exposed, examples:
  - shales, claystone.
  - but also foliated metamorphic rock such as phyllites.
  - and igneous rocks with deep weathering profiles such as basalts.

Table 6.16 Rock degradation (Walkinshaw and Santi, 1996).

Test	*Strong and durable*	*Weak and non-durable*
Point load index (MPa)	>6 MPa	<2 MPa
Free swell (%)	≤4%	>4%

6.17 Material use

- Rocks In-situ can perform differently when removed and placed in earthworks.
- Its behaviour as a soil or rock will determine its slope and compaction characteristics.

Table 6.17 Rock degradation (Strohm et al., 1978).

Test	*Rock like*	*Intermediate*	*Soil like*
Slake durability test (%)	>90	60–90	<60
Jar slake test	6	3–5	≤2
Comments	Unlikely to degrade with time		Susceptible to weathering and long term degradation

Chapter 7

Soil properties and state of the soil

7.1 Soil behaviour

- A geotechnical model is often based on its behaviour as a sand (granular) or a clay (fine grained), with many variations in between these 2 models.
- A soil with a fine content of greater than 30% (depending on the gradation and size of the coarse material) may behave as fine grained material, although it has over 50% granular material.
- The table provides the likely behaviour for these 2 models.

Table 7.1 Comparison of behaviour between sands and clays.

Property	*Sands*	*Clays*	*Comments*
Permeability (k)	High k. Drains quickly (assumes <30% fines).	Low k. Drains slowly (assumes non-fissured or no lensing in clay).	Permeability affects the long term (drained) and short term (undrained) properties.
Effect of time	Drained and undrained responses are comparable.	Drained and undrained response needs to be considered separately.	Settlement and strength changes are immediate in sands, while these occur over time in clays.
Water	Strength is reduced by half when submerged.	Relatively unaffected by short term change in water.	In the long term the effects of consolidation, or drying and wetting behaviour may affect the clay.
Loading	Immediate response. Not sensitive to shape.	Slow response. 30% change in strength from a strip to a square/circular footing.	See Table 21.5 for N_c bearing capacity factor (shape influenced).
Strength	Frictional strength governs.	Cohesion in the short term often dominates, while cohesion and friction to be considered in the long term.	In clay materials both long term and short term analysis are required, while only one analysis is required for sands.
Confinement	Strength increases with confining pressure, and depth of embedment.	Little dependence on the confining pressure. However, some strain	If overburden is removed in sands a considerable loss in strength may occur at

(*Continued*)

Table 7.1 (Continued)

Property	*Sands*	*Clays*	*Comments*
		softening may occur in cuttings and softened strength (cohesion loss) then applies.	the surface. See Table 21.5 for N_q bearing capacity factor (becomes significant at $\phi > 30°$).
Compaction	Influenced by vibration. Therefore a vibrating roller is appropriate.	Influenced by high pressures. Therefore a sheepsfoot roller is appropriate.	Deeper lifts can be compacted with sands, while clays require small lifts. Sands tend to be self-compacting.
Settlement	Occurs immediately (days or weeks) on application of the load.	Has a short and long term (months or years) settlement period.	A self-weight settlement can also occur in both. In clays the settlement is made up of consolidation and creep.
Effect of climate	Minor movement for seasonal moisture changes.	Soil suction changes are significant with volume changes accompanying.	These volume changes can create heave, shrinkage uplift pressures. In the longer term this may lead to a loss in strength.

- In cases of uncertainty of clay/sand governing property, the design must consider both geotechnical models. The importance of simple laboratory classification tests becomes evident.
- Given the distinct behaviour of the two types of soils, then the importance of the soil classification process is self-evident. The requirement for carrying out laboratory classification tests on some samples to validate the field classification is also evident. Yet there are many geotechnical reports that rely only on the field classification due to cost constraints.

7.2 State of the soil

- The state of the soil often governs the soil properties. Therefore any discussion of soil property assumes a given state.

Table 7.2 Some influences of the state of the soil.

	State of soil			
Soil property	*Moisture*	*Compaction*	*OCR*	*Relative influence*
Strength	Dry	High	High	Higher strength
	Wet	Low	Low	Reduced strength
Colour	Dry			Lighter colour
	Wet			Dark colour
Suction	Dry	High	High	High suction
	Wet	Low	Low	Low suction
Density	Dry	High	High	High density
	Wet	Low	Low	Lower density

- OCR – Over-consolidation ratio.
- The above is a relative comparison only for a given soil as clay in a wet state can still have higher soil suction than sand in a dry state.

7.3 Soil weight

- The soil unit weight varies depending on the type of material and its compaction state.
- Rock in its natural state has a higher unit weight than when used as fill (Refer chapters 9 and 12).
- The unit weight for saturated and dry soils varies.

Table 7.3 Representative range of dry unit weight.

Type	*Soil description*	*Unit weight range (kN/m³)*	
		Dry	*Saturated*
Cohesionless Compacted broken rock	Soft sedimentary (chalk, shale, siltstone, coal)	12	18
	Hard sedimentary (Conglomerate, sandstone)	14	19
	Metamorphic	18	20
	Igneous	17	21
Cohesionless Sands and gravels	Very loose	14	17
	Loose	15	18
	Medium dense	17	20
	Dense	19	21
	Very dense	21	22
Cohesionless Sands	Loose – Uniformly graded	14	17
	Loose – Well graded	16	19
	Dense – Uniformly graded	18	20
	Dense – Well graded	19	21
Cohesive	Soft – organic	8	14
	Soft – non organic	12	16
	Stiff	16	18
	Hard	18	20

- Use saturated unit weight for soils below the water table and within the capillary fringe above the water table.
- Buoyant unit weight = Saturated unit weight – unit weight of water (9.81 kN/m³).
- The compacted rock unit weight shown is lower than the in situ unit weight.

7.4 Significance of colour

- The colour provides an indication of likely soil properties.

Table 7.4 Effect of colour.

Colour effect	*Significance*
Light to dark	Increasing moisture content. Dry soils are generally lighter than a wet soil
Black, dark shades of brown and grey	Organic matter likely
Bright shades of brown and grey. Red, yellow and whites	Inorganic soils
Mottled colours	Poor drainage, residual soils
Red, yellow – brown	Presence of iron oxides

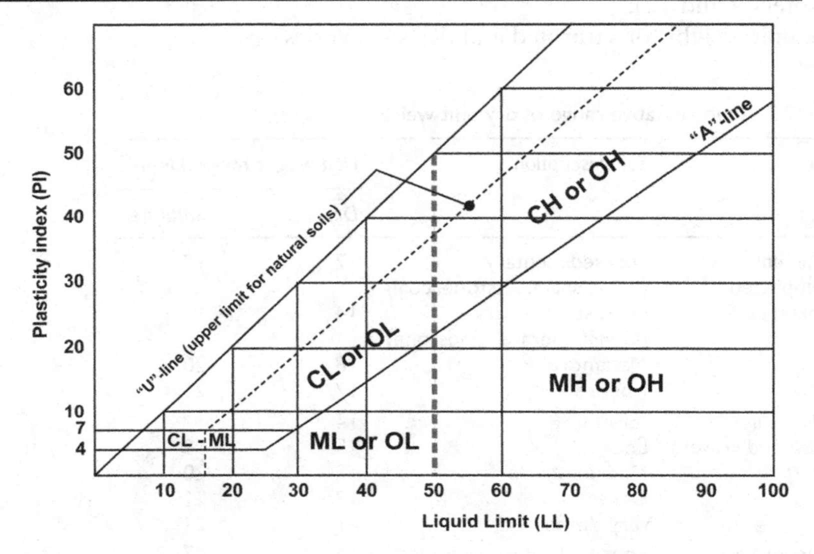

Figure 7.1 Soil plasticity chart.

7.5 Plasticity characteristics of common clay minerals

- Soils used to develop the plasticity chart tended to plot parallel to the A-Line (Refer Figure).
- A-Line divides the clays from the silt in the chart.
- A-Line: PI = 0.73 (LL – 20).
- The upper limit line U-line represents the upper boundary of test data.
- U-Line: PI = 0.9 (LL – 8).

Table 7.5 Plasticity characteristics of common clay minerals (from Holtz and Kovacs, 1981).

Clay mineral	*Plot on the plasticity chart*
Montmorillonites	Close to the U-Line. LL = 30% to Very High LL > 100%
Illites	Parallel and just above the A-Line at LL = 60% ± 30%
Kaolinites	Parallel and at or just below the A-Line at LL = 50% ± 20%
Halloysites	In the general region below the A-Line and at or just above LL = 50%

- Volcanic and bentonite clays plot close to the U Line at very high LL.

7.6 Weighted plasticity index

- The plasticity index by itself can sometimes be a misleading indicator of movement potential.
- The Atterberg test is carried out on the % passing the 425 micron sieve, i.e. any sizes greater than 425 μm is discarded. There have been cases when a predominantly "rocky/granular" site has a high PI test results with over 75% of the material discarded. This is a common occurrence in residual soils.
- The weighted plasticity index (WPI) considers the % of material used in the test.
- WPI = PI × % passing the 425 micron sieve.

Table 7.6 Weighted plasticity index classification (modified from Look, 1994).

Volume change classification	*Weighted plasticity index*
Very low	<1200
Low	1200–2200
Moderate	2200–3200
High	3200–5000
Very high	>5000

7.7 Effect of grading

- The grading affects the strength, permeability and density of soils.
- Different grading requirements apply to different applications.

Table 7.7 Effect of grading.

Grading	*Benefits*	*Application*	*Comments*
Well graded	Low porosity with a low permeability	Structural concrete, to minimize cement content	Well graded U > 5 and C = 1 to 3
Uniformly graded	Single sized or open – graded aggregate has high porosity with a high permeability	Preferred for drainage	Uniform grading U < 2 Moderate grading: 2 < U < 5. Open graded identified by their nominal size through which all of nearly all of material (D_{90})
$P (\%) = (D/D_{max})^n \times 100$ P – % passing size D (mm)	Maximum density	Road base/sub-base specification grading	n = 0.5 (Fuller's curves) D_{max} = maximum particle size
Well graded	Increased friction angle	Higher bearing capacity	Most common application

- D_{90} = 19 mm is often referred to as 20 mm drainage gravel.
- D_{90} = 9.5 mm is often referred to as 10 mm drainage gravel.

7.8 Effective friction of granular soils

- The friction depends on the size and type of material, its degree of compaction and grading.

Table 7.8 Typical friction angle of granular soils.

Type	*Description/state*	*Friction angle (degrees)*
Cohesionless Compacted broken rock	Soft sedimentary (Chalk, shale, siltstone, coal)	30–40
	Hard sedimentary (Conglomerate, sandstone)	35–45
	Metamorphic	35–45
	Igneous	40–50
Cohesionless Gravels	Very loose/loose	30–34
	Medium dense	34–39
	Dense	39–44
	Very dense	44–49
Cohesionless Sands	Very loose/loose	27–32
	Medium dense	32–37
	Dense	37–42
	Very dense	42–47
Cohesionless Sands	Loose – uniformly graded	27–30
	Loose – well graded	30–32
	Dense – uniformly graded	37–40
	Dense – well graded	40–42

- Particle shape (rounded vs. angular) also has an effect, and would change the above angles by about 4 degrees – angular has a high friction as compared to rounded grains.
- Fine, medium or coarse sizes determine the friction angle. The coarser grain sizes have a higher friction angle.
- When the percentage fines exceed 30%, then the fines govern the strength.
- Refer Figure 5.1.

7.9 Effective strength of cohesive soils

- The typical peak strength is shown in the table.
- This should not be confused with the critical state strength which is significantly lower.
- Allowance should be made for long term softening of the clay, with loss of effective cohesion.
- Remoulded strength and residual strength values would have a reduction in both cohesion and friction.

Table 7.9 Effective strength of cohesive soils

Type	*Soil description/state*	*Effective cohesion (kPa)*	*Friction angle (degrees)*
Cohesive	Soft – organic	5–10	10–20
	Soft – non organic	10–20	15–25
	Stiff	20–50	20–30
	Hard	50–100	25–30

- Friction may increase with sand and stone content, and for lower plasticity clays. Refer Figure 5.2.
- When the percentage coarse exceeds 30%, then some frictional strength is present.

- In some cases (e.g. cuttings) the cohesion may not be able to be relied on for the long term. The softened strength then applies. It is not unusual to use much lower values than shown (typical 1 to 10 kPa) for even stiff to hard clays.
- Some practioners advocate a small c′ for over consolidated fissured clays as meaning zero cohesion. Chandler and Skempton (1974) reject that approach as unnecessarily conservative and recommend 1 to 2 kPa for the cohesion.
- Using zero cohesion can create numerical errors or inconsistencies in slope stability analysis software.

7.10 Over-consolidation ratio

- The Over-consolidation ratio (OCR) provides an indication of the stress history of the soil. This is the ratio of its maximum past overburden pressure to its current overburden pressure.
- Material may have experienced higher previous stresses due to water table fluctuations or previous overburden being removed during erosion.

Table 7.10 Over-consolidation ratio.

Overconsolidation ratio (OCR)	$OCR = P'_c/P'_o$
Preconsolidation pressure = Maximum stress ever placed on soil	P'_c
Present effective overburden	$P'_o = \Sigma\gamma' z$
Depth of overlying soil	z
Effective unit weight	γ'
Normally consolidated	OCR ~ 1 but < 1.5
Lightly over-consolidated	OCR = 1.5–4
Heavily over-consolidated	OCR > 4

- For aged glacial clays OCR = 1.5–2.0 for PI > 20% (Bjerrum, 1972).
- Normally consolidated soils can strengthen with time when loaded.
- Over-consolidated soils can have strength loss with time when unloaded (a cutting or excavation) or when high strains apply.

7.11 Pre-consolidation stress from cone penetration testing

- The pre-consolidation stress is the maximum stress that has been experienced in its previous history.
- Current strength would have been based on its past and current overburden.

Table 7.11 Pre-consolidation pressure from net cone tip resistance (from Mayne et al., 2002).

Net cone stress	$q_T - P'_o$	*kPa*	*100*	*200*	*500*	*1000*	*1500*	*3000*	*5000*
Pre-consolidation pressure	P'_c	kPa	33	67	167	333	500	1000	1667
Excess pore water pressure	Δu_1	kPa	67	133	333	667	1000	2000	3333

- For intact clays only.
- For fissured clays $P'_c = 2000$ to 6000 with $\Delta u_1 = 600$ to 3000 kPa.
- The electric piezocone (CPTu) only is accurate for this type of measurement. The mechanical CPT is inappropriate.

7.12 Pre-consolidation stress from Dilatometer

- The Dilatometer should theoretically be more accurate than the CPTu in measuring the stress history. However, currently the CPTu is backed by greater data history with resulting greater prediction accuracy.

Table 7.12 Pre-consolidation pressure from net cone tip resistance (from Mayne et al., 2002).

Net contact pressure	$P_o - u_0$	*kPa*	*100*	*200*	*500*	*1000*	*1500*	*3000*	*5000*
Preconsolidation pressure	P'_c	kPa	50	100	250	500	750	1500	2500

- For intact clays only.
- For fissured clays $P'_c = 1000$ to 5000 with $P_o - u_0 = 600$ to 4000 kPa.

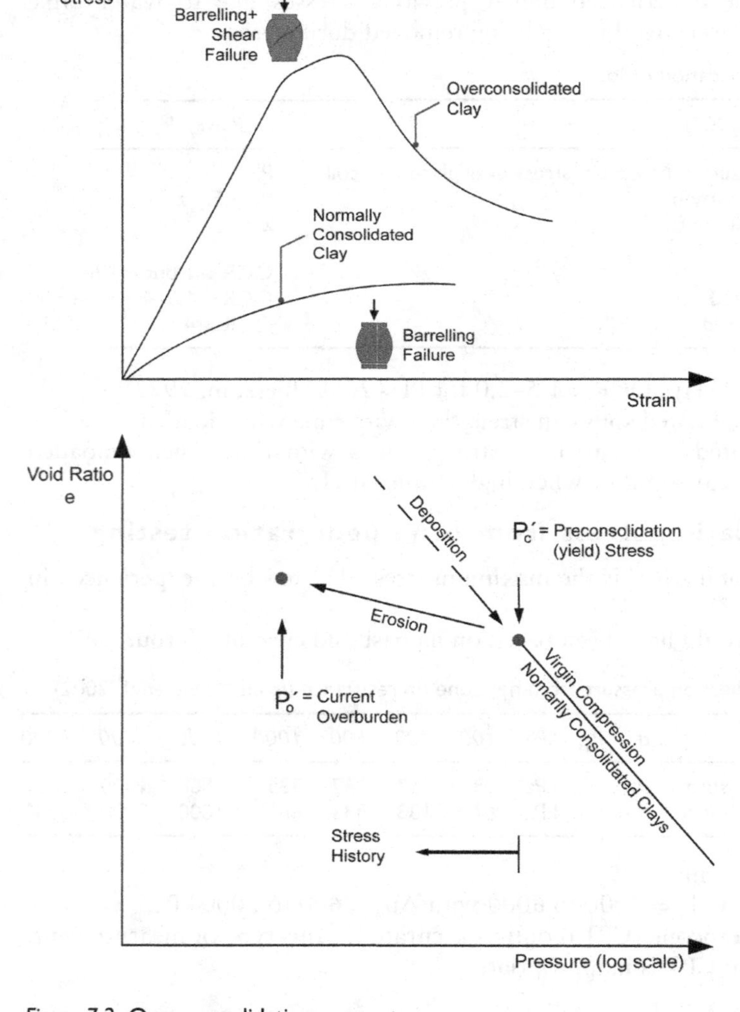

Figure 7.2 Over-consolidation concept.

7.13 Pre-consolidation stress from shear wave velocity

- The shear wave velocity for low pre-consolidation pressures would require near surface (Rayleigh) waves to be used.

Table 7.13 Pre-consolidation pressure from shear wave velocity (from Mayne et al., 2002).

Shear wave velocity	V_s	*m/s*	*20*	*40*	*70*	*100*	*150*	*250*	*500*
Pre-consolidation pressure	P'_c	kPa	9	24	55	92	168	355	984

- For intact clays only.
- For fissured clays $P'_c = 2000$ to 4000 with $V_s = 150$ to 400 m/s.

7.14 Over-consolidation ratio from Dilatometer

- Many correlation exists for OCR to dilatometer measurement of K_D.
- $K_D = 1.5$ for a naturally deposited sand (Normally Consolidated).
- $K_D = 2$ for a normally consolidated clays.
- $OCR = (0.5\ K_D)^{\cdot 56}$ (Kulhawy and Mayne, 1990).
- Table is for insensitive clays only.

Table 7.14 Over-consolidation from Dilatometer testing using the above relationship.

K_D =	*1.5–3.0*	*2.5–6*	*3–8*	*5–10*	*8–20*	*12–35*	*20–50*
OCR	1	2	3	5	10	20	30

- For intact clays only.
- For fissured clays OCR = 25 to 80 with $K_D = 7$ to 20.

7.15 Lateral soil pressure from Dilatometer test

- The Dilatometer is useful to determine the stress history and degree of over consolidation of a soil.

Table 7.15 Lateral soil pressure from Dilatometer test (Kulhawy and Mayne, 1990).

Type of clay	*Empirical parameter β_o*	*Over consolidation ratio (OCR)*				
		Formulae	*2*	*5*	*10*	*15*
Insensitive clays	0.5	$(K_D * 0.5)^{1.56}$	1.0	4.2	12	23
Sensitive clays	0.35	$(K_D * 0.35)^{1.56}$	N/A	2.4	7	13
Glacial till	0.27	$(K_D * 0.27)^{1.56}$	N/A	1.6	4.7	9
Fissured clays	0.75	$(K_D * 0.75)^{1.56}$	1.9	7.9	23	44

- $K_D \sim 2$ or less then the soil is normally consolidated. A useful indicator in determining the slip zones in clays.
- Parameter β_o used in the formulae shown.

7.16 Over consolidation ratio from undrained strength ratio and friction angles

- The friction angle of the soil influences the OCR of the soil.
- Sensitive CH clays are likely to have a lower friction angle.
- CL sandy clays are likely to have the 30 degree friction angles.
- Clayey sands are likely to have the higher friction angles.

Table 7.16 Over-consolidation from undrained strength ratio (after Mayne et al., 2001).

C_u/σ'_v	0.2	0.22	0.3	0.4	0.5	0.7	1.0	1.25	1.5	2.0
Friction angle	*Over-consolidation ratio*									
20°	1.5	1.7	2.3	3.1	3.8	5	8	10	11	15
30°	1.0	1.0	1.4	1.9	2.4	3.3	5	6	7	10
40°	1.0	1.0	1.0	1.4	1.7	2.4	3.5	4	5	7

- Applies for unstructured and uncemented clays.
- Value of 0.22 is the most common value typically adopted, but applies to normally consolidated soils.

7.17 Over-consolidation ratio from undrained strength ratio

- The undrained strength ratio is dependent on the degree of over consolidation.

Table 7.17 Over-consolidation from undrained strength ratio (after Ladd et al., 1977).

Over-consolidation ratio	C_u/σ'_v		
	OH clays	*CH clays*	*CL clays/silts*
1	0.25 to 0.35	0.2 to 0.3	0.15 to 0.20
2	0.45 to 0.55	0.4 to 0.5	0.25 to 0.35
4	0.8 to 0.9	0.7 to 0.8	0.4 to 0.6
8	1.2 to 1.5	0.9 to 1.2	0.7 to 1.0
10	1.5 to 1.7	1.3 to 1.5	0.8 to 1.2

7.18 Sign posts along the soil suction pF scale

- Soil suction occurs in the unsaturated state. It represents the state of the soil's ability to attract water.
- Units are pF or KPa (negative pore pressure). $PF = 1 + \text{Log}\, S$ (kPa).

Table 7.18 Soil suction values (Gay and Lytton, 1972; Hillel, 1971).

Soil suction		State	Soil–plant–atmosphere continuum
pF	kPa		
1	1	Liquid limit	
2	10	Saturation limit of soils in the field	15 kPa for lettuce
3	100	Plastic limit of highly plastic clays	Soil/stem
4	1,000	Wilting point of vegetation (pF = 4.5)	Stem/leaf: 1500 kPa for citrus trees
5	10,000	Tensile strength of water	Atmosphere; 75% relative humidity (pF = 5.6)
6	100,000	Air dry	45% relative humidity
7	1,000,000	Oven dry	

- Equilibrium moisture condition is related to equilibrium soil suction. Refer to section 13.
- Soil suction contributes to strength in the soil. However, this strength cannot be relied upon in the long term and is often not directly considered in the analysis.

7.19 Soil suction values for different materials

- The soil suction depends on the existing moisture content of the soil. This soil–water retention relationship (soil water characteristic curve) does vary depending on whether a wetting or a drying cycle.

Table 7.19 Typical soil suction values for various soils (Braun and Kruijne, 1994).

Volumetric moisture content (%)	Soil suction (pF)		
	Sand	Clay	Peat
0	7.0	7.0	7.0
10	1.8	6.3	5.7
20	1.5	5.6	4.6
30	1.3	4.7	3.6
40	0.0	3.7	3.2
50		2.0	2.8
60		0.0	2.2
70			0.3

- Volumetric moisture content is the ratio of the volume of water to the total volume.
- Soils in its natural state would not experience the soil suction pF = 0, as this is an oven dried condition. Thus for all practical purpose the effect of soil suction in sands are small.
- Greater soil suction produces greater moisture potential change and possible movement/swell of the soil.

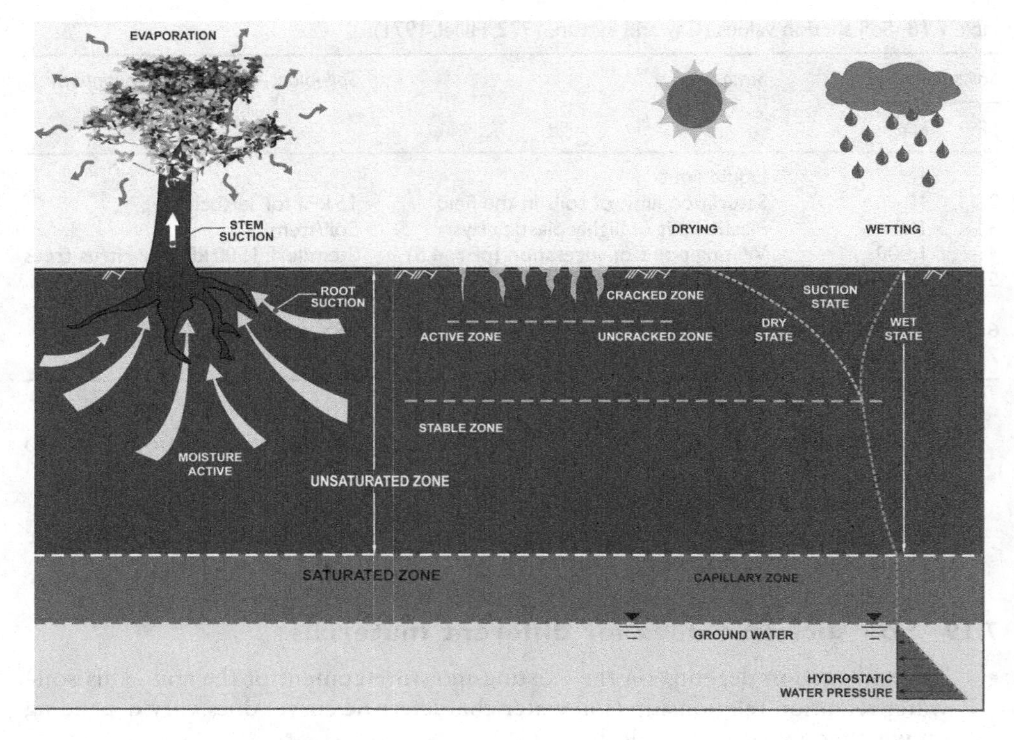

Figure 7.3 Saturated and unsaturated zones.

7.20 Capillary rise

- The capillary rise depends on the soil type, and whether it is in a drying or wetting phase.
- The table presents a typical capillary rise based on the coefficient of permeability and soil type.

Table 7.20 Capillary rise based on the soil type (Vaughan et al., 1994).

Type of soil	*Coefficient of permeability m/s*	*Approximate capillary rise*
Sand	10^{-4}	0.1–0.2 m
Silt	10^{-6}	1–2 m
Clay	10^{-8}	10–20 m

7.21 Equilibrium soil suctions in Australia

- The equilibrium soil suction depends on the climate and humidity.

Table 7.21 Equilibrium soil suctions in Australia (NAASRA, 1972, Australian Bureau of Meteorology).

Location	*Equilibrium soil suction (pF)*	*Climatic environment*	*Annual average rainfall (mm)*
Darwin	2 to 3	Tropical	1666
Sydney	3 to 4	Wet coastal	1220
Brisbane	3 to 4	Wet coastal	1189
Townsville		Tropical	1136
Perth	2 to 3	Temperate	869
Melbourne	2 to 3	Temperate	661
Canberra		Temperate	631
Adelaide	2 to 3	Temperate	553
Hobart	2 to 3	Temperate	624
Alice Springs	>4.0	Semi-arid	274

7.22 Effect of climate on soil suction change

- The larger soil suction changes are expected in the drier climates.

Table 7.22 Soil suction based on climate (AS 2870, 1996).

Climate description	*Soil suction change (Δu, pF)*	*Equilibrium soil suction, pF*
Alpine/wet coastal	1.5	3.6
Wet temperate	1.5	3.8
Temperate	1.2–1.5	4.1
Dry temperate	1.2–1.5	4.2
Semi-arid	1.5–1.8	4.4

7.23 Effect of climate on active zones

- The deeper active zones are expected in drier climates.
- Thornwaithe Moisture Index (TMI) based on rainfall and evaporation rates.

Table 7.23 Active zones based on climate (Walsh et al., 1998).

Climate description	*H_s (metres)*	*Thornwaithe Moisture Index (TMI)*
Alpine/west coastal	1.5	>40
Wet temperate	1.8	10 to 40
Temperate	2.3	−5 to 10
Dry temperate	3.0	−25 to −5
Semi-arid	4.0	<−25

7.24 Compaction concepts

- By compacting at or near the OMC more efficient use of compaction equipment was achieved while checking the effectiveness of the compaction by relating field density to the MDD achieved in the laboratory test for a given level of compaction.
- Regrettably the words "Optimum" and "Maximum" have been translated literally as if it was the "best" condition to target. The words refer to a peak on a graph and the corresponding moisture content – not a best condition. It is however a very useful reference point.

- MDD & OMC is dependent on a specific compactive effort and method of compaction. It does not necessarily reflect what can be achieved in the field. It is not a fundamental soil property. It is a laboratory compaction model. Refer Figure 7.4.

Table 7.24 Compaction concepts.

Key parameter	*Symbol*	*Comment*
Maximum dry density Optimum moisture content	MDD OMC	Can be standard (Proctor) or modified. While MDD is a repeatable parameter the OMC has high variability (Chapter 10). The MDD is a target, but the OMC is only a recommendation for construction efficiencies and workability. The same emphasis should not be placed on both. The target MDD can be achieved at MC other than OMC
Equilibrium moisture content	EMC (% OMC)	EMC is more important than OMC. The OMC does apply to places such as California where it was developed (as EMC ~ OMC). In varying climates in Australia, the OMC only applies at some locations – refer chapter 13
Dry of OMC	% OMC< OMC	Soil lumps hard and stiff, becoming brittle at low MC
Wet of OMC	% OMC> OMC	Soil lumps soft and easy to mold, becoming sticky at high MC
Target density	% MDD	Refer chapter 13. Target density varies with type of construction and material

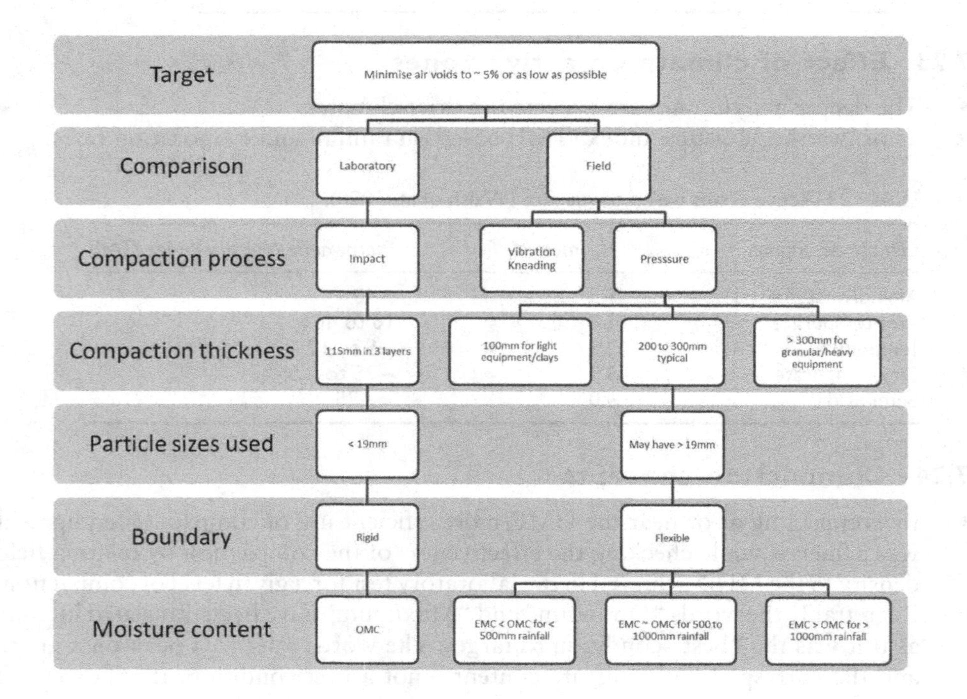

Figure 7.4 Comparison between laboratory and field compaction.

- Because OMC has been used successfully in countries where OMC = EMC, it has wrongly been applied in different climatic environments such as Australia which varies from arid to wet tropical climates.
- Soil suction then results in significant movements if expansive soils are compacted to OMC at such arid or wet tropical climates.

7.25 Effect of compaction on suction

- The compaction affects the soil suction.
- Soils compacted wet of optimum has less suction than those dry of optimum.
- Heavier compaction induces greater soil suction.

Table 7.25 Effect of compaction and suction (Bishop and Bjerrum, 1960; Dineen et al., 1999).

Soil type	*Compaction*	*Moisture content*	*Soil suction*
OMC = 9%–10%		2% Dry of OMC	150 kPa
MDD = 2.05 Mg/m^3	Standard	OMC	30 kPa
		2% Wet of OMC	<10 kPa
Bentonite enriched soil	Standard	% Dry of OMC	550 kPa
		OMC	200 kPa
		2% Wet of OMC	150 kPa
	Modified	% Dry of OMC	
		OMC	1000 kPa
		2% Wet of OMC	

Chapter 8

Permeability and its influence

8.1 Typical values of permeability

- The void spaces between the soil grains allow water to flow through them.
- Laminar flow is assumed.

Table 8.1 Typical values of coefficient of permeability (k).

Soil Type	*Description*		*k, m/s*	*Drainage*
Cobbles and Boulders	Flow may be turbulent, Darcy's law may not be valid		1	
Gravels	Coarse	Uniformly graded coarse aggregate	10^{-1}	Very good
	Clean		10^{-2}	
			10^{-3}	
Gravel Sand Mixtures	Clean	Well graded without fines	10^{-4}	
Sands	Clean, very fine	Fissured, desiccated, weathered	10^{-5}	Good
	Silty	clays	10^{-6}	
	Stratified clay / silts	Compacted clays – dry of optimum	10^{-7}	
Silts	Homogeneous below		10^{-8}	
Clays	zone of weathering	Compacted clays – wet of optimum	10^{-9}	
			10^{-10}	Poor
			10^{-11}	
Artificial	Bituminous, cement stabilized soil		10^{-12}	Practically impermeable
	Geosynthetic clay Liner / Bentonite enriched soil.			
	Concrete			

- Granular material is no longer considered free draining when the fines >15%.
- Granular material is often low permeability (if well compacted) when the fines >30%.

8.2 Permeability equivalents

- While the units typically used for k is m/s this value sometimes lacks meaning to the user. The equivalent values in metres/day or metres/year is tabulated below for ease of use in estimating orders of magnitudes of flows.

Table 8.2 Equivalent permeability.

k (m/s)	*k (m/day)*	*k (m/week)*	*k (m/month)*	*k (m/year)*
1	8.6×10^4	6.0×10^5	2.6×10^6	3.1×10^7
10^{-2}	8.6×10^2	6.0×10^3	2.6×10^4	3.1×10^5
10^{-4}	8.6	6.0×10^1	2.6×10^2	3.1×10^3
10^{-6}	8.6×10^{-2}	6.0×10^{-1}	2.6	3.1×10^1
10^{-8}	8.6×10^{-4}	6.0×10^{-3}	2.6×10^{-2}	3.1×10^{-1}
10^{-10}	8.6×10^{-6}	6.0×10^{-5}	2.6×10^{-4}	3.1×10^{-3}

8.3 Comparison of permeability with various engineering materials

- Material types have different densities.
- Materials with a higher density (for that type) generally have a lower permeability.

Table 8.3 Variability of permeability compared with other engineering materials (Cedergren, 1989).

Material	*Permeability relative to soft clay*
Soft clay	1
Soil cement	100
Concrete	1,000
Granite	10,000
High strength steel	100,000

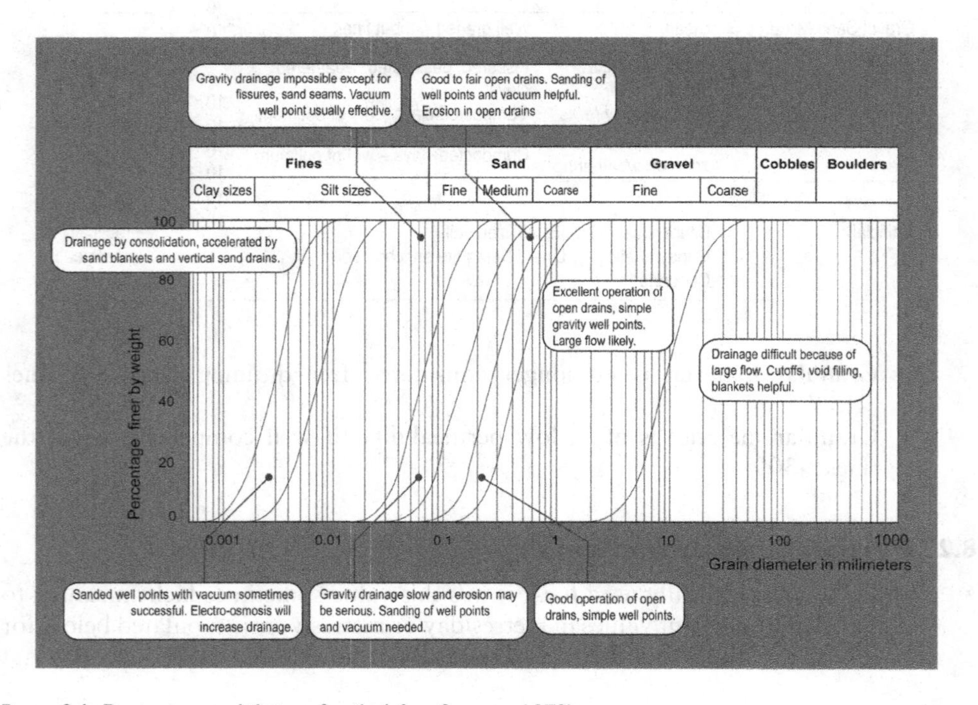

Figure 8.1 Drainage capabilities of soils (after Sowers, 1979).

8.4 Permeability based on grain size

- The grain size is one of the key factors affecting the permeability.
- Hazen Formula applied below is the most commonly used correlation for determining permeability.

Table 8.4 Permeability based on Hazen's relationship.

Coarse grained size	*>Fine sands*			*>Medium sands*				*>Coarse sands*		
Effective grain size d_{10}, mm	0.1	0.2	0.3	0.4	0.5	0.6	0.7	0.8	0.9	1.0
Permeability ($k = Cd_{10}^2$)	10^{-4} m/s				10^{-3} m/s				10^{-2} m/s	
C = 0.10 (above equation)	1	4	0.9	1.6	2.5	3.6	4.9	6.4	0.8	1.0
C = 0.15	1.5	6	1.4	2.4	3.8	5.4	7.4	9.6	1.2	1.5

- Hazen's formula is appropriate for coarse grained soils only (0.1 mm to 3 mm).
- Ideally for uniformly graded material with U < 5.
- Inaccurate for gap graded or stratified soils.

8.5 Permeability based on soil classification

- If the soil classification is known, this can be a first check on the permeability magnitude.

Table 8.5 Permeability based on soil classification.

Soil type	*Description*	*USC symbol*	*Permeability, m/s*
Gravels	Well graded	GW	10^{-3} to 10^{-1}
	Poorly graded	GP	10^{-2} to 10
	Silty	GM	10^{-7} to 10^{-5}
	Clayey	GC	10^{-8} to 10^{-6}
Sands	Well graded	SW	10^{-5} to 10^{-3}
	Poorly graded	SP	10^{-4} to 10^{-2}
	Silty	SM	10^{-7} to 10^{-5}
	Clayey	SC	10^{-8} to 10^{-6}
Inorganic silts	Low plasticity	ML	10^{-9} to 10^{-7}
	High plasticity	MH	10^{-9} to 10^{-7}
Inorganic clays	Low plasticity	CL	10^{-9} to 10^{-7}
	High plasticity	CH	10^{-10} to 10^{-8}
Organic	With silts/clays of low plasticity	OL	10^{-8} to 10^{-6}
	With silts/clays of high plasticity	OH	10^{-7} to 10^{-5}
Peat	Highly organic soils	Pt	10^{-6} to 10^{-4}

- Does not account for structure or stratification.
- Sometimes easier to relate to metres/day.
- For gravels GW – k ~ 864 m/day; GM – k ~ 0.86 m/day while for Silts ML – k ~ 0.008 m/day.
- Based on well compacted.

8.6 Permeability from dissipation tests

- The measurement of in situ permeability by dissipation tests is more reliable than the laboratory testing, due to the scale effects.
- The laboratory testing does not account for minor sand lenses, which can have significant effect on permeability.

Table 8.6 Coefficient of permeability from measured time to 50% dissipation (Parez and Fauriel, 1988).

Hydraulic conductivity, k (m/s)	10^{-3} to 10^{-5}	10^{-4} to 10^{-6}	10^{-6} to 10^{-7}	10^{-7} to 10^{-9}	10^{-8} to 10^{-10}
Soil type	Sand and gravel	Sand	Silty sand to sandy silt	Silt Silt	Clay
t_{50} (sec)	0.1 to 1	0.3 to 10	5 to 70	30 to 7000	>5000
t_{50} (min/hrs)	<0.2 min		0.1 to 1.2 min	0.5 min to 2 hrs	>1.5 hrs

- Pore water pressure u_2 measured at shoulder of piezocone.
- Soil mixtures would have intermediates times.

8.7 Effect of pressure on permeability

- The permeability of coarse materials are affected less by overburden pressure, as compared with finer materials.

Table 8.7 Permeability change with application of consolidation pressure (Cedergren, 1989).

Soil type	*Change in permeability with increase in pressure*		*Comment*
	0.1 kPa	*100 kPa*	
Clean gravel	50×10^{-2} m/s	50×10^{-2} m/s	No change
Coarse sand	1×10^{-2} m/s	1×10^{-2} m/s	
Fine sand	5×10^{-4} m/s	1×10^{-4} m/s	Some change
Silts	5×10^{-6} m/s	5×10^{-7} m/s	
Silty clay	1×10^{-8} m/s	1×10^{-9} m/s	
Fat clays	1×10^{-10} m/s	1×10^{-11} m/s	

8.8 Effect of fines on permeability

- For a washed filter aggregate (sand) example, a 7% fine sand or less material changes permeability by factor of 100 (after Cedergreen, 1989).

Table 8.8 Effect of fines on permeability.

% Passing = 0.15 mm	*Permeability coefficient (m/s)*	*Decrease in permeability*
0	$1.0–0.3 \times 10^{-3}$	–
2	$4.0–0.4 \times 10^{-4}$	~3
4	$2.0–0.1 \times 10^{-4}$	~6
6	$7.0–0.2 \times 10^{-5}$	~15
7	$1.0–0.1 \times 10^{-5}$	~100

8.9 Permeability of compacted clays

- Permeability is a highly variable parameter.
- At large pressure there is a small change in permeability. This minor change is neglected in most analysis.

Table 8.9 Laboratory permeability of compacted Cooroy clays – CH classification (Look, 1996).

Stress range (kPa)	*40–160*	*160–640*	*640–1280*	*1280–2560*
Typical soil depth (m)	2.0–8.0 m	8.0 m–32 m	32–64 m	>64 m
Permeability, k (m/s)	$0.4–70 \times 10^{-10}$	$0.4–6 \times 10^{-10}$	$0.2–0.7 \times 10^{-10}$	$0.1–0.4 \times 10^{-10}$
Median value, k (m/s)	2×10^{-10}	0.8×10^{-10}	0.4×10^{-10}	0.2×10^{-10}

8.10 Effect of moulding water content on permeability

- Compacting wet of optimum reduces permeability.
- Example in table is for a silty clay at modified compaction with maximum dry density of 1885 kg/m^3 and optimum moisture content of 15.0% (from Schor and Gray, 2007).
- Example shows a factor of 100 change in compacting dry of optimum to wet of optimum.

Table 8.10 Effect of moulding moisture content on permeability.

Degree of compaction (% of maximum dry density – Modified)	*Dry side compaction*		*Wet side compaction*	
	Water Content (%)	*Permeability (m/s)*	*Water Content (%)*	*Permeability (m/s)*
98	13	0.5×10^{-6}	16.0	1.0×10^{-8}
96	13	1.0×10^{-6}	17.0	0.8×10^{-8}
94	12.3	2.0×10^{-6}	18.5	0.3×10^{-8}
87	12.4	7.2×10^{-6}	22.5	0.6×10^{-8}

8.11 Permeability of untreated and asphalt treated aggregates

- Permeability of asphalt aggregates is usually high.

Table 8.11 Permeability of untreated and asphalt treated open graded aggregates (Cedergren, 1989).

Aggregate size	*Permeability (m/s)*	
	Untreated	*Bound with 2% Asphalt*
38 mm to 25 mm	0.5	0.4
19 mm to 9.5 mm	0.13	0.12
4.75 mm to 2.36 mm	0.03	0.02

8.12 Dewatering methods applicable to various soils

- The dewatering techniques applicable to various soils depend on its predominant soil type.

Table 8.12 Dewatering techniques (here from Hausmann, 1990; Somerville, 1986).

Predominant soil type	*Clay*	*Silt*	*Sand*	*Gravel*	Cobbles
Grain size (mm)	<0.002	0.06	2	60	>60
Dewatering method	Electro-osmosis	Wells and/or well points with vacuum	Gravity drainage	Subaqueous excavation or grout curtain may be required. Heavy Yield. Sheet piling or other cut off and pumping	
Drainage impractical	⇐	Gravity Drainage too slow	Sump pumping	Range may be extended by using large sumps with gravel filters	

- Refer to Figure 8.1 for the drainage capabilities of soils.
- Well points in fine sands require good vacuum. Typical 150 mm pump capacity: 60 L/s at 10 m ahead.

8.13 Radius of influence for drawdown

- The drawdown at a point produces a cone of depression. This radius of influence is calculated in the table.

Table 8.13 Radius of drawdown (Somerville, 1986).

Drawdown (m)	*Radius of influence (metres) for various soil types and permeability (m/s)*		
	Very fine sands	*Clean sand and gravel mixtures*	*Clean gravels*
	10^{-5} m/s	*10^{-4} m/s*	*10^{-3} m/s*
1	9	30	95
2	19	60	190
3	28	90	285
4	38	120	379
5	47	150	474
7	66	210	664
10	95	300	949
12	114	360	1138
15	142	450	1423

- There is an increase in effective pressure of ground within cone of depression.
- Consolidation of clays if depression is for a long period.
- In granular soils, settlement takes place almost immediately with drawdown.

8.14 Typical hydrological values

- Specific Yield is the % volume of water that can freely drain from rock.

Table 8.14 Typical hydrological values (Waltham, 1994).

Material	*Permeability*		*Specific yield (%)*
	m/day	*m/s*	
Granite	0.0001	1.2×10^{-9}	0.5
Shale	0.0001	1.2×10^{-9}	1
Clay	0.0002	2.3×10^{-9}	3
Limestone (Cavernous)	Erratic		4
Chalk	20	2.3×10^{-4}	4
Sandstone (Fractured)	5	5.8×10^{-5}	8
Gravel	300	3.5×10^{-3}	22
Sand	20	2.3×10^{-5}	28

- An aquifer is a source with suitable permeability that is suitable for groundwater extraction.
- Impermeable rock $k < 0.01$ m/day.
- Exploitable source $k > 1$ m/day.

8.15 Relationship between coefficients of permeability and consolidation

- The coefficient of consolidation (c_v) is dependent on both the soil permeability and its compressibility.
- Compressibility is a highly stress dependent parameter. Therefore c_v is dependent on stress level.

Table 8.15 Relationship between coefficients of permeability and consolidation.

Parameter	*Symbol and relationship*
Coefficient of vertical consolidation	$c_v = k/(m_v \gamma_w)$
Coefficient of permeability	k
Unit weight of water	γ_w
Coefficient of compressibility	m_v
Coefficient of horizontal consolidation	$c_h = 2$ to $10c_v$
Coefficient of vertical permeability	k_v
Coefficient of horizontal permeability	$k_h = 2$ to $10k_v$

- Permeability can be determined from the coefficient of consolidation. This is from a small sample size and does not account for overall mass structure.

8.16 Typical values of coefficient of consolidation

- The smaller value of the coefficient of consolidation produces a longer time for consolidation to occur.

Table 8.16 Typical values of the coefficient of consolidation (Carter and Bentley, 1991).

Soil	*Classification*	*Coefficient of consolidation, c_v, m^2/yr*
Boston blue clay	CL	12±6
Organic silt	OH	0.6–3
Glacial lake clays	CL	2.0–2.7
Chicago silty clays	CL	2.7
Swedish medium	CL–CH	0.1–1.2 (Laboratory)
Sensitive clays		0.2–1.0 (Field)
San Francisco bay mud	CL	0.6–1.2
Mexico city clay	MH	0.3–0.5

8.17 Variation of coefficient of consolidation with liquid limit

- The coefficient of consolidation is dependent on the liquid limit of the soil.
- c_v decrease with strength improvement, and with loss of structure in remoulding.

Table 8.17 Variation of coefficient of consolidation with liquid limit (NAVFAC, 1988).

Liquid Limit, %	*30*	*40*	*50*	*60*	*70*	*80*	*90*	*100*	*110*
	Coefficient of consolidation, c_v, m^2/yr								
Undisturbed – virgin compression	120	50	20	10	5	3	1.5	1.0	0.9
Undisturbed – recompression	20	10	5	3	2	1	0.8	0.6	0.5
Remoulded	4	2	1.5	1.0	0.6	0.4	0.35	0.3	0.25

- LL > 50% is associated with a high plasticity clay/silt.
- LL < 30% is associated with a low plasticity clay/silt.

8.18 Coefficient of consolidation from dissipation tests

- The previous sections discussed the measurement of permeability and the dissipation tests carried out with the piezocone. This also applies to testing for the coefficient of consolidation. The measurement of in situ coefficient of permeability by dissipation tests is more reliable than laboratory testing.
- Laboratory testing does not account for minor sand lenses, which can have a significant effect on permeability.

Table 8.18 Coefficient of consolidation from measured time to 50% dissipation (Mayne, 2002).

Coefficient of Consolidation, C_h	*cm^2/min*	*0.001 to 0.01*	*0.01 to 0.1*	*0.1 to 1*	*1 to 10*	*10 to 200*
	m^2/yr	*0.05 to 0.5*	*0.5 to 5.3*	*5.3 to 53*	*53 to 525*	*525 to 10,500*
t_{50} (mins)		400 to 20,000	40 to 2000	4 to 200	0.4 to 20	0.1 to 2
t_{50} (hrs)		6.7 to 330 hrs	0.7 to 33 hrs	0.1 to 3.3 hrs	<0.3 hrs	

- Pore water pressure u_2 measured at shoulder of $10\,cm^2$ piezocones.
- Multiply by 1.5 for $15\,cm^2$ piezocones.
- Soil mixtures would have intermediates times.

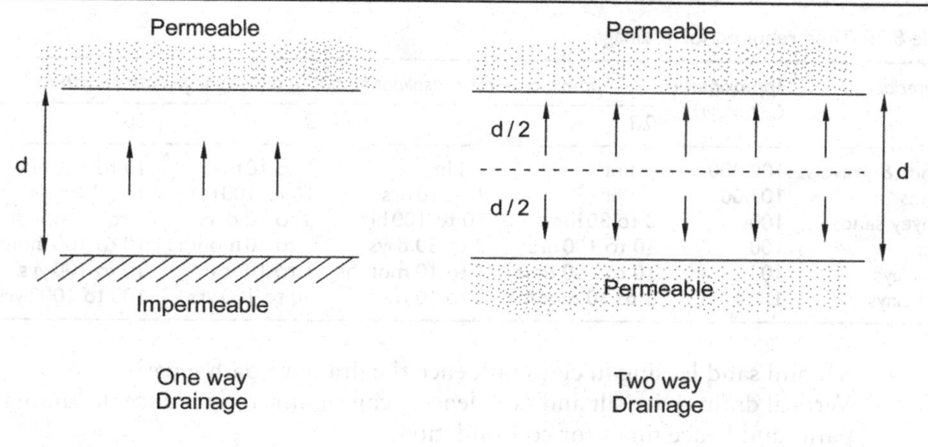

Figure 8.2 Drainage paths.

8.19 Time factors for consolidation

- The time to achieve a given degree of consolidation = t = $T_v d^2/c_v$.
- Time Factor = T_v.
- D = maximum length of the drainage path = ½ layer thickness for drainage top and bottom.
- Degree of consolidation = U = Consolidation settlement at a given time (t)/Final consolidation settlement.
- $\alpha = u_0$ (top)/u_0 (bottom), where u_0 = initial excess pore pressure.

Table 8.19 Time factor values (From NAVFAC DM 7-1, 1982).

Degree of consolidation	*Time factor T_v*		
	$\alpha = 1.0$ (two way drainage)	*$\alpha = 0$ (one way drainage – bottom only)*	*$\alpha = \infty$ (one way drainage – top only)*
10%	0.008	0.047	0.003
20%	0.031	0.100	0.009
30%	0.071	0.158	0.024
40%	0.126	0.221	0.048
50%	0.197	0.294	0.092
60%	0.287	0.383	0.160
70%	0.403	0.500	0.271
80%	0.567	0.665	0.440
90%	0.848	0.940	0.720

8.20 Time required for drainage of deposits

- The drainage time depends on the coefficient of consolidation, and the drainage path.
- t_{90} – time for 90% consolidation to occur.

Table 8.20 Time required for drainage.

Material	*Approximate C_v (m^2/yr)*	*Approx. time for consolidation based on Drainage Path Length (m)*			
		0.3	*1*	*3*	*10*
Sands & gravels	100,000	<1 hr	<1 hr	1 to 10 hrs	10 to 100 hrs
Sands	10,000	<1 hr	1 to 10 hrs	10 to 100 hrs	1 to 10 days
Clayey sands	1000	3 to 30 hrs	10 to 100 hrs	3 to 30 days	1 to 10 months
Silts	100	10 to 100 hrs	3 to 30 days	1 to 10 months	10 to 100 months
CL clays	10	10 to 100 days	1 to 10 months	1 to 10 yrs	10 to 100 yrs
CH clays	1	3 to 30 months	1 to 10 yrs	30 to 100 yrs	100 to 1000 yrs

- Silt and sand lensing in clays influence the drainage path length.
- Vertical drains with silt and sand lensing can significantly reduce the drainage paths and hence times for consolidation.
- Conversely without some lensing wick drains are likely to be ineffective for thick layers, with smearing of the wicks during installation, and possibly reducing the permeability.

8.21 Estimation of permeability of rock

- The primary permeability of rock (intact) condition is several orders less than in situ permeability.
- The secondary permeability is governed by discontinuity frequency, openness and infilling.

Table 8.21 Estimation of secondary permeability from discontinuity frequency (Bell, 1992).

Rock mass description	*Term*	*Permeability (m/s)*
Very closely to extremely closely spaced discontinuities	Highly permeable	10^{-2}–1
Closely to moderately widely spaced discontinuities	Moderately permeable	10^{-5}–10^{-2}
Widely to very widely spaced discontinuities	Slightly permeable	10^{-9}–10^{-5}
No discontinuities	Effectively impermeable	$<10^{-9}$

8.22 Effect of joints on rock permeability

- The width of joints, its openness, and the joint sets determine the overall permeability.
- The likely permeability for various joints features would have most of the following characteristics.

Table 8.22 Effect of joint characteristics on permeability.

Joint openness	*Typical joint characteristics*			*Permeability m/s*
	Filling	*Width*	*Fractures*	
Open	Sands and gravels	>20 mm	≥3 interconnecting Joint sets	$>10^{-5}$
Gapped	Non plastic fines	2–20 mm	1 to 3 interconnecting Joint sets	10^{-5} to 10^{-7}
Closed	Plastic clays	<2 mm	≤1 Joint Sets	$<10^{-7}$

8.23 Lugeon tests in rock

- The Lugeon test (also known as a Packer Test) is a water pressure test, where a section of the drill hole is isolated and water is pumped into that section until the flow rate is constant.
- A Lugeon is defined as the water loss of 1 litre/minute/length of test section at an effective pressure of 1 MPa.
- 1 Lugeon $\sim 10^{-7}$ m/s.

Table 8.23 Indicative rock permeability from the Lugeon test.

Lugeon	*Joint condition*
<1	Closed or no Joints
1–5	Small joint openings
5–50	Some open joints
>50	Many open joints

- Permeability less than 1 Lugeon suggests not enough joints / openings to allow grouting

Chapter 9

Rock properties

9.1 General engineering properties of common rocks

The engineering characteristics are examined from 3 general conditions:

- Competent rock – Fresh, unweathered and free of discontinuities, and reacts to an applied stress as a solid mass.
- Decomposed rock – Weathering of the rock affecting its properties, with increased permeability, compressibility and decrease in strength.
- Non intact rock – Defects in the rock mass governing its properties. Joint spacing, opening, width, and surface roughness are some features to be considered.

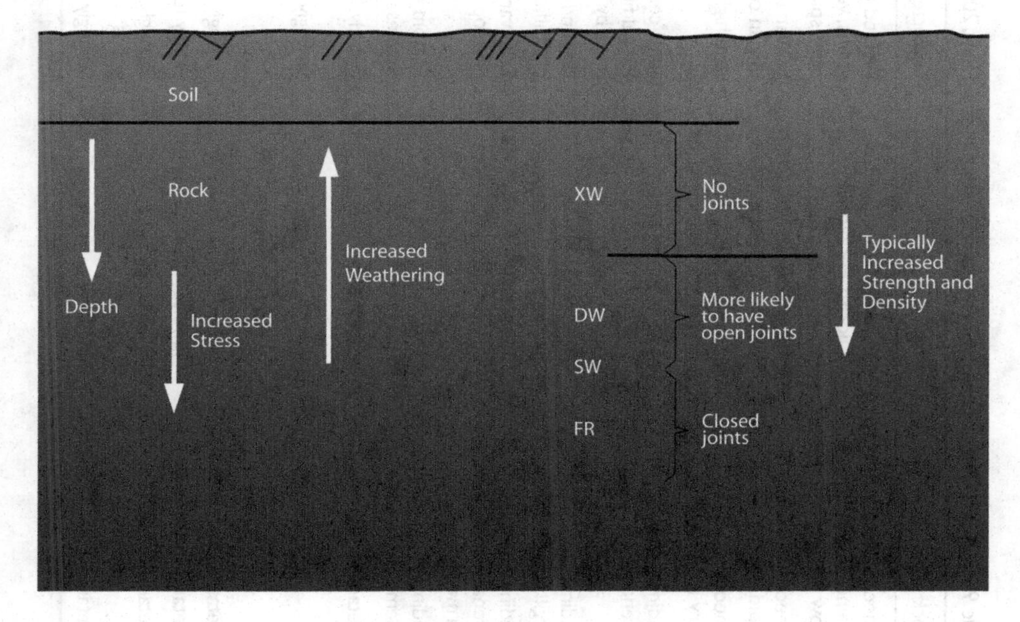

Figure 9.1 Typical changes in rock properties with depth.

Table 9.1 General Engineering properties of common rocks (Hunt, 2005).

Rock origin	*Type*	*Characteristics*	*Permeability*	*Deformability*	*Strength*
Igneous coarse to medium grained – very slow to slow cooling	Granite, Granadiorite, Diorite, Peridiotite	Welded interlocking grains, very little pore space	Essentially impermeable	Very low	Very high
Igneous fine grained – rapid cooling	Rhyolite, Trachyte, Quartz, Dacite, Andesite, Basalt	Similar to above, or can contain voids	With voids can be highly permeable	Very low to low	Very high to high
Igneous Glassy – Very rapid chilling	Pumice, Scoria, Vesicular Basalt	Very high void ratio	Very high	Relatively low	Relatively low
Sedimentary – Arenaceous Clastic	Sandstones	Voids cement filled. Partial filling of voids by cement coatings	Low Very high	Low Moderate to high	High Moderate to low
Sedimentary – Argillaceous Clastic	Shales	Depends on degree of lithification	Impermeable	High to low, can be highly expansive	Low to high
Sedimentary – Arenaceous Chemically formed	Limestone	Pure varieties normally develop caverns	High through caverns	Low except for cavern arch	High except for cavern arch
Sedimentary – argillaceous chemically formed	Dolomite	Seldom develops cavities	Impermeable	Lower than limestone	Higher than limestone
Metamorphic	Gneiss	Weakly foliated	Essentially Impermeable	Low	High
		Strongly foliated	Very low	Moderate normal to foliations. Low parallel to foliations	High normal to foliations Low parallel to foliations
Metamorphic	Schist	Strongly foliated	Low	As for Gneiss	
Metamorphic	Phyllite	Highly foliated	Low	Weaker than Gneiss	
Metamorphic	Quartzite	Strongly welded grains	Impermeable	Very Low	Very High
Metamorphic	Marble	Strongly welded	Impermeable	Very Low	Very High

- The table is for fresh intact condition only.
- Basalts cool rapidly, while Granites cool slowly. The rapid cooling produces temperature induced cracks, which acts as the pathway for deep weathering.

9.2 Rock weight

- The rock unit weight would vary depending on its type, and weathering.
- This table is for intact rock only. Compacted rock would have reduced values.

Table 9.2 Representative range of dry unit weight.

Origin	*Rock type*	*Unit weight range (kN/m³)*			
	Weathering	*XW*	*DW*	*SW*	*Fr*
Sedimentary	Shale	20–22	21–23	22–24	23–25
	Sandstone	18–21	20–23	22–25	24–26
	Limestone	19–21	21–23	23–25	25–27
Metamorphic	Schist	23–25	24–26	25–27	26–28
	Gneiss	23–26	24–27	26–28	27–29
Igneous	Granite	25–27	26–27	27–28	28–29
	Basalt	20–23	23–26	25–28	27–30

- Specific gravity, $G_s = 2.70$ typically, but varies from 2.3 to 5.0.

9.3 Rock minerals

- The rock minerals can be used as a guide to the likely rock properties.
- Rock minerals by itself do not govern strength.
- For example, Hornfels (non foliated) and schists (foliated) are both metamorphic rocks with similar mineralogical compositions, but the UCS strengths can vary by a factor of 4 to 12. Hornfels would be a good aggregate, while schist would be poor as an aggregate.

Table 9.3 Typical predominant minerals in rocks (after Waltham, 1994).

Origin	*Rock type*	*Approximate primary mineralogical composition (secondary minerals not shown to make up 100% of composition)*							
		Quartz	*Feldspar*	*Micas*	*Mafics*	*Calcite*	*Kaolinite*	*Illite*	*Chlorite*
Sedimentary	Sandstone	80%	>10%						
	Limestone					95%			
	Mudstone						20%	60%	
Metamorphic	Schist	25%		35%					20%
	Hornfels	30%		30%					
Igneous	Granite	25%	50%		10%				
	Basalt	<10%	50%		50%				

- ◦ Quartz is resistant to chemical weathering.
- ◦ Feldspar weathers easily into clay minerals.
- ◦ Biotite, Chlorite produces planes of weaknesses in rock mass.

9.4 Silica in igneous rocks

- Silica has been used to distinguish between groups as it is the most important constituent in igneous rocks.

Table 9.4 Silica in igneous rocks (Bell, 1992).

Igneous rock group	*Silica*
Acid/Silicic	>65%
Intermediate	55–65%
Basic/mafic	45–55%
Ultra-basic/ultramafic	<45%

9.5 Hardness scale

- The rock hardness is related to drillability, but is not necessarily a strength indicator.
- Each mineral in scale is capable of scratching those of a lower order.
- Attempts to deduce hardness by summing hardness of rock minerals by its relative proportion has not proved satisfactory.

Table 9.5 Moh's hardness values.

Material	*Hardness*	*Common objects scratched*
Diamond	10	–
Corundum	9	Tungsten Carbide
Topaz	8	
Quartz	7	Steel
Orthoclase	6	Glass
Apatite	5	Penknife scratches up to 5.5
Fluorspar	4	
Calcite	3	Copper coin
Gypsum	2	Fingernail scratches up to 2.5
Talc	1	

9.6 Rock hardness

- Rock hardness depends on mineral presents.

Table 9.6 Typical main mineral hardness values of various rock types (after Waltham, 1994).

Hardness	*Mineral*	*Specific Gravity*	*Origin*		
			Sedimentary	*Metamorphic*	*Igneous*
7	Quartz	2.7	√	√	√
6	Feldspar	2.6		√	√
6	Hematite	5.1	√		
6	Pyrite	5.0	√		
6	Epidote	3.3		√	
5.5	Mafics	>3.0			√
5.0	Limonite	3.6	√	√	
3.5	Dolomite	2.8	√		
3.0	Calcite	2.7	√	√	
2.5	Muscovite	2.8	√	√	√
2.5	Biotite	2.9		√	√
2.5	Kaolinite	2.6	√	√	
2.5	Illite	2.6	√		
2.5	Smectite	2.6	√		
2.0	Chlorite	2.7		√	
2.0	Gypsum	2.3	√		

9.7 Influence of properties on bored pile

- Rate of pile installation depends on rock material properties as well as type of equipment.
- The compressive strength, while the key design parameter needs to be considered with the tensile strength (to break the rock) and the content of abrasive material (wear and tear and production effects) for pile installation.

Table 9.7 Indicative characteristics of rock types and effect on bored pile drilling (Larisch, 2013).

Rock type	*Compressive strength (MPa)*		*Tensile strength (MPa)*		*Content of abrasive mat'l (%)*		*Characteristics*
	Avg.	*min–max*	*Avg.*	*min–max*	*Avg.*	*min–max*	
Quartz	280	240–340	19.3	15.3–24	75	80–100	Very abrasive, homogeneous
Andesite	190	150–240	17.9	12.0–24	27	24–30	High strength, fine grained
Sandstone	170	25–320	17.3	5.1–31	60	30–90	High abrasiveness, massive
Granite	170	85–285	13.8	7.8–23	45	30–60	Hard and massive few cracks
Basalt	145	50–280	14.0	7.3–23	24	17–30	Little abrasiveness, fissures
Limestone	130	75–270	11.9	6.0–19	5	<5–15	Non-abrasive, brittle
Slate or Schist	110	55–250	13.6	9.3–31	39	15–75	Fissures and cracks present
Siltstone	95	30–180	8.5	4.1–19	15	5–33	Inhomogeneous

9.8 Mudstone–shale classification based on mineral proportion

- Shale is the commonest sedimentary rock – characterised by its laminations.
- Mudstones are similar grain size as shales – but non-laminated.
- Shale may contain significant quantities of carbonates.

Table 9.8 Mudstone–shale classification (Spears, 1980).

Quartz content	*Fissile*	*No Fissile*
>40%	Flaggy (parting planes 10–50 mm apart) Siltstone	Massive siltstone
30–40%	Very coarse shale	Very coarse mudstone
20–30%	Coarse shale	Coarse mudstone
10–20%	Fine shale	Fine mudstone
<10%	Very fine shale	Very fine mudstone

9.9 Relative change in rock property due to discontinuity

- The discontinuities in a rock have a significant effect on its engineering properties.
- Rock mass strength = intact strength factored for discontinuities. Similarly for other properties.

Table 9.9 Relative change in rock property.

Rock property	*Change in intact property due to discontinuity*	
	Typical range	*Typical magnitude change*
Strength	1 to 10	5
Deformation	2 to 20	10
Permeability	10 to 1000	100

9.10 Rock strength due to failure angle

- The confining stress affects the rock strength but is not as significant a factor as with the soil strength.
- The table is for zero confining stress.

Table 9.10 Relative strength change due to discontinuity inclination (after Brown et al., 1977).

Angle between failure plane and major principal stress direction	*Major principal Stress at Failure (Relative Change)*	*Comments*
0°	100%	Horizontal
15°	70%	Sub-horizontal
30°	30%	
45°	15%	
60°	20%	
75°	40%	Sub-vertical
90°	70%	Vertical

9.11 Rock defects and rock quality designation

- The RQD is an indicator of the rock fracturing.
- RQD measurement methods do vary. Measure according to the methods described in Chapter 3.

Table 9.11 Correlation between Rock Quality Designation (RQD) and discontinuity spacing.

RQD (%)	*Description*	*Fracture frequency per metre*	*Typical mean discontinuity spacing (mm)*
0–25	Very poor	>15	<60 mm
25–50	Poor	15–8	60–120
50–75	Fair	8–5	120–200
75–90	Good	5–1	200–500
90–100	Excellent	≤1	>500

9.12 Rock laboratory to field strength

- The RQD does not take into account the joint opening and condition.

Table 9.12 Design values of strength parameters (Bowles, 1996).

RQD (%)	*Rock description*	*Field/Laboratory compressive strength*
0–25	Very poor	0.15
25–50	Poor	0.20
50–75	Fair	0.25
75–90	Good	0.3–0.7
>90	Excellent	0.7–1.0

9.13 Rock shear strength and friction angles of specific materials

- The geologic age of the rock may affect the intact strength for sedimentary rocks.
- The table assumes fresh to slightly weathered rock.
- More weathered rock can have significantly reduced strengths.

Table 9.13 Typical shear strength of intact rock.

Origin	*Rock type*	*Shear strength*	
		Cohesion (MPa)	*Friction angle°*
Sedimentary – soft	Sandstone (Triassic), coal, chalk, shale, Limestone (Triassic)	1–20	25–35
Sedimentary – hard	Limestone, dolomite, Greywacke Sandstone (Carborniferous), Limestone (Carborniferous)	10–30	35–45
Metamorphic – non foliated	Quartzite, Marble, Gneiss	20–40	30–40
Metamorphic – foliated	Schist, Slate, Phyllite	10–30	25–35
Igneous – acid	Granite	30–50	45–55
Igneous – basic	Basalt	30–50	30–40

9.14 Rock shear strength from RQD values

- The rock strength values from RQD can be used in rock foundation bearing capacity assessment.

Table 9.14 Rock mass properties (Kulhaway and Goodman, 1988).

RQD (%)	*Rock mass properties*		
	Design compressive strength	*Cohesion*	*Angle of friction*
0–70 (Very poor to fair)	$0.33q_u$	$0.1q_u$	30°
70–100 (Good to excellent)	$0.33–0.8q_u$	$0.1q_u$	30–60°

- q_u = UCS = Uniaxial Compressive Strength of intact rock core.
- When applied to bearing capacity equations for different modes of failure (refer later chapters), the design compressive strength seems to be high. Chapter 22 provides comparative values.

9.15 Rock shear strength and friction angles based on geologic origin

- The geology determines the rock strength.
- Values decrease as the weathering increases.

Table 9.15 Likely shear strength of intact fresh to slightly weathered rock.

Origin	*Grain type*	*Rock type*	*Shear strength*	
			Cohesion (MPa)	*Friction angle(°)*
Sedimentary	Rudaceous (>2 mm)	Clastic	30	45
		Chemically formed	20	40
		Organic remains	10	40
	Arenaceous (0.06–2 mm)	Clastic	15	35
		Chemically formed	10	35
		Organic remains	5	35
	Argillaceous (>2 mm)	Clastic	5	25
		Chemically formed	2	30
		Organic remains	1	30
Metamorphic	Coarse	Foliated	20	35
		Non-foliated	30	40
	Medium	Foliated	10	30
		Non-foliated	15	35
	Fine	Foliated	2	25
		Non-foliated	5	30
Igneous	Coarse (Large intrusions)	Pyroclastic	20	40
		Non pyroclastic	40	50
	Medium (Small intrusions)	Pyroclastic	10	35
		Non pyroclastic	30	45
	Fine (Extrusions)	Pyroclastic	5	30
		Non pyroclastic	20	40

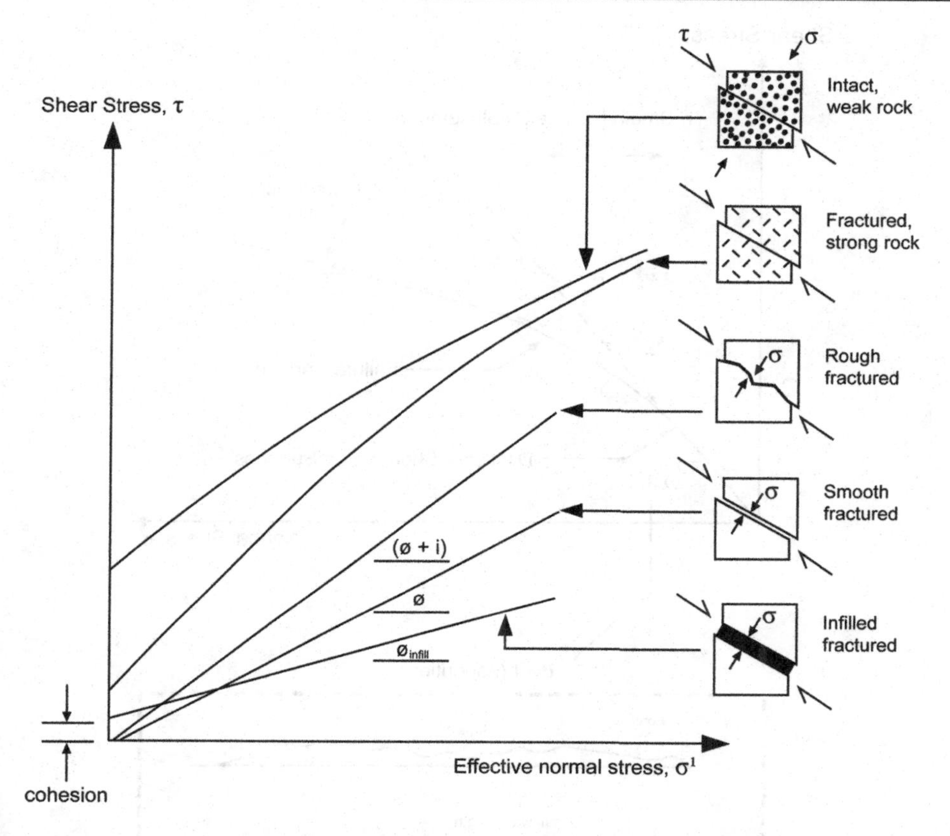

Figure 9.2 Variation of rock strength for various geological conditions (TRB, 1996).

9.16 Friction angles of rocks joints

- At rock joints the friction angle is different from the intact friction angles provided in the previous tables.

Table 9.16 Typical range of friction angles (TRB, 1990).

Rock class	*Friction angles range (degrees)*	*Typical rock types*
Low Friction	20 to 27	Schists, Shale
Medium Friction	27 to 34	Sandstones, Siltstone, Chalk, Gneiss, Slate
High Friction	34 to 40	Basalt, Granite, Limestone, Conglomerate

- Effective rock friction angle = Basic friction angle (ϕ) + Roughness angle (i).
- Above table assumes no joint infill is present.

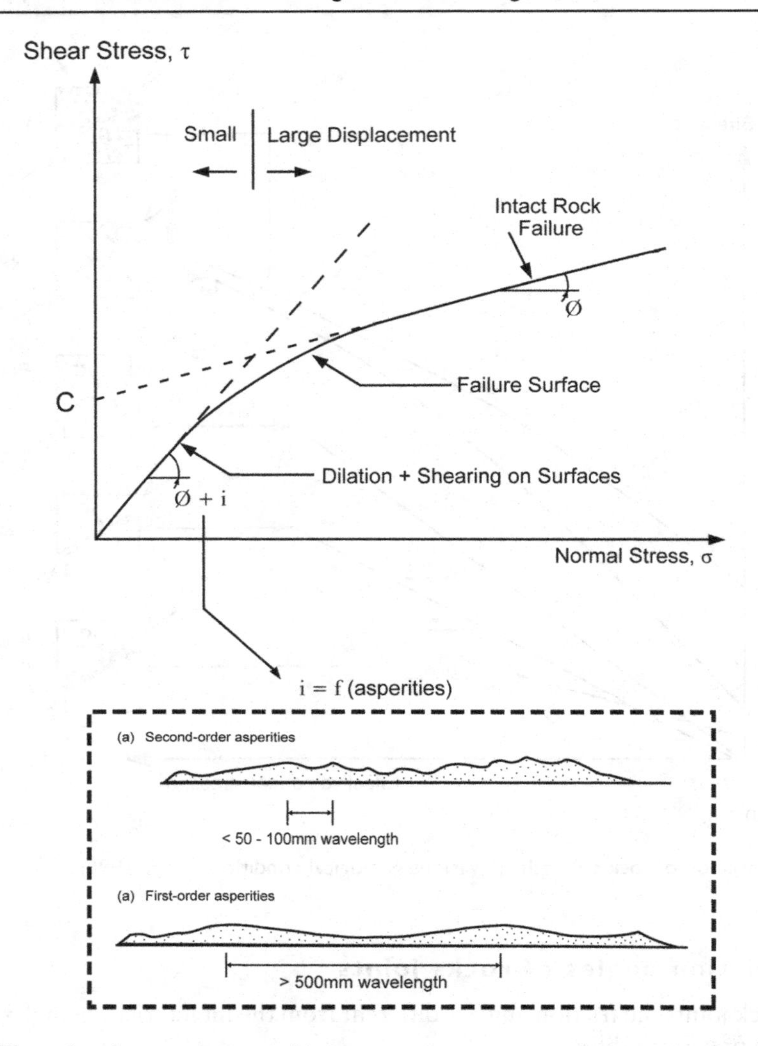

Figure 9.3 Effect of surface roughness on friction.

9.17 Asperity rock friction angles

- The wavelength of the rock joint determines the asperity angle.

Table 9.17 Effect of asperity on roughness angles, (Patton, 1966).

Order of asperities	*Wavelength*	*Typical asperity angle (i°)*
First	500 mm	10 to 15
Second	<50 to 100 mm	20 to 30

9.18 Shear strength of filled joints

- The infill of the joints can affect the friction angle.
- If movements in clay infill have occurred then the residual friction angle is relevant.

Table 9.18 Shear strength of filled joints (Barton, 1974).

Material	*Description*	*Peak*		*Residual*	
		c (kPa)	$\phi°$	c_r *(kPa)*	$\phi_r°$
Granite	Clay filled joint	0–100	24–45		
	Sand-filled joint	50	40		
	Fault zone jointed	24	42		
Clays	Overconsolidated clays	180	12–18	0–30	10–16

Chapter 10

Material and testing variability with risk assessment

10.1 Variability of materials

- Nature offers a significantly larger variability of soil and rock than man-made materials.
- A structural engineer can therefore predict with greater accuracy the performance of the structural system.

Table 10.1 Variability of materials (Harr, 1996).

Material	*Coefficient of variation*	*Comments*
Structural steel – tension members	11%	Man made
Flexure of reinforced concrete – Grade 60	11%	
Flexure of reinforced concrete – Grade 40	14%	
Flexure strength of wood	19%	Nature resistance
Standard penetration test	26%	Field testing
Soils – unit weight	3%	Nature
Friction angle – sand	12%	
Natural water content (silty clay)	20%	
Undrained shear strength, C_u	40%	
Compression index, C_c	30%	

- Coefficient of variation (%) = Standard deviation/Mean.
- For a wind loading expect COV > 25%.

10.2 Variability of soils

- The variability of the soil parameters must always be at the forefront in assessing its relevance, and emphasis to be placed on its value.

Table 10.2 Variability of soils (Kulhawy, 1992).

Property	*Test*	*Mean COV without outliers*
Index	Natural moisture content, w_n	17.7
	Liquid Limit, LL	11.1
	Plastic Limit, PL	11.3
	Initial void ratio, e_o	19.8
	Unit weight, γ	7.1

(Continued)

Table 10.2 (Continued)

Property	*Test*	*Mean COV without outliers*
Performance	Rock uniaxial compressive strength, q_u	23.0
	Effective stress friction angle, ϕ'	12.6
	Tangent of ϕ'	11.3
	Undrained shear strength C_u	33.8
	Compression index C_c	37.0

- Greater confidence can be placed on index parameters than strength and deformation parameters.
- This does not mean that strength correlations derived from index parameters are more accurate, as another correlation variable is now introduced.

10.3 Variability of in-situ tests

- The limitations of in-situ test equipment needs to be understood.
- The likely measurement error needs to be considered with the inherent soil variability.
- The SPT is a highly variable in-situ test.
- Electric cone penetrometer and Dilatometer has the least variability.
- The table shows cumulative effect of equipment, procedure, random variability.

Table 10.3 Variability of in-situ tests (Phoon and Kulhawy, 1999).

Test	*Coefficient of variation (%)*
Standard penetration test	15–45
Mechanical cone penetration test	15–25
Self-boring pressure meter test	15–25
Vane shear test	10–20
Pressure meter test, prebored	10–20
Electric cone penetration test	5–15
Dilatometer test	5–15

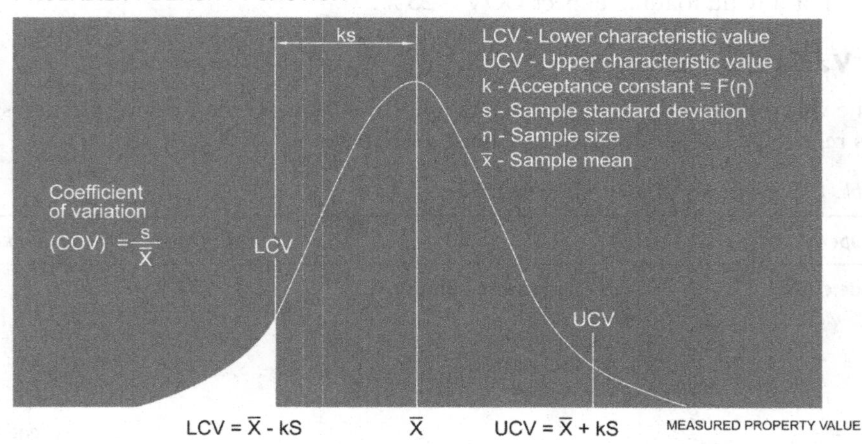

Figure 10.1 Normal distribution of properties.

10.4 Soil variability from laboratory testing

- The density of soils can be accurately tested.
- There is a high variability on the shear strength test results of clays and the Plasticity Index.

Table 10.4 Variability from laboratory testing (Phoon and Kulhawy, 1999).

Test	Property	Soil type	Coefficient of variation (%)	
			Range	Typical mean
Atterberg tests	Plasticity index	Fine grained	5–51	24
Triaxial compression	Effective angle of friction	Clay, silt	7–56	24
Direct shear	Shear strength, C_u	Clay, silt	19–20	20
Triaxial compression	Shear strength, C_u	Clay, silt	8–38	19
Direct shear	Effective angle of friction	Sand	13–14	14
Direct shear	Effective angle of friction	Clay	6–22	14
Direct shear	Effective angle of friction	Clay, silt	3–29	13
Atterberg tests	Plastic limit	Fine grained	7–18	10
Triaxial compression	Effective angle of friction	Sand, silt	2–22	8
Atterberg tests	Liquid limit	Fine grained	3–11	7
Unit weight	Density	Fine grained	1–2	1

10.5 Guidelines for inherent soil variability

- Variability is therefore the sum of natural variability and the testing variability.

Table 10.5 Guidelines for inherent soil variability (Phoon and Kulhawy, 1999).

Test type		Property	Soil type	Coefficient of variation (%)	
				Range	Estimated mean
Lab strength	UC	Shear strength, C_u	Clay	20–55	40
	CIUC			20–40	30
	UU			10–30	20
Lab strength		Effective angle of friction	Clay & sand	5–15	10
Standard penetration test		N-value		25–50	40
Dilatometer		A	Clay	10–35	25
		B			
		A	Sand	20–50	35
		B			
		I_D	Sand	20–60	40
		K_D		20–60	
		E_D		15–65	
Pressuremeter		P_L	Clay	10–35	25
			Sand	20–50	35
		E_{PMT}	Sand	15–65	40
Cone penetrometer test		q_c	Clay	20–40	30
			Sand	20–60	40
Vane shear test		Shear strength, C_u	Clay	10–40	25
Lab index		Natural moisture content	Clay and silt	8–30	20
		Liquid limit		6–30	
		Plastic limit		6–30	

10.6 Compaction testing

- In a compaction specification, the density ratio has less variation than the moisture ratio.
- The density ratio controls can be based on a standard deviation of 3% or less (Hilf, 1991).

Table 10.6 Precision values (MTRD, 1994).

Conditions	*Maximum dry density*	*Optimum moisture content*	
		Granular materials	*Clay*
Repeatability	1% of mean	10% of mean	13% of mean
Reproducibility	2.5% of mean	12% of mean	19% of mean

- ○ The placement moisture is therefore only a guide to achieving the target density, and one should not place undue emphasis on such a variable parameter.

10.7 Guidelines for compaction control testing

- Clays tend to be more variable than granular materials.
- At higher moisture contents, the variation in densities is reduced.

Table 10.7 Guidelines for compaction control testing.

Test control	*Coefficient of variation*		
	Homogeneous conditions	*Typical*	*Highly variable*
Maximum dry density	1.5%	3%	5%
Optimum moisture content	15%	20%	30%

10.8 Subgrade and road material variability

- Testing for road materials is the more common type of tests.

Table 10.8 Coefficient of variations for road materials (extracted from Lee et al., 1983).

Test type	*Test*	*Coefficient of variation*
Strength	Cohesion (undrained)	20–50%
	Angle of friction (clays)	12–50%
	Angle of friction (sands)	5–15%
	CBR	17–58%
Compaction	Maximum dry density	1–7%
	Optimum moisture content	20–30%
Durability	Absorption	25%
	Crushing value	8–14%
	Flakiness	13–40%
	Los Angeles abrasion	31%
	Sulphate soundness	92%
Deformation	Compressibility	18–73%
	Consolidation coefficient	25–100%
	Elastic modulus	2–42%
Flow	Permeability	200–300%

10.9 Deflection testing for pavements

- Deflection readings along roads are taken with Benkelman beams and deflectometers. The latter is gradually replacing the former which was the original "standard" of measurement.

Table 10.9 Coefficient of variation of deflection testing (Lay, 1990).

Construction	*Coefficient of variation*
Very uniform	<15%
Fair	20%–30%
Non-uniform	>40%
Consider remedial	>50%

10.10 Distribution functions

- Variability can be assessed by distribution functions.
- The normal distribution is the taught fundamental distribution, in maths and engineering courses. It is the simplest distribution to understand, but is not always directly relevant to soils and rocks.
- The tables following illustrate this discrepancy in selecting the normal distribution as the default.

Table 10.10 Appropriate distribution functions

Distribution type	*Application on geotechnical engineering*	*Comments*
Probability distribution function (PDF)	Various as below	Relative likelihood a random variable will assume a particular value. The area under a PDF is unity
Cumulative distribution function (CDF)	CBR selection. Alternative way of presenting the PDF	Probability value will have a value less than or equal to a particular value. CDF is the integral of the corresponding PDF
Pearson VI, Lognormal, Gamma, Weibull, Beta	Soil (cohesion) and rock strength (UCS). Especially for strength indices such as Point Load index and CBR.	These distributions avoid the negative values that sometimes occur when a normal distribution is applied. Use Lognormal as simplicity in its approach.
Exponential	Earthquakes, lengths of joints	A negative (decreasing) distribution. Applies mainly for the time needed to wait before an event occurs with a "rate" parameter as a descriptor. Also useful for a constant probability per unit distance
Normal	Density	Most used in industry due to familiarity but applies where small coefficient of variations. Not generally applicable to strength as negative values may result at low characteristic values

10.11 Distribution functions for rock strength

- When applied to soil or rock strength properties, negative values can result at say lower 5 percentile if a normal distribution is used (Look and Griffiths, 2004).
- The assumed distribution can affect the results considerably. For example the probability of failure of a slope can vary by a factor of 10 if a normally distributed or gamma distribution used.

Table 10.11 Appropriate distribution functions in a rock strength property assessment (Look and Griffiths, 2004).

Distribution type rank	*Overall*	*Typical application outside of geotechnical engineering*
Pearson VI	1	Time to perform a task.
Lognormal	2	Measurement Errors. Quantities that are the product of a large number of other quantities. Distribution of physical quantities such as the size of an oil field.
Gamma	3	Time to complete some task, such as building a facility, servicing a request.
Weibull	4	Lifetime of a service for reliability index.
Beta	5	Approximate activity time in a PERT network. Used as a rough model in the absence of data.
Normal	11	Distribution characteristics of a population (height, weight); size of quantities that are the sum of other quantities (because of central limit theorem).

- Above rank are based on various goodness-of-fit tests for 25 distribution types on point load index results.
- Due to non-normality of distribution, the median is recommended instead of mean in characterisation of a site.
- Overall the log normal provides a reasonable best fit (and well ranked in most geotechnical applications) where strength is measured as shown in the examples below.

10.12 Effect of distribution functions on rock strength

- An example of the effect of the distribution type on a design value obtained from point load index results.
- Typically a characteristic value at the lower 5% adopted for design in limit state codes.
- Using an assumption of a normal distribution resulted in negative values.
- Mean values are similar in these distributions.

Table 10.12 Effect of distribution type on statistical values (Look and Griffiths, 2004).

Rock		*Distribution applied to Point Load Index test results*								
Type	*Weathering*	*Normal*			*Lognormal*			*Weibull*		
		5%	Mean	95%	5%	Mean	95%	5%	Mean	95%
Argillite/	DW	(−0.4)	1.0	2.4	0.1	1.0	2.6	0.2	1.1	3.1
Greywacke	SW	(−0.8)	2.0	4.8	0.2	2.0	5.2	0.3	2.1	6.3
Sandstone/	DW	(−0.3)	0.6	1.5	0.1	0.6	1.7	0.1	0.7	2.1
Siltstone	SW	(−1.1)	1.1	3.2	0.0	1.1	3.3	0.1	1.1	3.1
Tuff	DW	(−0.1)	0.4	0.8	0.1	0.4	0,9	0.1	0.4	1.2
	SW	(−1.5)	3.3	8.0	0.3	3.3	8.5	0.6	3.2	8.7
Phyllites	DW	(−0.3)	0.9	2.0	0.1	0.9	2.2	0.1	0.9	2.7
	SW	(−0.4)	1.0	2.5	0.1	1.0	2.6	0.2	1.0	2.8

- A lognormal distribution is recommended for applications in soils and rock. Although, depending on the application different distributions may be more accurate (refer Figure 10.2). However the lognormal distribution is highly ranked overall and offers simplicity in its application that is not found in more rigorous distribution functions.

10.13 CBR values for a linear (transportation) project

- Variation occurs both in testing at a specified location (previous Tables) and spatially for a given project
- This example is for linear project of 13 km length and the results of soaked CBR testing in the preliminary design phase. The lower characteristic value (LCV) is compared for various reliability levels and distribution functions.

Table 10.13 Results of distribution models at 10% and 25% risk for the various CBR zones.

Chainage	*All (13 km)*	*3.5 km*	*1.5 km*	*3.5 km*	*4.5 km*
No. of values	96	31	6	38	21
COV	90%	72%	50%	101%	71%
10% LCV		CBR (%) value			
Best fit	2.9	3.0	15.0	2.3	4.0
Log-normal	2.8	3.0	14.8	2.2	4.0
Normal	(−1.7)	0.7	10.6	(−2.8)	1.3
25% LCV		CBR (%) value			
Best Fit	4.6	4.2	18.8	3.8	6.5
Log-normal	4.6	4.2	18.7	3.7	6.5
Normal	4.5	4.4	19.6	3.2	7.1

- A COV of 90% over the route is significantly higher than the previous tables but at a given location a COV of 50% was the lowest for the design segments.
- At the 10 percentile defects, a negative design value occurs if the normal distribution is applied. The best fit and log normal distribution provides comparable values.

- At the 25 percentile defects, the best fit, the log normal and normal are comparable for the LCV.

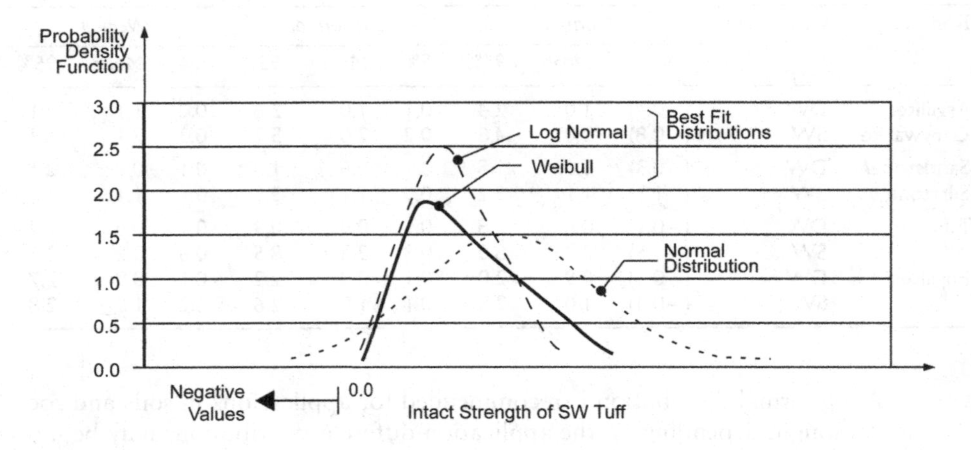

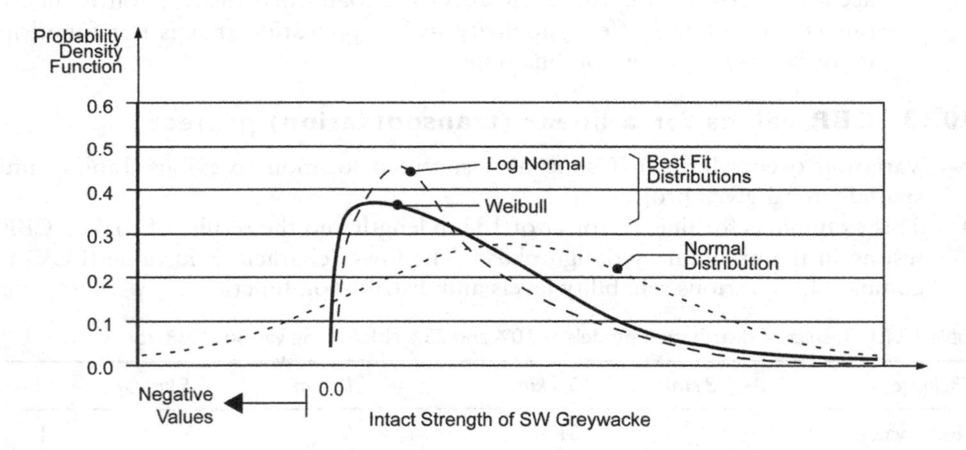

Figure 10.2 Typical best fit distribution functions for rock strength compared with the normal distribution.

10.14 Point load index values for a vertical linear (bridge) project

- Point Load strength Index results (330 No) in a DW interbedded sandstone-siltstone were carried out for 24 piles at a given bridge pier (Look and Wijeyakulasuriya, 2009).
- For that bridge pier the I_s (50) COV was 91% but varied from approximately 40% to 150% for the individual pile location.
- Use of the normal distribution at the characteristic values would result in negative values at some piles, and near zero overall for this bridge pier.
- The LCV (10%) based on a better fit normal distribution would be 7 times that predicted by the normal distribution.

Table 10.14 Results of distribution models at 10% for point load index rock strength values at bridge pier.

Pier 6	*Diametral I_s (50) statistics*			*10% Characteristic (MPa)*	
Pile #	*Mean (MPa)*	*COV*	*No. of points*	*Normal*	*Log Normal*
P6-5	0.85	39%	10	0.46	n/a
P6-6	1.01	151%	10	0.26	0.43
P6-7	0.57	56%	15	0.15	0.19
P6-8	0.74	68%	15	0.12	0.30
P6-21	0.94	37%	16	0.48	0.51
P6-22	0.81	113%	17	(−0.13)	0.20
P6-23	0.81	40%	13	0.40	0.45
P6-24	0.61	87%	18	(−0.12)	0.12
P6-ALL	0.82	91%	330	0.03	0.24

- The overall results are shown in Figure 10.3. This illustrates the 5% to 10% values can be misleading when a normal distribution is applied but approximately comparable at the 20% to 30% value.

10.15 Variability in design and construction process

- Section 5 provided comment on the errors involved in the measurement of soil properties.
- The table shows the variation in the design and construction process.

Table 10.15 Variations in design and construction process based on fundamentals only (Kay, 1993).

Variability component	*Coefficient of variation*
Design model uncertainty	0–25%
Design decision uncertainty	15–45%
Prototype test variability	0–15%
Construction variability	0–15%
Unknown unknowns	0–15%

- Natural variation over site (state of nature) is 5 to 15% typically.
- Sufficient statistical samples should be obtained to assess the variability in ground conditions.
- Ground profiling tools (boreholes, CPT) provide only spatial variability. Use of broad strength classification systems (Chapters 2 and 3) are of limited use in an analytical probability model.
- Socially acceptable risk is outside the scope of this text, but the user must be aware that voluntary risks (Deaths from smoking and alcohol) are more acceptable than involuntary risks (e.g. death from travelling on a construction project), and the following probability of failures should not be compared with non-engineering risks.

10.16 Prediction variability for experts compared with industry practice

- This is an example of the variability in prediction in practice.
- Experts consisted of 4 eminent engineers to predict the performance characteristic, including height of fill required to predict the failure of an embankment on soft clays. 30 participants also made a prediction.
- Table shows the variation in this prediction process.

Table 10.16 Variations in prediction of height difference at failure (after Kay, 1993).

Standard of prediction	*No. of participants*	*Coefficient of variation*
Expert level	4	14%
Industry practice	30	32%

- ○ A much lower variation of experts also relates to the effort expended, which would not normally occur in the design process.
- ○ The experts produced publications, detailed effective stress and finite element analyses, including one carried out centrifuge testing. These may not be cost effective in industry where many designs are cost driven.

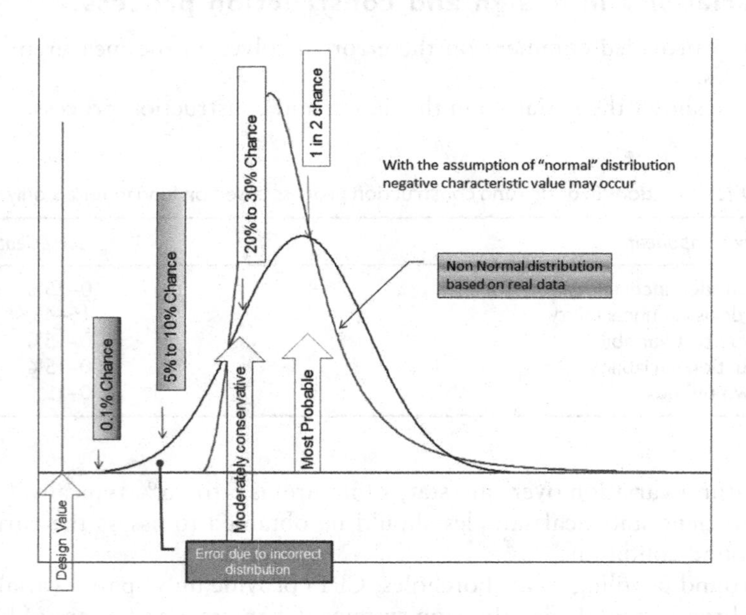

Figure 10.3 Normal distribution does not always apply (illustration from 330 point load index results at a large bridge pier) – Look and Campbell (2013).

10.17 Variability in selecting design values

- Selection of design values is often subjective unless statistical approaches are adopted.

- In a survey of 112 participants for the case study of a local (low traffic) road in Queensland, the variability in response is shown in the Table for selecting a design value between CBR of 2% to 15%.
- Although the geotechnical engineer was the least likely to make a poor judgement on the design value, only 4 out of 10 (39%) selected a given value, and suggests a high variability in how the design value is selected.

Table 10.17 Variability in selecting design values (Look and Campbell, 2013).

Participant *No in survey/%*	*Geotechnical engineer* *55/52%*	*Project manager* *18/17%*	*Civil + structural engineer* *21/20%*	*Other* *12/11%*
Preferred CBR (% by professional background	5% (39%)	5% (47%)	3% and 5% (24% each)	5% (33%)
Judgement	Best engineering "judgement"	Within expectations	Disproportionate amount of poor engineering "judgement"	
Comment		Most consistent	Lowest selection	

  - Other interesting findings, were
    - When participants were informed of relative cost of each CBR selection decision, 47% of participants increased their design value selection. This suggests cost is not necessarily a considered part of the design decision unless explicitly presented.
    - Professionals with less than 5 years of experience were the most likely to show poor judgement in selecting a design value. The converse was not true i.e. those with greater than 20 years was not the least likely to show poor judgement. That belonged to those with 5 to 10 years of experience, although the 20 yrs+ was the most consistent in 46% selecting the same value as compared to 32% for the 5 to 10 yr experience level.

10.18 Tolerable risk for new and existing slopes

- The probabilities of failure are more understandable to other disciplines and clients than factors of safety. A factor of safety of 1.3 does not necessarily mean that system has a lower probability of failure than a factor of safety of 1.4.
- Existing and new slopes must be assessed by different criteria.

Table 10.18 Tolerable risks for slopes (AGS, 2000).

Situation	*Tolerable risk probability of failure*	*Loss of life*
Existing slope	10^{-4}	Person most at risk
	10^{-5}	Average of persons at risk
New slopes	10^{-5}	Person most at risk
	10^{-6}	Average of persons at risk

10.19 Probability of failures of rock slopes

- A guidance on catastrophic versus minor failures probabilities are provide in the Table.

Table 10.19 Probability of failure in rock slope analysis (Skipp, 1992).

Failure category	*Annual probability*	*Comment*
Catastrophic	0.0001 (1×10^{-4})	
Major	0.0005 (5×10^{-4})	
Moderate	0.001 (1×10^{-3})	
Minor	0.005 (5×10^{-3})	For unmonitored permanent urban slopes with free access

10.20 Qualitative risk analysis

- Qualitative risk analysis (QRA) is one of the tools used in landslide risk management. It involves and assessment of probability of failure and its consequences.
- Other considerations affect the risk decision beyond cost of property (Look and Thorley, 2011). A few of these that include
  - The cost of rectification in many cases may outweigh the property value
  - Societal impacts
  - Available funding with due consideration to the pecking order as compared to other hazards
  - Public perception.

Table 10.20 Qualitative risk analysis matrix – risk to property (AGS, 2007).

Level	*Descriptor*	*Approx. annual probability*	*Consequences*				
			Catastrophic 200%	*Major 60%*	*Medium 20%*	*Minor 5%*	*Insignificant 0.5%*
A	Almost certain	10^{-1}	VH	VH	VH	H	M or L
B	Likely	10^{-2}	VH	VH	H	M	L
C	Possible	10^{-3}	VH	H	M	M	VL
D	Unlikely	10^{-4}	H	M	L	L	VL
E	Rare	10^{-5}	M	L	L	VL	VL
F	Barely credible	10^{-6}	L	VL	VL	VL	VL

- VH – Very high; H – High: M – Moderate; L – Low; VL – Very low.
- Refer to following tables for detailed assessment of probability and consequences and risk level implications.

10.21 Qualitative measure of likelihood

- Likelihood is a qualitative description of probability of frequency of occurrence.
- Probability is often uncertain for exteme events.

Table 10.21 Qualitative measure of likelihood (AGS, 2007).

Approximate annual probability	*Implied recurrence landslide interval*	*Description*	*Descriptor*	*Level*
10^{-1}	10 years	The event is expected to occur over design life	Almost certain	A
10^{-2}	100 years	The event will probably occur under adverse conditions over the design life	Likely	B
10^{-3}	1000 years	The event could occur under adverse circumstances over the design life	Possible	C
10^{-4}	10,000 years	The event might occur under very adverse circumstances over the design life	Unlikely	D
10^{-5}	100,000 years	The event is conceivable but only under exceptional circumstances over the design life	Rare	E
10^{-6}	1,000,000 years	The event is inconceivable or fanciful over the design life	Barely credible	F

- The table should be used from left to right.

10.22 Qualitative measure of consequences to property

- The approximate cost of damage is expressed as a percentage of market value, being the cost of the improved value of the unaffected property which includes the land plus the unaffected structures.
- The approximate cost is to be an estimate of the direct cost of the damage, such as the cost of reinstatement of the damaged portion of the property (land plus structures), stabilisation works required to render the site to tolerable risk level for the landslide which has occurred and professional design fees, and consequential costs such as legal fees, temporary accommodation.

Table 10.22 Qualitative measure of consequences to property (AGS, 2007).

Approximate cost of damage	*Description*	*Description*	*Level*
200%	Structure(s) completely destroyed and/or large scale damage requiring major engineering works for stabilisation. Could cause at least 1 adjacent property major consequence damage	Catastrophic	1
60%	Extensive damage to most of structure, and/or extending beyond site boundaries requiring significant stabilisation works. Could cause at least 1 adjacent property medium consequence damage	Major	2
20%	Moderate damage to some of structure, and/or significant part of site requiring large stabilisation works. Could cause at least 1 adjacent property minor consequence damage	Medium	3
5%	Limited damage to part of structure and/or part of site requiring some reinstatement stabilisation works	Minor	4
1%	Little damage	Insignificant	5

- The table should be used from left to right.

10.23 Risk level implications

- The implications for a particular situation are to be determined by all parties to the risk assessment and may depend on the nature of the property at risk; these are only given as a general guide.

Table 10.23 Risk level implications (AGS, 2007).

Risk level	*Description*	*Example implication*
VH	Very high risk	Unacceptable without treatment. Extensive detailed investigation and research, planning and implementation of treatment options essential to reduce risk to Low: may be too expensive and not practical. Work likely to cost more than value of the property
H	High risk	Unacceptable without treatment. Detailed investigation planning and implementation of treatment options essential to reduce risk to Low. Work would cost a substantial sum in relation to the value of the property
M	Moderate risk	May be tolerable in certain circumstances (subject to regulator's approval) but requires investigation planning and implementation of treatment options essential to reduce risk to Low. Treatment options to reduce to low risk should be implemented as soon as possible
L	Low risk	Usually acceptable to regulators. Where treatment has been required to reduce risk to this level, ongoing maintenance is required

10.24 Acceptable probability of slope failures

- The acceptable probability depends on its effect on the environment, risk to life, cost of repair, and cost to users.

Table 10.24 Slope stability – acceptable probability of failure (Santamarina et al., 1992).

Conditions	*Risk to life*	*Costs*	*Probability of failure (P_f)*
Unacceptable in most cases			$<10^{-1}$
Temporary structures	No potential life loss	Low repair costs	10^{-1}
Nil consequences of failure Bench slope, open pit mine	No potential life loss	High cost to lower p_f	1 to 2×10^{-1}
Existing slope of riverbank at docks. available alternative docks	No potential life loss	Repairs can be promptly done	5×10^{-2}
to be constructed: same condition		Do – nothing attractive idea	$<5 \times 10^{-2}$
Slope of riverbanks at docks no alternative docks	No potential life loss	Pier shutdown threatens operations	1 to 2×10^{-2}
Low consequences of failure	No potential life loss	Repairs can be done when time permits. repair costs < costs to lower p_f	10^{-2}
Existing large cut – interstate highway	No potential life loss	Minor	1 to 2×10^{-2}
To be constructed: same condition	No potential life loss	Minor	$<10^{-2}$
Acceptable in most cases	No potential life loss	Some	10^{-3}
Acceptable for all slopes	Potential life loss	Some	10^{-4}
Unnecessarily low			$<10^{-5}$

10.25 Probabilities of failure based on lognormal distribution

- The factor of safety can be related to the probability of failure based on different coefficients of variations (COV).
- A lognormal distribution is used.
- The factor of safety is the most likely value.

Table 10.25 Probability of failure based on lognormal distribution (Duncan and Wright, 2005).

Factor of safety	*Probability of failures (%) based on COV*				
	COV = 10%	*20%*	*30%*	*40%*	*50%*
1.2	3.8	21	32	39	44
1.3	0.5	11	23	31	37
1.4	0.04	5.5	16	25	32
1.5	$\sim 10^{-3}$	2.6	11	20	27
2.0	$< 10^{-3}$	0.03	1.3	5	11
2.5		$\sim 10^{-3}$	0.15	1.4	4.4
3.0		$< 10^{-3}$	0.02	0.39	1.8

- For layered soils, different COVs are likely to apply to each layer.

10.26 Project reliability

- Reliability is based on the type of project and structure.
- Lowest value of strength is not used in design unless only limited samples available (3 or less).
- Design values are referenced to a normal distribution as this is what is applied to steel and concrete design, and many codes apply this normality concept also to soil and rock. As commented above non-normality of soils and rock applies, to avoid negative values.
- Ultimate conditions (strength criteria) and serviceability (deformation criteria) requires a different acceptance criterion. The literature is generally silent on this issue and a suggested criterion is provided in the table.

Table 10.26 Ground conditions acceptance based on type of project.

Type of project	*Typical design values*		*Comment*
	Ultimate	*Serviceability*	
Structure	1%	5%	5% for a normal distribution is likely to be 10% to 20% for a lognormal distribution.
Road	5%	10%	10% for a normal distribution is likely to be 30% to 50% for a lognormal distribution: 20 % to 30% is typically close to the median value.

- Correct distribution needs to be applied, i.e. non normal.
- At interfaces such as embankments next to a bridge structure then tighter controls would be required. This would be 1% to 5% serviceability for major to minor roads, respectively.

- If the above is translated into a physical criteria, then in terms of absolute conditions, e.g. if 10% design is used then no more than 1 m in 10 m of road length would be above a criteria of say 50 mm acceptable movement.

10.27 Road reliability values

- The desired road reliability level is based on the type of road.
- A normal distribution is assumed, and comments on the non-normality of soil and rocks apply.

Table 10.27 Typical road reliability levels.

Road class	*Traffic*	*Project reliability (typical)*
Highway	Lane AADT > 2000	90–97.5% (95%)
	Lane AADT ≤ 2000 (Rural)	85–95% (90%)
Main roads	Lane AADT > 500	85–95% (90%)
Local roads	Lane AADT ≤ 500	80–90 % (85%)

- These values do vary between road authorities.
- Reliability has a higher cost of construction implication to project that should be understood by owner of asset.

10.28 Reliability index

- The reliability index β indicates the probability of failure p_f.
- Assumptions on distribution type may affect result on characteristic value.
- Larger $\beta \rightarrow$ smaller p_f.

Table 10.28 Reliability index relationship with probability of failure (US Corps of Engineers, 1997; FHWA, 2001).

Reliability index β		*Probability of failure p_f*	*Probability of unsatisfactory performance*	*Other indicators*
Normal	*Lognormal distribution*			
1.0	1.96	0.16	Hazardous	
1.28		10^{-1}		
1.5		0.07	Unsatisfactory	
2.0	2.5	2.3×10^{-2}	Poor	Systems with a high degree of redundancy
2.33		10^{-2}		
2.5		6×10^{-3}	Below average	
3		1×10^{-3}	Above average	
3.09	3.03	10^{-3}		Systems with a low level of redundancy
3.71	3.57	10^{-4}		
4	4.10	3×10^{-5}	Good	
4.26		10^{-5}		
4.75	4.64	10^{-6}		
5		3×10^{-7}	High	
5.19		10^{-7}		
5.62		10^{-8}		
5.99		10^{-9}		

- Probability of unsatisfactory performance is the probability that the value of the performance function will approach the limit state, or that an unsatisfactory performance event will occur.

10.29 Concrete quality

- The COV of manufactured material fall within a relatively narrow range as compared to soils and rock

Table 10.29 Quality of concrete (Rathati, 1988, here from Phoon and Kulhawy, 2008).

Quality of concrete	*Excellent*	*Good*	*Satisfactory*	*Bad*
Coefficient of variation	≥10%	10%–15%	15%–20%	≥20%

10.30 Soil property variation for reliability calibration

- Three ranges of soils variability are sufficient to achieve uniform reliability levels (Phoon and Kulhawy, 2008).

Table 10.30 Ranges of soil property variation for reliability calibration (Phoon and Kulhawy, 2008).

Geotechnical parameter	*Property variability*	*Coefficient of variation (%)*	*Comment*
Undrained shear strength	Low	10–30	Good quality direct lab or field measurement
	Medium	30–50	Indirect correlations with good field data except for the SPT
	High	50–70	Indirect correlations with SPT filed data and with strictly empirical correlations
Effective stress friction angle	Low	5–10	Good quality direct lab or field measurement
	Medium	10–15	Indirect correlations with good field data except for the SPT
	High	15–20	Indirect correlations with SPT filed data and with strictly empirical correlations
Horizontal stress coefficient	Low	30–50	Good quality direct lab or field measurement
	Medium	50–70	Indirect correlations with good field data except for the SPT
	High	70–90	Indirect correlations with SPT filed data and with strictly empirical correlations

10.31 Testing, spatial and temporal variation

- There is the natural change in ground profile with depth that can be expected
- However, field testing within a few metres of each other would not provide an identical result numerically even for the same ground profile. This is testing variation.
- By shifting by 10 m even in a "uniform" site then spatial variation is added to the testing variation.

- If one carries out the same tests on the same site and at similar locations, but over different periods of time, then temporal variation is added to the above.
- This table shows the results of a dynamic cone penetrometer at a uniform residual soil site over 3 different periods with the resulting coefficient of variation.

Table 10.31 Testing, spatial and temporal variation (Mellish, 2013).

Variable parameter	*Testing variation*	*Testing and spatial variation*	*Testing, spatial and temporal variation*
Uniformity	Same location. Tests within 1 metre	Uniform site. All tests within 15 m	Same test and site location as before, but over 3 periods of time.
Coefficient of variation	24%	38%	46%

Chapter 11

Deformation parameters

11.1 Modulus definitions

- The stiffness of a soil or rock is determined by its modulus value. The modulus is the ratio of the stress versus strain at a particular point or area under consideration.
- Materials with the same strength can have different stiffness values.
- The applicable modulus is dependent on the strain range under consideration.
- The long term and short term modulus is significantly different for fine grained soils, but slightly different for granular soils. The latter is considered approximately similar for all practical purposes.
- Additional modulus correlations with respect to roads are provided in Chapter 13 for subgrades and pavements.

Table 11.1 Modulus definitions.

Modulus type	*Definition*	*Strain*	*Comment*
Initial tangent modulus	Slope of initial concave line	Low	Due to closure in micro-cracks from sampling stress relief (laboratory) or existing discontinuities (in-situ).
Elastic tangent modulus	Near linear slope prior to yield	Medium	Also elastic modulus. Can be any specified point on the stress strain curve, but usually at a specified stress levels such as 50% of maximum or peak stress.
Deformation modulus	Slope of line between zero and maximum or peak stress	Medium to high	Also secant modulus.
Constrained modulus	Slope of line between zero and constant volume stress	High	This is not detailed in the literature. But values are lower than a secant modulus, and it is obtained from odeometer tests where the sample is prevented from failure, therefore sample has been taken to a higher strain level.
Recovery modulus	Slope of unload line	High	In-situ tests seldom stressed to failure, and unload line does not necessarily mean peak stress has been reached. Usually concave in shape.
Reload modulus	Slope of reload Line	High	Following unloading the reload line takes a different stress path to the unload line. Usually convex in shape. Also resilient modulus.

(Continued)

Table 11.1 (Continued)

Modulus type	*Definition*	*Strain*	*Comment*
Cyclic modulus	Average slope of unload/reload line	High	Strain hardening can occur with increased number of cycles.
Equivalent modulus	A combination of various layers into on modulus	Various	A weighted average approach is usually adopted.

- Modulus usually derived from strength correlations. The 2 most common are:
  - Secant modulus is usually quoted for soil – structure interaction models.
  - Resilient modulus applies for roads.
- Typically vertical modulus $E_V \approx 2$ horizontal modulus E_H

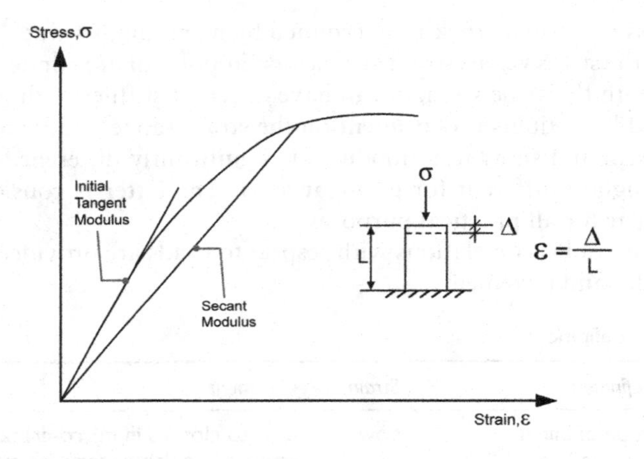

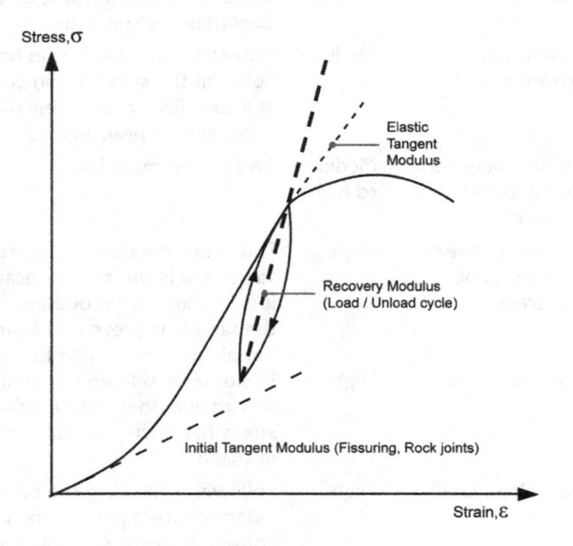

Figure 11.1 Stress strain curve showing various modulus definitions.

11.2 Small strain shear modulus

- The small strain shear modulus is significantly higher than at high strains.
- The table provides small – strain typical values.

Table 11.2 Typical values of small strain – shear modulus (Sabatani et al., 2002).

Shear modulus, G	*Small – strain shear modulus G_0 (MPa)*
Soft clays	3 to 15
Firm clays	7 to 35
Silty sands	30 to 140
Dense sands and gravels	70 to 350

- For large strains $G_{ls} = E/2.5$.
- For small strains $G_{ss} = 2E = 5\ G_{ls}$.

11.3 Comparison of small to large strain modulus

- The applicable modulus is dependent on the strain level.
- The table provides the modulus values at small and large strains.

Table 11.3 Stiffness degradation range for various materials (summarised from Heymann, 1998).

Strain level comparison	*Stiffness ratio*
$E_{0.01}/E_0$	0.8 to 0.9
$E_{0.1}/E_0$	0.4 to 0.5
$E_{1.0}/E_0$	0.1 to 0.2

- Modulus at 0% strain = E_0.
- Modulus at 0.01% strain = $E_{0.01}$ (small strain).
- Modulus at 1.0% strain = $E_{1.0}$ (large strain).
- Materials tested were intact chalk, London clay and Bothkennar clay.
- Figure 11.2 (from Sabatani et al. 2002) shows the types of tests appropriate at various strain levels.

11.4 Strain levels for various applications

- The applicable modulus value below a pavement is different from the modulus at a pile tip even for the same material.
- Different strain level produces different modulus values.
- Jardine et al. (1986) found shear strain levels for excavations to be <0.1% for walls and as low as 0.01% if well restrained.
- The modulus value for the design of a pavement is significantly different from the modulus values used for the support of a flexible pipe in a trench.

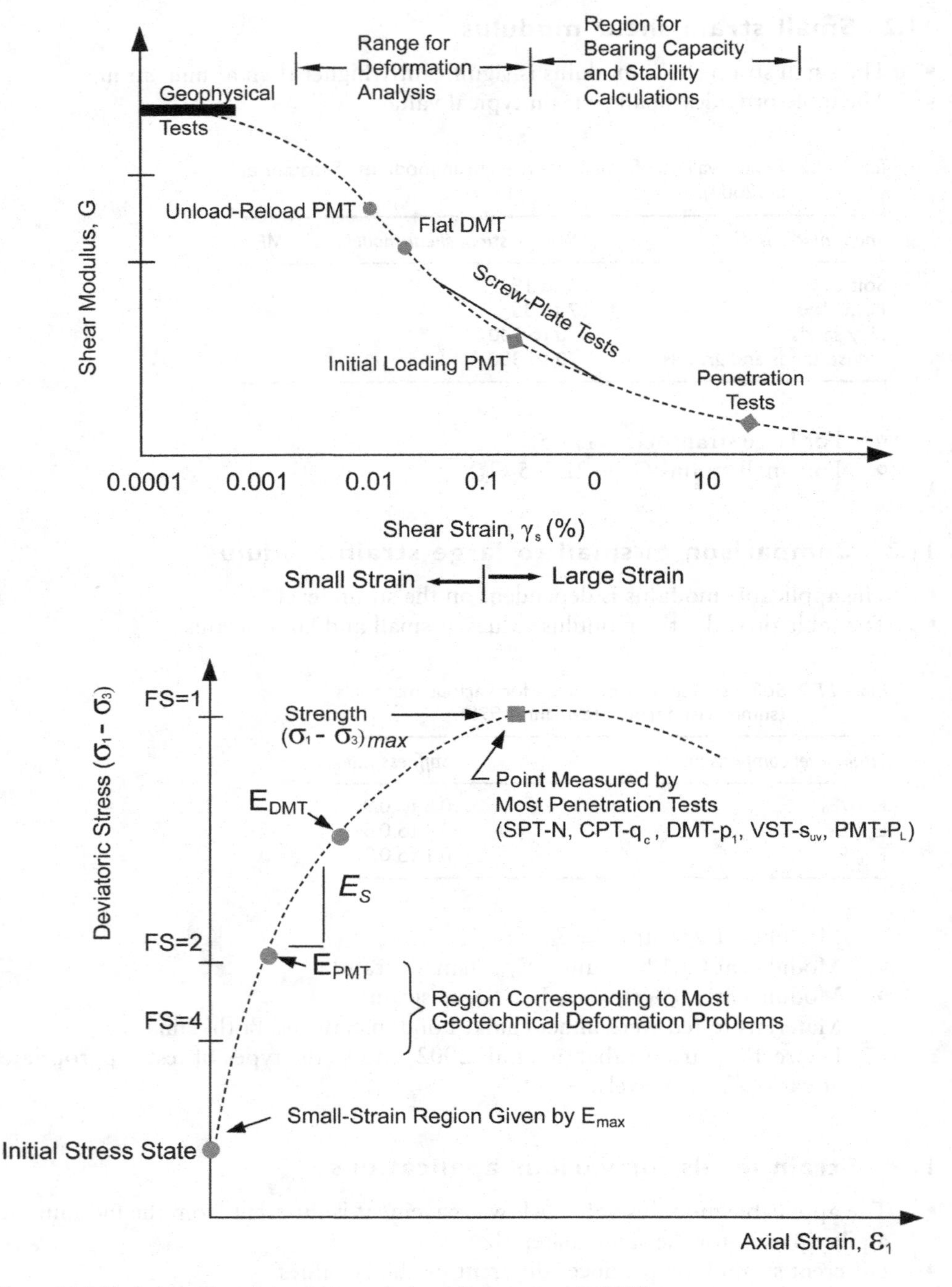

Figure 11.2 Variation of modulus with strain level (Sabatani et al., 2002).

Table 11.4 Strain levels.

Application	*Type*	*Strain Level*	*Typical Movement*	*Shear Strain*	*Applicable Testing*
Pavement	Rigid	Very small	5–10 mm	<0.001%	Dynamic methods
	Flexible base	Large	5–30 mm	>0.1%	Dynamic methods/ local gauges
	Sub base	Small/large	5–20 mm	0.01–0.1%	
	Subgrade	Small/very small	5–10 mm	0.001–0.01%	
	Haul/access	Very large	50–200 mm	>0.5%	Conventional soil testing
	Unpaved road	Large	25–100 mm	>0.1%	
Foundations	Pile shaft	Small	5–20 mm	0.01–0.1%	Local gauges
	Pile tip	Small/medium	10–40 mm		
	Shallow	Small/large	10–50 mm	0.05–0.5%	Local gauges
	Embankments	Large/very large	>50 mm	>0.1%	Conventional soil testing
Retention systems	Retaining wall	Active – small	10–50 mm	0.01–0.1%	Local gauges
		Passive – large	>50 mm	>0.1%	
	Tunnel	Large	10–100 mm	>0.1%	Conventional soil testing

- Retention systems and tunnels have both horizontal and vertical movements.
- Horizontal movement typically 25% to 50% of vertical movement.
- Different modulus values also apply for plane strain versus axisymmetric conditions.
- The modulus values for fill can be different (less) for in situ materials for the same soil description.

11.5 Modulus applications

- There is much uncertainty on the modulus values, and its application.
- The table provides a likely relative modulus ranking. Rank is 1 for smallest values and increasing in number to larger modulus. However this can vary between materials. For example, an initial tangent modulus without micro cracks in clay sample could have a higher modulus than the secant modulus at failure, which is different from the rank shown in the table.
- The relative values depend on material type, state of soil and loading factors.
- Some applications (e.g. pavements) may have a high stress level, but a low strain level. In such cases a strain criteria applies. In other applications, such as foundations, a stress criterion applies in design.
- In most cases, only 1 modulus is used in design although the soil may experience wide modulus ranges.
- Modulus values between small strain and large stain applications can vary by a factor of 5 to 10.
- The dynamic modulus can be greater than 2, 5 and 10 times that of a static modulus value for granular, cohesive material and rock, respectively.

Table 11.5 Modulus applications.

Rank	*Modulus type*	*Application*	*Comments*
1 (Low value)	Initial tangent modulus	Fissured clays. At low stress levels. Some distance away from loading source, e.g. at 10% $q_{applied}$ Low height of fill	Following initial loading and closing of micro-cracks, modulus value then increases significantly. For intact clay, this modulus can be higher than the secant modulus.
2	Constrained modulus	Wide loading applications such as large fills Wide embankments	Used where the soil can also fail, i.e. exceed peak strength.
3	Deformation (secant) Modulus	Spread footing Pile tip	Most used "average" condition, with secant value at ½ peak load (i.e. working load).
4	Elastic tangent modulus	Movement in incremental loading of a multi-storey building Pile shaft	The secant modulus can be 20% the initial elastic tangent modulus for intact clay.
5	Reload (Resilient) Modulus	Construction following excavation Subsequent loading from truck/train	Difficult to measure differences between reload/unload or cyclic. Resilient modulus term interchangeably used for all of them.
6	Cyclic modulus	Machine foundations Offshore structures/ wave loading earthquake/blast loading	Also called dynamic modulus of elasticity.
7	Recovery (Unload) Modulus	Heave at the bottom of an excavation After loading from truck/train Excavation in front of wall and slope	
Varies	Equivalent Modulus	Simplifying overall profile, where some software can have only 1 input modulus	Uncertainty on thickness of bottom layer (infinite layer often assumed). Relevant layers depend on stress influence.

11.6 Typical values for elastic parameters

- The strength of metals is significantly higher than the ground strength. Therefore movements from the ground tend to govern the performance of the structure.

Table 11.6 Typical values for Young's Modulus of various materials (after Gordon, 1978).

Classification	*Material*	*Young's Modulus, E (MPa)*
Human	Cartilage	24
	Tendon	600
	Fresh bone	21,000
Timber	Wallboard	1,400
	Plywood	7,000
	Wood (along grain)	14,000

(Continued)

Table 11.6 (Continued)

Metals	Magnesium	42,000
	Aluminium	70,000
	Brasses and bronzes	120,000
	Iron and steel	210,000
	Sapphire	420,000
	Diamond	1,200,000
Construction	Rubber	7
	Concrete	20,000
Soils	Soft Clays	5
	Stiff Clays, loose sands	20
	Dense sands	50
Rocks	Extremely weathered, soft	50
	Distinctly weathered, soft	200
	Slightly weathered, fresh, hard	50,000

- Modulus values of 30,000 MPa for industrial concrete floors would apply.

11.7 Elastic parameters of various soils

- Secant modulus values are used for foundations. This can be higher or lower depending on strain levels.

Table 11.7 Elastic parameters of various soils.

Type	*Strength of soil*	*Elastic modulus, E (MPa)*	
		Short term	*Long term*
Gravel	Loose	25–50	
	Medium	50–100	
	Dense	100–200	
Medium to coarse sand	Very loose	<5	
	Loose	3–10	
	Medium dense	8–30	
	Dense	25–50	
	Very dense	40–100	
Fine sand	Loose	5–10	
	Medium	10–25	
	Dense	25–50	
Silt	Soft	<10	<8
	Stiff	10–20	8–15
	Hard	>20	>15
Clay	Very soft	<3	<2
	Soft	2–7	1–5
	Firm	5–12	4–8
	Stiff	10–25	7–20
	Very stiff	20–50	15–35
	Hard	40–80	30–60

- These modulus values should not be used in a different application, i.e. non foundations.
- For example, the modulus values of similar soils in a trench as backfill surrounding a pipe would be significantly less than the above values.

11.8 Typical values for coefficient of volume compressibility

- The coefficient of volume compressibility (m_v) is used to compute settlements for clay soils.
- The m_v value is obtained from the consolidation (odeometer) test. This test is one dimensional with rigid boundaries, i.e. the Poisson Ratio $\upsilon' = 0$ and $E' = 1/m_v$.
- The elastic modulus obtained by this approach is referred to as the constrained modulus and is based on the assumption that negligible lateral strain occurs (in odeometer), so that Poisson's ratio is effectively zero.
- One-dimensional settlements $= \rho_{od}$. Refer Table 23.4 for correction factors

Table 11.8 Typical values for coefficient of volume compressibility (after Carter, 1983).

Type of clay	*Descriptive term*		*Coefficient of volume compressibility,*	*Constrained modulus,*
	Strength	*Compressibility*	m_v *(10^{-3} kPa^{-1})*	*$1/m_v$, (MPa)*
Heavily overconsolidated Boulder clays, weathered mudstone.	Hard	Very low	<0.05	>20
Boulder Clays, tropical red clays, moderately over consolidated.	Very stiff	Low	0.05 to 0.1	10–20
Glacial outwash clays, lake deposits, weathered marl, lightly to normally consolidated clays.	Firm	Medium	0.1–0.3	3.3–10
Normally consolidated alluvial clays such as estuarine and delta deposits, and sensitive clays.	Soft	High	0.3–1.0 (non sensitive) 0.5–2.0 (organic, sensitive)	0.7–3.3
Highly organic alluvial clays and peat.	Very soft	Very high	>1.5	<0.7

11.9 Coefficient of volume compressibility derived from SPT

- The m_v value is inversely proportional to the strength value. The correlation with the SPT N-value is provided in the table for clays with varying plasticity index.
- The table was based on data for stiff clays.

Table 11.9 Coefficient of volume compressibility derived from SPT N-value (after Stroud and Butler, 1975).

Plasticity index (%)	*Conversion factor (f_2)*	*m_v (10^{-3} kPa^{-1}) based on N-value: $m_v = 1/(f_2 N)$*				
		N = 10	*20*	*30*	*40*	*50*
10%	800	0.12	0.06	0.04	0.03	0.02
20%	525	0.19	0.09	0.06	0.05	0.04
30%	475	0.21	0.10	0.07	0.05	0.04
40%	450	0.22	0.11	0.07	0.06	0.04

11.10 Deformation parameters from CPT results

- The Coefficient of volume change and the constrained modulus (i.e. large strain condition) values can be derived from the CPT results.

Table 11.10 Deformation parameters from CPT results (Fugro, 1996; Meigh, 1987).

Parameter	*Relationship*	*Comments*
Coefficient of volume change, m_v	$m_v = 1/(\alpha q_c)$	For normally and lightly over consolidated soils $\alpha = 5$ for classifications CH, MH, ML $\alpha = 6$ for classifications CL, OL $\alpha = 1.5$ for classifications OH with moisture >100% for over consolidated soils $\alpha = 4$ for classifications CH, MH, CL, ML $\alpha = 2$ for classifications ML, CL with $q_c > 2$ MPa
Constrained modulus, M	$M = 3q_c$	$M = 1/m_v$
Elastic (Young's) modulus, E	$E = 2.5q_c$ $E = 3.5q_c$	Square pad footings – axisymetric Strip footings – plane strain

11.11 Drained soil modulus from cone penetration tests

- The approximate relationship between CPT value and drained elastic modulus for sands is provided.

Table 11.11 Preliminary drained elastic modulus of sands from cone penetration tests.

Relative density	*Cone resistance, q_c (MPa)*	*Typical drained elastic modulus, E′ (MPa)*
V. loose	<2.5	<10
Loose	2.5–5.0	10–20
Med dense	5.0–10.0	20–30
Dense	10.0–20.0	30–60
V. dense	>20.0	>60

11.12 Soil modulus in clays from SPT values

- The modulus varies significantly between small strain and large strain applications.

Table 11.12 Drained E′ and undrained E_u modulus values with SPT N-value (CIRIA, 1995).

Material	*E′/N (MPa)*	*E_u/N (MPa)*
Clay	0.6 to 0.7 0.9 for $q/q_{ult} = 0.4$ to 0.1	1.0 to 1.2 6.3 to 10.4 for small strain values ($q/q_{ult} < 0.1$)
Weak rocks	0.5 to 2.0 for N_{60}	

  - $E_u/N = 1$ is appropriate for footings.
  - For rafts, where smaller movements occur $E_u/N = 2$.
  - For very small strain movements for friction piles $E_u/N = 3$.

11.13 Drained modulus of clays based on strength and plasticity

- The drained modulus of soft clays is related to its undrained strength C_u and its plasticity index.

Table 11.13 Drained modulus values (from Stroud et al., 1975).

Soil plasticity, %	*E′/C_u*
10–30	270
20–30	200
30–40	150
40–50	130
50–60	110

11.14 Undrained modulus of clays for varying over consolidation ratios

- The undrained modulus E_u depends on the soil strength, its plasticity and over consolidation ratio (OCR).
- The table below is for a secant modulus at a Factor of safety of 2, i.e. 50% of the peak strength.

Table 11.14 Variation of the undrained modulus with over consolidation ratio (Jamiolkowski et al., 1979).

Over consolidation ratio	*Soil plasticity*	*E_u/C_u*
<2	PI < 30%	600–1500
2–4		400–1400
4–6		300–1000
6–10		200–600
<2	PI = 30–50%	300–600
2–4		200–500
4–10		100–400
<2	PI > 50%	100–300
2–10		50–250

- The E_u/C_u value is dependent on the strain level.
- For London Clays (Jardine et al., 1985) found a E_u/C_u ratio of 1000 to 500 for foundations but a larger ratio for retaining walls, when smaller strains apply.

11.15 Soil modulus from SPT values and plasticity index

- These values correlate approximately with previous tables for large strain applications.
- This applies to rigid pavements.
- Do not use for soft clays.

Table 11.15 Modulus values (Industrial Floors and Pavements Guidelines, 1999).

E_s/N	*Material*
3.5	Sands, gravels and other cohesionless soils
2.5	Low PI (<12%)
1.5	Medium PI (12% < PI < 22%)
1.0	High PI (22% < PI < 32%)
0.5	Extremely high PI (PI > 32%)

11.16 Short and long term modulus

- For granular materials the long term and short term strength and modulus values are often considered similar. However for these materials there can still be minor change between the long and short term state.
- Short term Young's modulus E_s = Long term modulus $E_l = \beta\ E_s$.

Table 11.16 Long term vs. short term (Industrial floors and pavements guidelines, 1999).

β	*Material*
0.9	Gravels
0.8	Sands
0.7	Silts, silty clays
0.6	Stiff clays
0.4	Soft clays

11.17 Poisson ratio in soils

- Clay in an undrained state has a Poisson Ration of 0.5. Not applicable in analysis – use 0.49.
- In the Odeometer test with negligible (near zero) lateral strain the Poisson Ratio is effectively 0.0.

Table 11.17 Poisson's ratio for soils (adapted from Industrial Floors and Pavements Guidelines, 1999).

Material	*Short term*	*Long term*
Sands, gravels and other cohesionless soils	0.30	0.30
Low PI (<12%)	0.35	0.25
Medium PI (12% < PI < 22%)	0.40	0.30
High PI (22% < PI < 32%)	0.45	0.35
Extremely high PI (PI > 32%)	0.45	0.40

11.18 Resilient modulus

- Resilient modulus (M_r) values vary with plasticity of clay, moisture content, compaction level, age, confining and deviatoric stress.
- $M_R = F_0 \times F_1 \times F_2 \times F_3 \times F_4 \times F_5 \times F_6$.
- $F_0 = 67.6$ MPa.
- The relative influence is shown in the following table which shows the high sensitivity to the moisture content.

Table 11.18 Resilient modulus correction factors (Pezo and Hudson, 1994).

Influence	*Correction factor*					*M_R, MPa*	
Moisture Content, %	10	15	20	25		Low MC	High MC
F_1	4.0	2.0	1.0	0.5		270	34
% Maximum dry density	100	95	90	85		High	Low % MDD
F_2	1.0	0.90	0.80	0.70		68	47
Plasticity index, %	10	20	30	≥40		Low PI	High PI
F_3	1.0	1.5	2.0	2.5		68	169
Sample age, days	2	10	20	≥30		Low age	High age
F_4	1.0	1.1	1.15	1.2		68	81
Confining stress, kPa	13.8	27.6	41.4			Low	High confining
F_5	1.0	1.05	1.10			68	75
Deviatoric stress, kPa	13.8	27.6	41.4	55.2	69.0	Low	High deviatoric
F_6	1.0	0.98	0.96	0.94	0.92	68	62

11.19 Typical rock deformation parameters

- The higher density rocks have a larger intact modulus.
- This needs to be factored for the rock defects to obtain the in situ modulus.

Table 11.19 Rock deformation based on rock description (adapted from Bell, 1992).

Rock density (kg/m³)	*Porosity (%)*	*Deformability (10^3 MPa)*
<1800	>30	<5
1800–2200	30–15	5–15
2200–2550	15–5	15–30
2550–2750	5–1	30–60
>2750	<1	>60

11.20 Rock deformation parameters

- This table is for intact rock properties, and compares the Young's Modulus (E) to the unconfined strength (q_u).

Table 11.20 Rock modulus values (Deere and Miller, 1966).

E/q_u	*Material*	*Comments*
1000	Steel, concrete	Man-made materials
500	Basalts & other flow rocks (Igneous rocks) Granite (Igneous) Schist: low foliation (Metamorphic) Marble (Metamorphic)	*High modulus ratio – UCS > 100 MPa* Metasediments 730 in SE Qld Basalt in Brisbane was 300 and 680 in SE Qld Granite in Queensland was 640 Phyllite (foliated metamorphic) in Brisbane was 500 Tuff (Pyroclastic Igneous) in Brisbane was 150
200	Gneiss, Quartzite (Hard metamorphic rocks) Limestone (Sedimentary) Dolomite (Calcareous sedimentary: Coral)	*Medium modulus ratio – UCS = 60–100 MPa*
100	Shales, Sandstones (Sedimentary rocks) Schist: steep foliation	*Low modulus ratio – UCS < 60 MPa* Horizontal bedding: Lower the E values Sedimentary 370 in SE Qld

- Intact rock properties would vary from in-situ conditions depending on the defects.
- Rock modulus correlations and the above general relationship should be calibrated with local conditions.
- The Brisbane and Qld relationships are from laboratory measurements. Overall a modulus ratio value of 650 applies for rocks of SE Queensland.

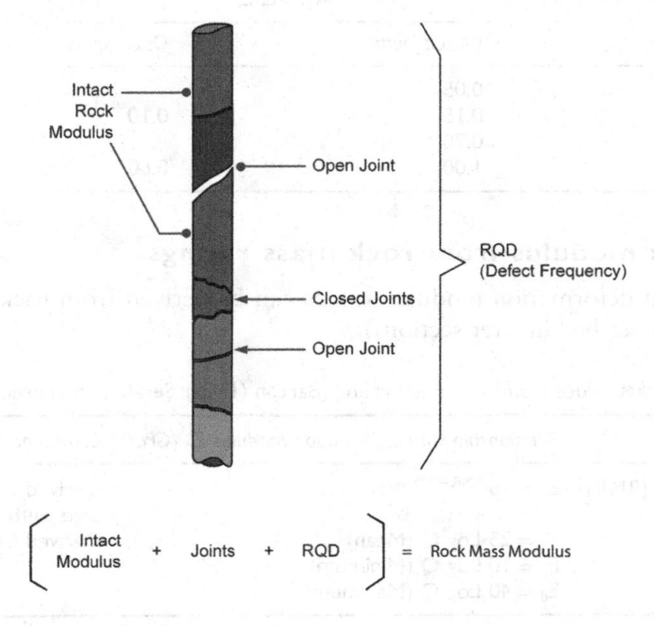

Figure 11.3 Rock mass modulus.

11.21 Rock mass modulus derived from the intact rock modulus

- Reduction factors needs to be applied to use the intact rock modulus in design.
- When the Young's modulus of the in-situ rock = E_r.

$$E_r = K_E E_i$$

where E_i = Intact rock modulus

Table 11.21 Modulus reduction ratio (after Bieniawski, 1984).

RQD (%)	*Modulus reduction ratio,* K_E
0–50	0.15
50–70	0.2
70–80	0.30
80–90	0.40
>90	0.70

11.22 Modulus ratio based on open and closed joints

- The modulus ratio (intact rock modulus/rock mass modulus) can be derived from the RQD combined with the opening of the rock joints, if known.
- Open joints have a higher reduction value at high RQD values.

Table 11.22 Estimation of the rock modulus based on the RQD values (after Carter and Kulhawy, 1988).

RQD (%)	$K_E = E_i/E_r$	
	Closed joints	*Open joints*
20	0.05	
50	0.15	0.10
70	0.70	
100	1.00	0.60

11.23 Rock modulus from rock mass ratings

- The in situ deformation modulus values can be derived from rock mass ratings systems (described in later sections).

Table 11.23 Modulus values from rock mass rating (Barton (1983); Serafim and Pereira (1983)).

Rock mass rating	*Relationship with deformation modulus* E_d *(GPa)*	*Comment*
Rock mass rating (RMR)	$E_d = 10^{(RMR - 10)/40}$	Derived from plate bearing tests with RMR = 25 to 85
Q – Index	$E_d = 25 \text{ Log } Q$ (Mean) $E_d = 10 \text{ Log } Q$ (Minimum) $E_d = 40 \text{ Log } Q$ (Maximum)	Derived from in-situ tests

11.24 Poisson ratio in rock

- These correlate approximately with the modulus ratios. Rocks with high modulus ratios tend to have lower Poisson's Ratio than rocks with low modulus ratios (see previous table).

Table 11.24 Poisson's ratio (rock).

Rock type	*Poisson's ratio*
Basalt	0.1 to 0.2
Granite	0.15 to 0.25
Sandstone	0.15 to 0.3
Limestone	0.25 to 0.35

  - Poison's ratio of concrete ~0.15.
  - Use a value of 0.15 for competent unweathered bedrock, and 0.3 for highly fractured and weathered bedrock.

11.25 Significance of modulus

- The relevant modulus value depends on the relative stress influence

Table 11.25 Significance of modulus (Deere et al., 1967).

Modulus ratios for rock	*Comments*
$E_d/E_{conc} > 0.25$	Foundation modulus has little effect on stresses generated within the concrete mass ($E_{conc} \sim 20{,}000$ MPa).
$0.06 < E_d/E_{conc} < 0.25$	Foundation modulus becomes significant with respect to stresses generated within the concrete mass.
$0.06 < E_d/E_{conc}$	Foundation modulus completely dominates the stresses generated within concrete mass.

Chapter 12

Earthworks

12.1 Earthworks issues

- The design and construction issues are covered in the table below.
- Issues related to pavements are discussed in the next chapter.
- Related issues on slopes and retaining walls are covered in later chapters.

Table 12.1 Earthworks issues.

Earthwork issues	*Comments*
Excavatability	Covered in this chapter. The material parameter is only 1 indicator of excavatability. Type of excavation and plant data also affects process.
Compaction characteristics	Covered in this chapter. Depends on material, type of excavation/operating space and plant. Lab compaction process is different from field compaction.
Bulk up	Covered in this chapter. Depends on material.
Pavements	Refer chapter 13
Slopes	Refer chapter 14
Retaining walls	Refer chapter 20
Drainage and erosion	Refer chapter 15
Geosynthetics	Refer chapter 16

12.2 Excavatability

- The excavatability depends on the method used as well as the material properties.

Table 12.2 Controlling factors.

Factor	*Parameter*
Material	Degree of weathering Strength Joint spacing Bedding spacing Dip direction
Type of excavation	Large open excavation Trench excavation Drilled shaft Tunnels
Type of plant	Size Weight
Space	Run direction Run up distance

- Some of these are not mutually exclusive, i.e. strength may be affected by degree of weathering, and run direction is relevant mainly for large open excavations, and when dip direction is an issue.
- Geological definition of rock is different form the contractual definition, where production rates are important.
- Different approaches often produce different conclusions on excavatability.

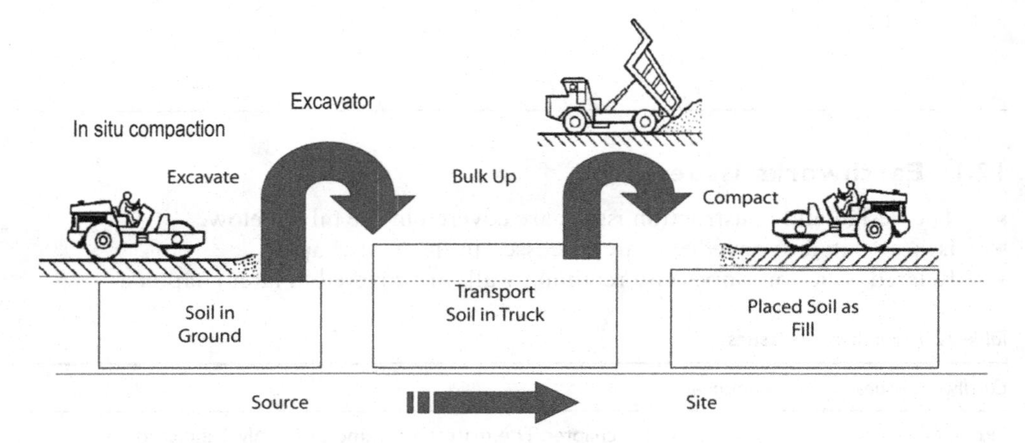

Figure 12.1 Earthworks process.

12.3 Excavation requirements

- The strength of the material is one of the key indicators in assessing the excavation requirements.
- Even this key indicator is likely to represent only 30% of the consideration.
- The table provides a preliminary assessment of the likely excavation requirements.

Table 12.3 Preliminary assessment of excavation requirements.

Material type	*Excavation requirements*
Very soft to firm clays Very loose to medium dense sands	Hand tools
Stiff to hard clays Dense to very dense sands Extremely low strength rocks – typically XW	Power tools
Very low to low strength rocks – typically XW/DW	Easy ripping
Medium to high strength rocks – typically DW	Hard ripping
Very high to extremely high – typically SW/Fr	Blasting

- Blasting term refers to the difficulty level and can include rock breakers, or expanding grouts.
- Significant cost factor (~ X 20) from easy excavation to blasting.

12.4 Excavation characteristics

- The excavatability characteristics based on rock hardness and strength.
- The above is combined with its bulk properties (seismic velocity) and joint spacing.

Table 12.4 Excavation characteristics (Bell, 1992).

Rock hardness description	*Unconfined compressive strength (MPa)*	*Seismic wave velocity (m/s)*	*Spacing of joints (mm)*	*Excavation characteristics*
Very soft	1.7–3.0	450–1200	<50	Easy ripping
Soft	3.0–10	1200–1500	50–300	Hard ripping
Hard	10–20	1500–1850	300–1000	Very hard ripping
Very hard	20–70	1850–2150	1000–3000	Extremely hard ripping or blasting
Extremely hard	>70	>2150	>3000	Blasting

- ○ Table below combines both factors of strength and fractures into one assessment.

12.5 Excavatability assessment

- The excavatability data shown are extracted from charts. It is therefore approximate values only.
- Higher strengths combined with closer discontinuity spacing shifts the excavatability rating.

Table 12.5 Excavatability assessment (Franklin et al., 1971 with updates from Walton and Wong, 1993).

Parameter	*Easy digging*	*Marginal digging without blasting*	*Blast to loosen*	*Blast to fracture*
Strength, I_s (50) (MPa)	<0.1	<0.3	>0.3	
Discontinuity spacing (m)	<0.02	<0.06	0.2 to 0.6	>0.6
RQD (%)	<10%	<90%	>90%	

- ○ Blast to loosen can be approximately equated to using a rock breaker.
- ○ Ripping involves using a tine attached to the rear of the bulldozer.

12.6 Excavatability assessment for heavy ripping equipment

- An updated chart by Pettifer and Fookes, 1994 considers the type of excavation equipment in addition to the rock parameters such as the discontinuity spacing index and point load strength index.

Table 12.6 Excavatability assessment updated (Pettifer and Fookes, 1994).

Parameter		*Digging*	*D6 to D7*	*D8*	*D9 to D11*	*Blasting required*
Strength, I_s (50) (MPa) and Discontinuity spacing (m)	From To	<0.2 MPa	0.3 MPa/0.6 m 10 MPa/0.02 m	0.2 MPa/0.6 m 30 MPa/0.02 m	0.2 MPa/2 m 30 MPa/0.06 m	0.2 MPa/3 m 30 MPa/0.2 m

12.7 Excavatability assessment based on seismic wave velocities

- Caterpillar handbooks provide assessment of excavatability by seismic wave velocities by type of equipment.

Table 12.7 Excavatability based on seismic wave velocities (Caterpillar, 2012).

Excavator model	*7E*	*D8R*	*D9R*	*D10T*	*D11T*
Flywheel power (kW)	175	228	302	433	634
Operating weight (t)	26	38	49	67	105
Ground pressure (kPa)	44	94	125	136	162
Rippable (typical)	NP	1700	2000	2300	2600
Unrippable (typical)	NP	2300	2600	3000	3500

- Rippability varies with rock type in Caterpillar charts.
- NP – not provided.

12.8 Excavatability production rates

- The various rating methods provided poor correlations (Macgregor et al., 1995).
- Quantification of productivity equations was developed based on rock type and various rock properties to produce productivity for the mass of the equipment.
- Allows for turning/pushing and 20% of time for dozers to push scrapers and/or doze loose material.
- To convert to values quoted in Caterpillar handbook, multiply by ~50%.

Table 12.8 Production rates (Macgregor et al., 1995).

Productivity grouping	*Ease of ripping*
0–250 m^3/hr	Very difficult/blasting
250–750 m^3/hr	Difficult
750–2000 m^3/hr	Medium
2000–3000 m^3/hr	Easy
>3000 m^3/hr	Very easy

12.9 Diggability index

- The rock weathering term is another term incorporated in this table as well as the type of equipment (backhoe or excavator).
- This table classifies the diggability only. The following table provides the implication for the type of equipment.

Table 12.9 Diggability index rating (adapted from, Scoble and Muftuoglu, 1984).

Parameter	*Symbol*	*Ranking*				
Weathering	W	Complete	High	Moderately	Slight	Fresh
	Rating	0	5	15	20	25
Strength (MPa): UCS	S	<20	20–50	40–60	60–100	>100
Is (50)		<0.5	0.5–1.5	1.5–2.0	2–3.5	>3.5
	Rating	0	5	15	20	25
Joint Spacing (m)	J	<0.3	0.3–0.6	0.6–1.5	1.5–2	>2
	Rating	5	15	30	45	50
Bedding Spacing (m)	B	<0.1	0.1–0.3	0.3–0.6	0.6–1.5	>1.5
	Rating	0	5	10	20	30

12.10 Diggability classification

- The diggability in terms of the type of plant required uses the index obtained from the previous table.

Table 12.10 Diggability classification for excavators (adapted from, Scoble and Muftuoglu, 1984).

Class	*Ease of digging*	*Index (W+S+J+B)*	*Typical plant which may be used without blasting*	
			Type	*Example*
I	Very easy	<40	Hydraulic backhoe <3 m³	CAT 235D
II	Easy	40–50	Hydraulic shovel or backhoe <3 m³	CAT 235FS, 235 ME
III	Moderately	50–60	Hydraulic shovel or backhoe >3 m³	CAT 245FS, 245 ME
IV	Difficult	60–70	Hydraulic shovel or backhoe >3 m³: Short boom of a backhoe	CAT 245, O&K RH 40
V	Very difficult	70–95	Hydraulic shovel or backhoe >4 m³	Hitachi EX 100
VI	Extremely difficult	95–100	Hydraulic shovel or backhoe >7 m³	Hitachi EX 1800, O&K RH 75

12.11 Excavations in rock

- The assessment of open excavations is different from excavations in limited space, such as trenches or drilled shafts.

Table 12.11 Excavation in rock (part data from Smith, 2001).

Type of excavation	*Parameter*	*Dig*	*Rip*	*Break/blast*
	Relative cost	1	2 to 5	5 to 25
Large open excavations	N-value RQD SWV	N ≤ 50 to 70 RQD < 25% < 1500 m/s		N = 100/100 mm, Use N* = 300 RQD > 50% 1850–2750 m/s
Trench excavations	SWV	750–1200 m/s Using backhoe		1850–2750 m/s Excavators in large excavations, rock breakers
Drilled shafts	N-value UCS SWV	N < 100/75 mm Use N* < 400 UCS < 20 MPa < 1200 m/s		N* > 600 UCS > 30 MPa >1500 m/s
Tunnels	UCS	UCS < 3 MPa		UCS > 70 MPa

- Seismic Wave Velocity – SWV
- Unconfined Compressive Strength – UCS
- For drilled shafts (AKA bored piers):
  - Limit of earth auger is 15cm penetration in a 5 – minute period –> Replace with rock auger.
  - Rock auger to down-the-hole hammers (break).
  - Refer chapter 22 for additional data table.
- For tunnelling shields:
  - Backhoes mounted inside tunnel shields must give way to road headers using drag pick cutters (similar to rock auger teeth for drilled shafts). Occurs at about UCS = 1.5 MPa.
  - Road headers —> drill and blast or TBM with disk cutters at about UCS = 70 to 80 MPa. Specialist road headers can excavate above that rock strength.

12.12 Rippability rating chart

- Weaver's charts combine concepts of strength, discontinuity, plant and joint characteristics.
- A bit dated as larger equipment currently available.

Table 12.12 Rippability rating chart (after Weaver 1975).

Rock class	*I*	*II*	*III*	*IV*	*V*
Description	Very good rock	Good rock	Fair rock	Poor rock	Very poor rock
Seismic velocity (m/s)	>2150	2150–1850	1850–1500	1500–1200	1200–450
Rating	26	24	20	12	5

(Continued)

Table 12.12 (Continued)

Rock class	*I*	*II*	*III*	*IV*	*V*
Rock hardness	Extremely hard rock	Very hard rock	Hard rock	Soft rock	Very soft rock
Rating	10	5	2	1	0
Rock weathering	Unweathered	Slightly weathered	Weathered	Highly weathered	Completely weathered
Rating	9	7	5	3	1
Joint spacing	>3000	3000–1000	1000–300	300–50	<50
Rating	30	25	20	10	5
Joint continuity	Non-continuous	Slightly continuous	Continuous – no gouge	Continuous – some gouge	Continuous – with gouge
Rating	5	5	3	0	0
Joint gouge	No separation	Slight separation	Separation <1 mm	Gouge <5 mm	Gouge >5 mm
Rating	5	5	4	3	1
*Strike and dip orientation	Very unfavourable	Unfavourable	Slightly unfavourable	Favourable	Very favourable
Rating	15	13	10	5	3
Total rating	100–90	90–70+	70–50	50–25	<25
Rippability assessment	Blasting	Extremely hard ripping and blasting	Very hard ripping	Hard ripping	Easy ripping
Tractor selection	–	DD9G/D9G	D9/D8	D8/D7	D7
Horsepower	–	770/385	385/270	270/180	180
Kilowatts	–	575/290	290/200	200/135	135

- Original strike and dip orientation now revised for rippability assessment.
- + Ratings in excess of 75 should be regarded as unrippable without pre-blasting.

12.13 Bulking factors

- The bulking factor for excavation to transporting to placement and compaction:
  - 0%–10% soils and soft rocks.
  - 5%–20% hard rocks.
- Typically wastage is ~5%.
  - This "transportation" bulking is not the same as bulking factor between placed and in situ material which is typically in the range of 5% to 25% for soft to hard rock material, respectively.

Table 12.13 Bulking factors for excavation to transporting (from Horner, 1988).

Material	*Bulk density (in-situ t/m³)*	*Bulk up on excavation (%)*
Granular soils		
– Uniform sand	1.6–2.1	
– Well graded sand	1.7–2.2	10–15
– Gravels	1.7–2.3	
Cohesive		
– Clays	1.6–2.1	
– Gravelly clays	1.7–2.2	20–40
– Organic clays	1.4–1.7	
Peat/topsoil	1.1–1.4	25–45
Rocks		
– Igneous	2.3–2.8	50–80
– Metamorphic	2.2–2.7	30–60
– Sedimentary	2.1–2.6	40–70
– Soft rocks	1.9–2.4	30–40

12.14 Practical maximum layer thickness

- The practical maximum layer thickness for compaction depends on the material to be compacted and equipment used.
- The table below is for large equipment in large open areas.

Table 12.14 Practical maximum layer thickness for different roller types (Forssblad, 1981).

Roller type static weight (drum module weight in brackets)		*Practical maximum layer thickness (m)*					
		Embankment				*Pavement*	
Type	*Weight (ton)*	*Rock fill*	*Sand/gravel*	*Silt*	*Clay*	*Subbase*	*Base*
Towed vibratory rollers	6	0.75	+0.60	+0.45	0.25	−0.40	+0.30
	10	+1.50	+1.00	+0.70	−0.35	−0.60	+0.40
	15	+2.00	+1.50	+1.00	−0.50	−0.80	–
	6 Padfoot	–	0.60	+0.45	+0.30	0.40	–
	10 Padfoot	–	1.00	+0.70	+0.40	0.60	–
Self propelled roller	7 (3)	–	+0.40	+0.30	0.15	+0.30	+0.25
	10 (5)	0.75	+0.50	+0.40	0.20	+0.40	+0.30
	15 (10)	+1.50	+1.00	+0.70	+0.35	+0.60	+0.40
	8 (4) padfoot	–	0.40	+0.30	+0.20	0.30	–
	11 (7) padfoot	–	0.60	+0.40	+0.30	0.40	–
	15 (10) padfoot	–	1.00	+0.70	+0.40	0.60	–
Vibratory tandem rollers	2	–	0.30	0.20	0.10	0.20	+0.15
	7	–	+0.40	0.30	0.15	+0.30	+0.25
	10	–	+0.50	+0.35	0.20	+0.40	+0.30
	13	–	+0.60	+0.45	0.25	+0.45	+0.35
	18 Padfoot	–	0.90	+0.70	+0.40	0.60	–

- Most suitable applications marked +.
- Thickness in confined areas should be 200 mm maximum loose lift thickness.
- For small sized equipment (<1.5 ton) the applicable thickness is $\frac{1}{2}$ to $\frac{1}{3}$ of the above.

12.15 Large compaction equipment

- Static rollers have a shallower depth of influence than vibratory or impact rollers. The latter rollers can therefore compact to a greater thickness—but this is dependent on material type
- However, in situ control testing typically uses a range of 150 mm to 300 mm and often governs placement thickness.
- Vibration levels also limit the size and type of equipment in urban areas or near to or over culverts and services.

Table 12.15 Large compaction equipment.

Roller type	*Contact pressure (kPa)*	*Use & material type*	*Compaction by*
Smooth drum (with vibration)	300–400	Finishing of surface. All soils except rocky (granular soils)	Static weight (and vibratory)
Pneumatic (4–6 tires)	600–700	Both granular & fine grained soils of low plasticity	Static weight & kneading action
Sheepsfoot/ tamping	1400–7000	Clayey soils	Static weight & kneading action
Grid or mesh (vibratory)	1400–7000	Granular soils. Breaks down soft rock materials	Static weight and vibratory
Impact (non circular)	500–1000	Dry soils in arid regions. Deep lifts	Impact

- Smooth drums can be used with and without vibration
- Impact is also called rolling dynamic compaction. The roller can be 3 to 5 sided. For 8 sides, the roller approaches the application of a round roller.

12.16 Ease of compaction

- Granular materials are easier to compact than fine grained materials
- Clean material (with little to no fines) are best compacted saturated
- Increase in fines and plasticity also increases the difficulty in compaction

Table 12.16 Ease of compaction.

Properties	*Easy*	*Ease of compaction*			*Difficult*
Type	Coarse grained			Fine grained	
Gradation/ plasticity	Well graded	Poorly graded/with fines		Low plasticity	High plasticity
Material	GW/SW	GP/GM/GC	SP/SM/SC	CL/ML/OL	MH/CH/OH
Roller type	←Vibratory Tamping→				

- Material classified as Pt (highly organic) is not included in the table as unsuitable for compaction.

12.17 Compaction requirements for various applications

- The compaction levels should be based on the type of application.
- Compaction assumes a suitable material, as well as adequate support from the underlying material.
- A very high compaction on a highly expansive clay can have an adverse effect in increasing swelling potential.
- The subgrade thickness is typically considered to be 1.0 m, but this varies depending on the application. Refer Chapter 12.

Table 12.17 Compaction levels for different applications.

Class	*Application*	*Compaction level*
1	Pavements Upper 0.5 m of subgrade under buildings	Extremely high
2	Upper 1.5 m of subgrade under airport pavements Upper 1.0 m of subgrade under rail tracks Upper 0.75 m of subgrade under pavements Upper 3 m of fills supporting 1 or 2 story buildings	Very high ≥98% (standard)
3	Deeper parts to 3 m of fills under pavements Deeper areas of fills under buildings Lining for canal or small reservoir Earth dams Lining for landfills	High (95%)
4	All other fills requiring some degree of strength or incompressibility Backfill in pipe or utility trenches Drainage blanket or filter (Gravels only)	Normal ≥90% (standard)
5	Landscaping material Capping layers (not part of pavements) Immediately behind retaining walls (self-compacting material "Drainage gravel" typical)	Nominal (80% to 90%)

- The compaction level may be related to a specified value of CBR strength.

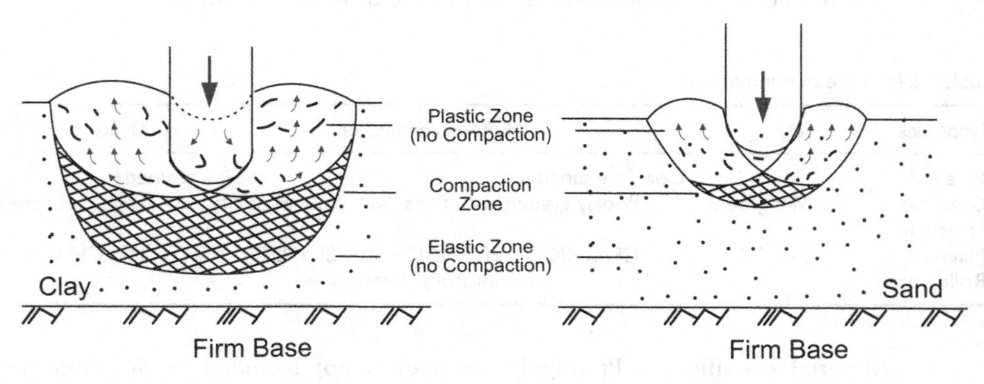

Figure 12.2 Effect of sheepsfoot roller on clays and sands (Here from Holts and Kovacs, 1981, Spangler and Handy, 1982).

12.18 Required compaction

- Relative compaction is the ratio of the field density with the maximum dry density.
- The relative compaction is required in an end product specification.
- Typically many specifications simply use 95% relative compaction. The table shows that this should vary depending on the application. The table is therefore a guide only. A movement sensitive building would require a higher level of compaction, than a less sensitive building such as a steel-framed industrial building.
- Refer previous table for compaction class

Table 12.18 Required compaction level based on various soil types (adapted and modified from Sower's 1979).

Soil type	*Soil classification*	*Required Compaction (% Standard MDD)*				
		Class 1	*Class 2*	*Class 3*	*Class 4*	*Class 5*
Rock sizes	>60 mm	Compaction standards do not apply				
Gravels	GW					
	GP	96	94			
	GM			–		–
	GC				90	
Sands	SW	98				
	SP		96			
	SM					
	SC			92		
Low plasticity	ML	100				
	CL					85
Fine grained	OL		98		92	
High plasticity	MH					
Fine grained	CH	–		96		
	OH		–			

- When the percentage of gravel sizes (>200 mm) exceeds 15%, and the percentage of cobble sizes (60 mm) exceeds 30%, then use a method specification.
- Method specifications require the type and weight of roller to be defined with the number of passes and the lift thickness.
- Important to differentiate between standard and modified compaction (see later sections).

12.19 Comparison of relative compaction and relative density

- The relative compaction applies to material with some fines content.
- The relative density applies to material that is predominantly granular.

Table 12.19 Approximation of relative density to relative compaction (Lee and Singh, 1971).

Granular consistency	*Relative density*	*Relative compaction*
Very dense	100	100
	90	98
	80	96
Dense	70	94
	60	92
Medium	50	98
	40	88
Loose	30	86
	20	84
Very loose	10	82
	0	80

- Relative compaction above 100% and below 70% can occur in practice.

12.20 Field characteristics of materials used in earthworks

- Different material types are required depending on the application.
- The table provides the typical field characteristics for different materials.

Table 12.20a Field characteristics of coarse materials used in earthworks (adapted from BS 6031 – 1981).

Material type	*Description*	*USC symbol*	*Drainage characteristics*	*Shrinkage or swelling properties*	*Value as a road foundation*	*Bulk density before excavation*		*Coefficient of bulking %*
						Dry or Moist Mg/m³	*Submerged Mg/m³*	
Boulders and cobbles	Boulder Gravels	—	Good	Almost none	Good to excellent	—	—	—
Other Materials	Hard broken rock	—	Excellent		Very good to excellent	—	—	20–60
	Soft rocks, Rubble	—	Fair to practically impervious	Almost none to slight	Good to excellent	1,10 to 2.00	0.65 to 1.25	40
Gravels and Gravelly Soils	Well graded	GW	Excellent	Almost none	Excellent	1.90 to 2.10	1.15 to 1.30	10–20
	Poorly graded	GP			Good	1.60 to 2.00	0.90 to 1.25	
	Silty	GM	Fair to practically impervious	Almost none to slight	Good to excellent	1.80 to 2.10	1.10 to 1.30	
	Clayey	GC	Practically impervious	Very slight	Excellent	2.00 to 2.25	1.00 to 1.35	

(Continued)

Table 12.20a (Continued)

Material type	*Description*	*USC symbol*	*Drainage characteristics*	*Shrinkage or swelling properties*	*Value as a road foundation*	*Bulk density before excavation*		*Coefficient of bulking %*
						Dry or Moist Mg/m³	*Submerged Mg/m³*	
Sands and sandy soils	Well graded	SW	Excellent	Almost none	Good to excellent	1.80 to 2.10	1.05 to 1.30	5 to 15
	Poorly graded	SP			Fair to good	1.45 to 1.70	0.90 to 1.00	
	Silty	SM	Fair to practically impervious	Almost none to medium		1.70 to 1.90	1.00 to 1.15	
	Clayey	SC	Practically impervious	Very Slight	Good to excellent	1.90 to 2.10	1.15 to 1.30	
	with silts/clays of high plasticity	OH		High	Very poor	1.50	0.50	—

Table 12.20b Field characteristics of coarse materials used in earthworks (adapted from BS 6031 – 1981).

Material type	*Description*	*USC symbol*	*Drainage characteristics*	*Shrinkage or swelling properties*	*Value as a road foundation*	*Bulk density before excavation*		*Coefficient of bulking %*
						Dry or Moist Mg/m³	*Submerged Mg/m³*	
Inorganic Silts	Low plasticity	ML	Fair to poor	Slight to Medium	Fair to poor	1.70 to 1.90	1.00 to 1.15	20 to 40
	High plasticity	MH	Poor	High	Poor	1.75	1.00	—
Inorganic Clays	Low plasticity	CL	Practically impervious	Medium	Fair to poor	1.60 to 1.80		20 to 40
	High plasticity	CH		High	Poor to very poor			—
Organic	With silts/clays of low plasticity	OL	Practically impervious	Medium to high	Poor	1.45 to 1.70	0.90 to 1.00	20 to 40
	With silts/clays high plasticity	OH		High	Very poor	1.50	0.50	—
Peat	Highly organic soils	Pt	Fair to poor	Very high	Extremely poor	1.40	0.40	—

12.21 Typical compaction characteristics of materials used in earthworks

- This table provides a guide to the use of different materials in method specifications.

Table 12.21 Compaction characteristics of materials used in earthworks (adapted from BS 6031 – 1981).

Material	*Suitable type of compaction plant*	*Minimum number of passes required*	*Maximum thickness of compacted layer*	*Remarks*
Natural rocks • Chalk • Other rock fills	Heavy vibratory roller – >1800 kg/m or Grid rollers – >8000 kg/m or Self-propelled tamping rollers	3 (for Chalk) 4 to 12	500 to 1500 mm depending on plant used	Maximum dimension of rock not to exceed 2/3 of layer thickness
Waste material • Burnt and unburnt colliery shale • Pulverised fuel ash • Broken concrete, bricks, steelworks slag	Vibratory roller, or Smooth wheeled rollers or Self-propelled tamping rollers Pneumatic tyred rollers for pulverised fuel ash only	4 to 12	300 mm	
Coarse grained soils • Well graded gravels and gravely soils • Well graded sands and sandy soils	Grid rollers >5400 kg/m or Pneumatic tyred rollers >2000 kg/wheel or Vibratory plate compactor >1100 kg/m^2 of baseplate Smooth wheeled rollers or Vibratory roller, or Self propelled tamping rollers	3 to 12	75 mm to 275 mm	
Coarse grained soils • Uniform sands and gravels	Grid rollers <5400 kg/m or Pneumatic tyred rollers < 1500 kg/wheel or Vibratory plate compactor Smooth wheeled rollers <500 Kg/m or Vibratory roller	3 to 16	75 mm to 300 mm	
Fine grained soils • Well graded gravels and gravely soils • Well graded sands and sandy soils	Sheepsfoot roller Pneumatic tyred rollers or Vibratory plate compactor >1400 kg/m^2 of baseplate Smooth wheeled rollers or Vibratory roller >700 kg/m	4 to 8	100 mm to 450 mm	High plasticity soils should be avoided where possible

- Thickness of compacted layers depends on type of plant used.
- Different plant types would need to be used for different materials and operating room.

12.22 Suitability of compaction plant

- Effective compaction requires consideration of the type of plant, materials being compacted and environment.

Table 12.22 Suitability of compaction plant (Hoerner, 1990).

Compaction plant	*Principal soil type*							
	Cohesive		*Granular*				*Rock*	
	Wet	*Others*	*Well graded*		*Uniform*		*Soft*	*Hard*
			Coarse	*Fine*	*Coarse*	*Fine*		
Smooth wheeled roller		√√	√√	√√			√√	
Pneumatic tyred roller	√√		√√	√√	√√	O	O	O
Tamping roller	√√	√√	O	√√	O		O	
Grid roller		√√	√√	√√	√√	O	√√	O
Vibrating roller	O	√√	√√	√√	√√	√√	O	√√
Vibrating plate		O	√√	√√	√√	√√	O	√√
Vibro-tamper		√√	√√	√√	√√	√√	O	√√
Power rammer	O	√√	√√	√√			O	O
Dropping weight		√√	√√	√√			√√	√√
Dynamic consolidation	O	√√	√√	√√			√√	√√

√√ Most suited.
O Can be used but les efficiently.

- Tamping rollers includes sheepsfoot and pad rollers.

12.23 Typical lift thickness

- The lift thickness is dependent on the type of material and the plant.
- In limited operating room (e.g. backfill of trenches) small plant are required and the thickness must be reduced from the typical (200 mm to 300 mm) to achieve the appropriate compaction level.
- Adjacent to area sensitive to load and/or vibration (e.g. over services, adjacent to buildings), then medium sized compaction equipment applies. The thickness levels would be smaller than in an open area, but not as small as in the light equipment application.

Table 12.23 Typical lift thickness.

Equipment weight	*Material type*	*Typical lift thickness*	*Comments*
Heavy ≥10 tonnes	Rock fill	750–2000 mm	Applies to open areas and assumes appropriate plant (e.g. vibration in granular materials)
	Sand & gravel	500–1200 mm	
	Silt	300–700 mm	
	Clay	200–400 mm	
Medium (1.5 to 10 tonnes)	Rock fill	400–1000 mm	Some controls required, e.g.
	Sand & gravel	300–600 mm	• Buildings are nearby
	Silt	200–400 mm	• Over service trenches
	Clay	100–300 mm	• Adjacent to walls
Small (<1.5 tonnes)	Rock fill	200–500 mm	In limited areas, e.g.
	Sand & gravel	150–400 mm	• In trenches
	Silt	150–300 mm	• Around instrumentation
	Clay	100–250 mm	• Adjacent to walls

12.24 Maximum size of equipment based on permissible vibration level

- Different weight rollers are required adjacent to buildings. This must be used with a suitable offset distance.
- The table is based on a permissible peak particle velocity of 10 mm/second. Commercial and industrial buildings may be able to tolerate a larger vibration level (20 mm/sec). Conversely, historical buildings and buildings with existing cracks would typically be able to tolerate significantly less vibration (2 to 4 mm/sec).

Table 12.24 Minimum recommended distance from vibrating rollers (Tynan, 1973).

Roller class	*Weight range*	*Minimum distance to nearest building*
Very light	<1.25 tonne	Not restricted for normal road use. 3 m
Light	1–2 tonnes	Not restricted for normal road use. 3 m
Light to medium	2–4 tonnes	5–10 m
Medium to heavy	4–6 tonnes	Not advised for city and suburban streets 10–20 m
Heavy	7–11 tonnes	Not advised for built up areas 20–40 m

- ○ The building criteria of PPV = 10 mm/s is likely to be painful for nearby residents. Refer Chapter 23 for vibration levels

12.25 Compaction required for different height of fill

- The height of fill should also determine the level of compaction, and number of passes.
- The table below shows an example of such a variation, assuming similar materials being used throughout the full height.

Table 12.25 Typical number of roller passes needed for 150 mm thick compacted layer.

Height of fill (m)	*Number of passes of appropriate heavy roller for material type*		
	Clayey Gravel (GC)	*Sandy Clay (CL), Clayey Sand (SC)*	*Clay, CH*
<2.5 m	3	3	4
2.5 to 5.0 m	4	5	6
5.0 to 10.0 m	5	7	8

- ○ The optimum compaction thickness depends on the type of equipment used.

12.26 Typical compaction test results

- Granular material tends to have a higher maximum dry density (MDD) and lower optimum moisture content.
- The optimum moisture content (OMC) increases with increasing clay content.

Table 12.26 Typical compaction test results (Hoerner, 1990).

Material	Type of compaction test	Optimum moisture content (%)	Maximum dry density (t/m³)
Heavy clay	Standard (2.5 kg hammer)	26	1.47
	Modified (4.5 kg hammer)	18	1.87
Silty clay	Standard	21	1.57
	Modified	12	1.94
Sandy clay	Standard	13	1.87
	Modified	11	2.05
Silty gravelly clay	Standard	17	1.74
	Modified	11	1.92
Uniform sand	Standard	17	1.69
	Modified	12	1.84
Gravelly sand/sandy gravel	Standard	8	2.06
	Modified	8	2.15
	Vibrating hammer	6	2.25
Clayey sandy gravel	Standard	11	1.90
	Vibrating hammer	9	2.00
Pulverised fuel ash	Standard	25	1.28
Chalk	Standard	20	1.56
Slag	Standard	6	2.14
Burnt shale	Standard	17	1.70
	Modified	14	1.79

12.27 Field compaction testing

- The sand cone replacement is a destructive test. For large holes or rock fill, water or oil of known density is used.
- The nuclear density gauge is a non-destructive test. Direct Transmission or Back Scatter Techniques used.

Table 12.27 Field compaction testing.

Equipment	Sand cone	Nuclear density gauge
Equipment cost	Low	High
Advantages	Large sample	Fast
	Direct measurement	Easy to redo
	Conventional approach	More tests can be done
Disadvantages	More procedural steps	No sample
	Slow	Radiation
	Less repeatable	Moisture content results unreliable
Potential problems	Vibration	Presence of trenches and objects within 1 m affects results

- Calibration required for nuclear density gauge:
  - Bi-annual manufacturer's certificate. Quarterly checks using standard blocks.
  - Material calibration as required.

- For nuclear density moisture content: Every tenth test should be calibrated with results of standard oven drying.
- For nuclear density measurement: Every 20 tests should be calibrated with results of sand cone.

12.28 Standard versus modified compaction

- Modified compaction developed after standard compaction to account for heavier equipment in use
- There is no direct conversion between modified and standard compactions.
- The table below is a guide, but should be checked for each local site material.
- In general modified compaction is applicable mainly to pavements. It should be avoided in subgrade materials, and especially in expansive clay materials.

Table 12.28 Equivalence of modified and standard compactions (MDD) (adapted from Ervin, 1993).

Material	*Standard/95% modified compactions*	*95% Modified/standard compaction*
Clays/silts	105–115%	85 to 95%
Sandy clays/clayey sands	107–100%	93 to 100%
Sands/gravels/crushed rock	102–97%	98 to 103%

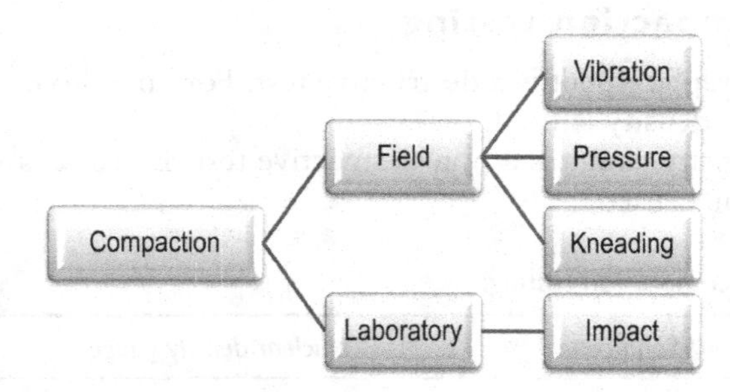

Figure 12.3 Compaction process.

12.29 Application of standard and modified compaction

- The modified test (1940s) was meant to be an improvement on the standard test (1930s) to account for larger equipment. This has been mistranslated to the modified being a better or more modern up to date test.
- 1940s equipment is not modern and history has shown no attempt to modify the test again in the1960s, 1980s or in the 21st century as equipment evolved.
- Modified test in itself does not "improve" compaction. It simply provided a different unit of measurement.
- MDD and OMC are not fundamentals of soils. They are reference units.

Table 12.29 Application of standard and modified compaction tests.

Type of test	*Standard compaction*	*Modified compaction*
Formation type	Subgrades, landscaping areas, "thin" layers	Pavements
Material types	Coarse gravel sizes <10% – see previous table. Expansive clays	Granular materials – but not materials prone to breaking down For significant coarse gravel sizes
Near structures	Adjacent to or over structures and services	Not applicable

- Modified (and high) over-compaction should be avoided on high WPI clays in wet environments
- High suctions induced at modified and high compactions.
- Figure 12.3 illustrates the additional variable that the process in the lab is different from that in the field.

12.30 Effect of excess stones

- The compaction tests are carried out for material passing the 20 mm sieve.
- If the stone fraction is included, it is likely that density and CBR would be higher, but with a lower OMC.
- The field density test that passes could be due to stone sizes influencing the results rather than an acceptable test result as compared to the laboratory reference density.
- The effect of stone size can be calculated, and depends on the quantity and type of material.

Table 12.30 Typical stone size effects.

% of stone sizes (% > 20 mm)	*Actual density compared with lab density*
<10%	Negligible
20%	~10% higher
40%	~20% higher

Chapter 13

Subgrades and pavements

13.1 Types of subgrades

- The subgrade is the natural material immediately below the pavement.
- The depth of "subgrade" varies depending on the type of load applications and the pavement type. Typically 1.0 m depth is used but this applies mainly to major roads.

Table 13.1 Depth of subgrades.

Application	*Type of load*	*Pavement type*	*Subgrade depth*
Airport	Dynamic/extra heavy	Flexible	2.0 m
		Rigid	1.5 m
Mine haul access	Dynamic/very heavy	Flexible	1.5 m
Rail	Dynamic/very heavy	Flexible/rigid	1.25 m
Major roads	Dynamic/heavy	Flexible	1.0 m
		Rigid	0.75 m
Industrial building	Dynamic/static/heavy	Rigid	0.75 m
Minor roads	Dynamic/medium	Flexible	0.75 m
		Rigid	0.5 m
Commercial and residential buildings	Static/medium	Rigid	0.5 m
Walkways/bike paths	Static/light	Rigid/flexible	0.25 m

- Contact pressures for flexible foundations on sands and clays approximately similar.
- Contact pressures for rigid foundations:
  - On sands, maximum pressure is at middle.
  - On clays, maximum pressure is at edge.
- Test location layout should reflect the above considerations.
- Subgrade refers to only direct bearing pressures, while material below the subgrade should also provide adequate support, although at reduced pressures. This underlying material can also affect movement considerations.
- Arguably for thick pavement designs/capping layers, the subgrade is now reduced to the top 0.5 m depth.

13.2 CBR laboratory model

- The California Bearing Ratio (CBR) was first used by the California State Highways by Porter in the 1930s.
- The CBR is a strength index and not an actual strength value.
- The "standard" crushed rock value of 100% applies.
- Key assumptions provided in Table and Figure 13.1, and expanded on in subsequent Tables.
- This illustrates many of the key assumptions do not apply directly. Other more current tests such as repeat load testing (which can provide a direct modulus value instead of a correlated value, and cyclic loading rather than a one off load) is significantly more expensive and with more time required relatively.
- Hence despite its many short comings (given current technology) the CBR remains the de facto "standard" used in industry.

Table 13.2 Assumptions in CBR test.

Parameter	*Assumptions*
Placement	Compacting to specified density and moisture content
Material used	20 mm maximum size – but 38 mm in large CBR mould
CBR value	100% – crushed rock is reference
Surcharge used	4.5 kg represents overburden
Swell value	Linked to overburden which can also affect CBR value itself
Confinement	Steel base and sides
Soaking	4 day soaked condition represents worst case condition
Failure	Quasi-static bearing failure

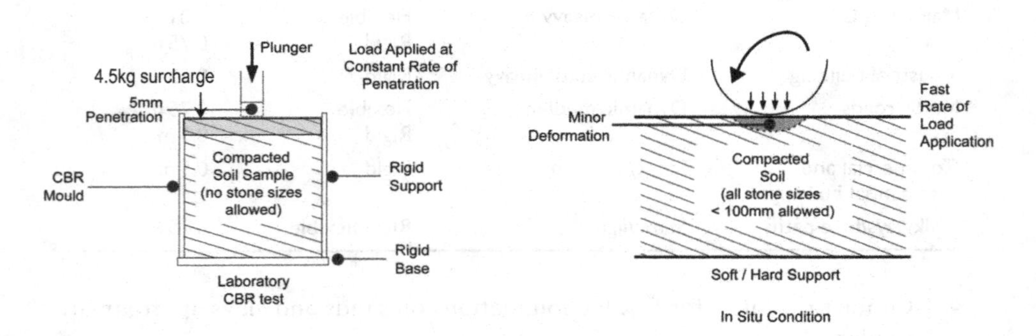

Figure 13.1 Laboratory CBR model versus field condition.

13.3 CBR tests in subgrade assessment

- The CBR test is the most common method of subgrade assessment.
- The lab CBR should not be directly translated as a design value. Refer also Figure 7.4.
- All key parameters shown in Table should be use from the lab report – there is a tendency to focus on the CBR % value only, when it may be an irrelevant value.

Table 13.3 Application of CBR test.

Parameter	*Typical condition*	*Comment on use of result*
Level of compaction – % MDD	At MDD	Test is misleading if specifications uses other than a 100% MDD e.g. 90%, 95%
Compaction moisture content – % OMC	At OMC	Equilibrium moisture content (EMC) more relevant. For rainfalls 500 to 1000 mm then EMC = OMC and results relevant
Percentage material above 38 mm discarded in test	All sample used	Test is misleading where significant oversize forms bulk of sample – yet discarded from test
CBR value	0 to 100%	Values can be above 100%. Unusual to use CBR > 20% for subgrades irrespective of test value. For CBRs < 10% swell may govern design and not strength
Swell value	0%–10%	For swell >3% movements likely to govern rather than CBR strength index
Surcharge used	Typically 4.5 kg	Especially for high swell soils, this surcharge should be modified as it affects both swell and CBR. 4.5 kg is representative of 125 mm of pavement only
Soaking	4 days	7 or 10 day soaked in tropical or sub-tropical climates for low permeability clays
Failure	Rutting	Design is a serviceability deformation (plastic strain) based on number of cycles (progressive strain) until the terminal rut criteria is reached and not a bearing capacity failure
Rigid Boundary	Flexible	Compatibility between underlying material and compacted material must be present to achieve required compaction or CBR

- Common misleading CBR results
  - construction specifications requires 95% standard compaction yet design value was based on lab testing at 100% standard compaction.
  - applying 100% MDD and 100% OMC as a target condition.
  - using CBR lab results without reviewing comments on percentage sample used or discarded.
  - using CBR value in design when swell governs.
  - using the CBR lab value as a design value without factoring for underlying material or climate.
  - Relevant surcharge not being applied.
  - Some of the above may increase the CBR value while others may decrease the value.

13.4 CBR reporting

- CBR should be reported to the nearest value as shown in the table. This reflects the degree of accuracy of this index testing
- CBRs < 8% have poor reproducibility.
- The laboratory density and moisture ratio is reported to the nearest 1%.
- The percentage of material retained on the 19 mm sieve and whether excluded or replaced is relevant to the test interpretation on its relevance to the field.

Table 13.4 CBR reporting (AS 1289).

CBR %	*Report value to the nearest*
<5	0.5
5–20	1
20–50	5
>50	10

13.5 CBR soaked and unsoaked tests

- Soaked or unsoaked tests are used depending on the site conditions.
- Soaked test is applicable for high water tables, floodways or where high inundation likely, poor drainage conditions or where seepage likely/base of cutting (through cut).
- The 4 day soaked is the "standard" test and was originally conceived as the worst case condition the soil may experience. In tropical areas or areas of high rainfall environments, flooding can occur longer than 4 days – use 7 or 10 day soaked. If the 4 day soaked is used then correction factors needs to be applied (refer Table 13.11).

Table 13.5 Soaked and unsoaked tests.

Rainfall (mm)	*Drainage condition*		*Compaction moisture content % (EMC)*	
	Excellent to good	*Fair to poor*	*WPI < 2200*	*WPI > 2200*
<500 mm	Unsoaked to 4 day soaked	1 to 4 day soaked	≤OMC	<OMC
500–1000	Unsoaked to 4 day soaked	4 to 7 day soaked	OMC	~OMC
>1000 mm	Unsoaked to 4 day soaked	4 to 10 day soaked	OMC	>0MC

13.6 Subgrade strength classification

- The subgrade strength is here defined in terms of the soaked CBR.
- Higher soaked time applies to low permeability soils.

Table 13.6 Subgrade strength classification.

Soaked CBR	*Strength classification*	*Comments*
<1%	Extremely weak	Geotextile reinforcement and separation layer with a working platform typically required.
1%–2%	Very weak	Geotextile reinforcement and/or separation layer and/or a working platform typically required.
2%–3%	Weak	Geotextile separation layer and/or a working platform typically required.
3%–10%	Medium	Geotextile separation layer and/or a working required if a high water table or poor drainage area.
10%–30%	Strong	Good subgrade to sub-base quality material.
>30%	Extremely strong	Sub-base to base quality material.

- Extremely weak to weak layers need a capping layer.
- Capping layer also referred to as a working platform.
- Design subgrade CBR values above 20% seldom used irrespective of test results.
- In-situ subgrade CBR = 6/subgrade deflection (mm) can be obtained from Benkeleman beam testing (Lay, 1990).

13.7 Damage from volumetrically active clays

- Volumetrically active materials are also called shrinkage clays, expansive clays, reactive clays, and plastic clays.

Table 13.7 Damage to roadways resulting from volumetrically active clays.

Mechanism	*Effect on roadway*
Swelling due to wetting/ Shrinkage due to drying	Longitudinal cracks on pavements and/or Unevenness of riding surface Culverts can rise out of ground unless suitably restrained
Swelling pressures where movement is prevented	Cracking of culverts High Pressures of retaining walls greater than at rest earth pressure coefficient
Loss of strength due to swelling or shrinkage	Localised failure of subgrade Slope failures of embankments

13.8 Subgrade volume change classification

- A subgrade strength criteria may be satisfied, but may not be adequate for volume change criteria, which must be assessed separately.
- The Weighted Plasticity Index (WPI) (Table 7.6) can be used for an initial assessment although the soaked CBR swell provides a better indicator of movement potential for design purposes.
- An approximate comparative classification is provided in this table.
- Swell is based on sample compacted to MDD (Standard Proctor) at its OMC and using a 4 day soak.

Table 13.8 Subgrade volume change classification for embankments.

Weighted plasticity index	*Soaked CBR swell*	*Subgrade volume change classification*	*Comments*
<1200	<1%	Very Low	Generally acceptable for base sub-base
1200–2200	1%–2%	Low	Applicable for capping layers
2200–3200	2%–3%	Moderate	Design for some movements
3200–5000	3%–5%	High	Unsuitable directly below pavements
>5000	>5%	Very High	Should be removed and replaced or stabilised

- Materials with a very low volume change potential tend to be CBR $\geq$ 10%.
- Clayey materials may still have swell after 4 days. Any WPI > 3200 should use a 7 day soaked test.

13.9 Minimising subgrade volume change

- Providing a suitable non-volumetrically active capping layer is the most cost effective way to minimise volume change.
- If sufficient non-reactive materials are unavailable then stabilisation of the subgrade may be required, for the thickness indicated.

Table 13.9 Typical improved subgrade to minimise volume change.

Subgrade volume change classification	*Thickness of non-reactive overlying layer*	
	Fills	*Cuts*
Very low/low	Subgrade strength governs pavement design	
Moderate	0.5 m to 0.75 m	0.25 m to 0.5 m
High	0.75 m–1.25 m	0.5 m to 0.75 m
Very High	1.0–2.0 m	0.75 m–1.5 m

- Thickness of overlying layer excludes pavement (125 mm assumed – i.e. "standard" 4.5 kg surcharge).
- Indicative thickness only. Depends also on climatic environment, which influences active zone.
- Pavement thickness (based on strength design) may be sufficient for no improved subgrade layer.
- Remoulded clays (fills) have a higher potential for movement (in its first few years of wet/dry cycles) than undisturbed clay subgrades (cuts). (Refer Figure 13.2).
- However the potential for rebound must also be checked for deep cuttings. Rebound is not a cyclic movement.
- Non-reactive material has WPI < 1200.

13.10 Subgrade moisture content

- The key to minimising initial volume change is to place the material as close as possible to its equilibrium moisture content and density.
- Optimum moisture content is a construction reference point, but should not be translated into a target or best condition
- Equilibrium moisture content (EMC) depends on its climatic environment as well the material properties itself.
- The data below was established for equilibrium conditions in Queensland, Australia.

Table 13.10 Equilibrium Moisture Conditions based on annual rainfall (Look, 2005).

Median annual rainfall (mm)	*Equilibrium moisture content*		
	WPI < 1200 (Low correlation)	*WPI = 1200–3200 (Medium correlation)*	*WPI > 3200 (High correlation)*
Median value for all rainfall	80% OMC	100% OMC	115% OMC
≤500	50%* to 90% OMC	70% to 100% OMC	50% to 80% OMC
500–1000			70% to 120% OMC
1000–1500	70% to 110% OMC	100% to 130% OMC	110% to 140% OMC
≥1500			130% to 160%* OMC

*Beyond practical construction limits

- The above equilibrium conditions also influence the strength of the subgrade.
- Use above EMC to obtain corresponding CBR value.
- Or apply correction factor to soaked CBR as in next section.
- The above can be summarised as:
  - For low WPI material, the EMC is dry or near OMC.
  - For medium WPI material, the EMC is near OMC.
  - For high WPI material, the EMC is sensitive to climate, and varies from dry of OMC for dry climates to wet of OMC for wet of climates.
- Using OMC and MDD as target condition applies to countries/states with rainfall of 500–1000 mm rainfall (e.g. UK/California). Above or below that rainfall (climate) the EMC applies.

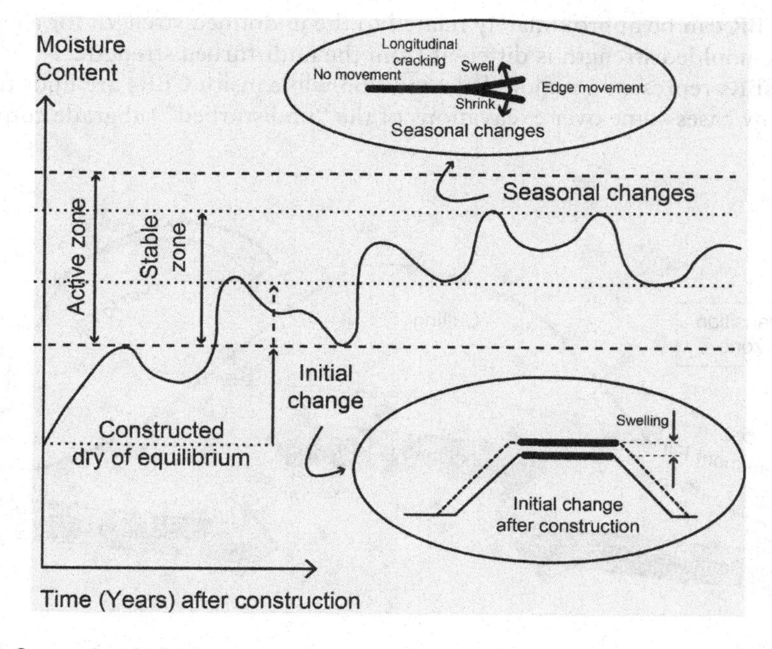

Figure 13.2 Seasonal and initial movements.

13.11 Subgrade strength correction factors to soaked CBR

- The CBR value needs to be factored to be used appropriately in its climatic environment.
- In many cases the soaked CBR may not be appropriate, and the unsoaked value should be used.
- For high rainfall environments, a correction factor is not required if a 7 day soaked value is obtained

Table 13.11 Correction factor to soaked CBR to estimate the equilibrium In Situ CBR (Mulholland et al., 1985).

Climatic zone	*Soil type*	
	Soil with PI < 11	*Soil with PI > 11*
Rainfall ≤ 600 mm	1.0–1.5	1.4–1.8
600 mm < Rainfall ≤ 1000 mm	0.6–1.1	1.0–1.4
Rainfall > 1000 mm	0.4–0.9	0.6–1.0

  - The 4 day soaked CBR models a 4 day flood condition (a worst case scenario at the location it was developed). In tropical climates and some locations in Australia 7 day soaked is more applicable (or a reduction as noted in table) to account for significantly longer durations of rainfall and flood events

13.12 Approximate CBR of clay subgrade

- The CBR can be approximately related to the undrained strength for a clay.
- The remoulded strength is different from the undisturbed strength.
- Lab CBRs represent a remoulded condition while insitu CBRs are undisturbed.
- In many cases some over excavations of the "undisturbed" subgrade cutting may occur.

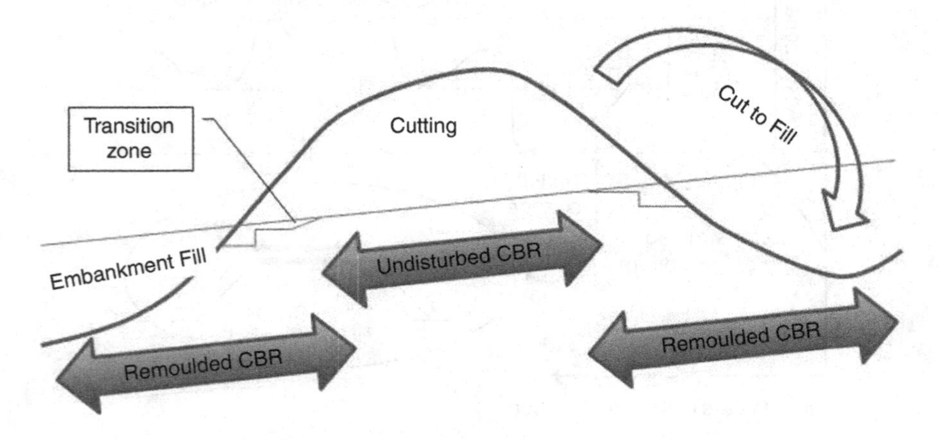

Figure 13.3 Undisturbed and remoulded CBR subgrades.

Table 13.12 Consistency of cohesive soil.

Term	*Field assessment*	*Undrained shear strength (kPa)*	*Approximate CBR %*	
			Undisturbed	*Remoulded*
Very soft	Exudes between fingers when squeezed	<12	≤1	≤1
Soft	Can be moulded by light finger pressure	12–25		1–2
Firm	Can be moulded by strong finger pressure	25–50	1–2	2–4
Stiff	Cannot be moulded by fingers Can be indented by thumb pressure	50–100	2–5	5–10
Very stiff	Can be indented by thumb nail	100–200	5–10	10–20
Hard	Difficult to indented by thumb nail	>200	>10	>20

13.13 Typical values of subgrade CBR

- The design subgrade CBR values depends on:
  - Site drainage.
  - Site rainfall/climate.
  - Soil classification.
  - Compaction level.
  - Confinement.

Table 13.13 Typical values of subgrade CBR.

Soil type	*USC symbol*	*Description*	*Drainage*	*CBR % (Standard)*
Competent broken rock, gravel sizes	GW, GP	e.g. Sandstone, granite, greywacke Well graded, poorly graded	All	20
Competent broken rock – some fines formed during construction gravel sizes, sands	GM, GC SW, SP	e.g. Phyllites, siltstones Silty, clayey, well graded, poorly graded	All	15
Weathered rock likely to weather or degrade during construction	ALL	e.g. Shales, mudstones	All	Treat as soil below
Sands	SM, SC	Silty, clayey	Good	10
Sands	SM, SC	Silty, clayey	Poor	7
Inorganic silts	ML	Low plasticity	Good	
Inorganic silts	ML	Low plasticity	Poor	5
Inorganic clays	CL	Low plasticity	Good	
Inorganic clays	CH	High plasticity	Good	
Inorganic silts	MH	High plasticity	Good	3
Inorganic clays	CL	Low plasticity	Poor	
Inorganic silts	MH	High plasticity	Poor	<3
Inorganic clays	CH	High plasticity	Poor	

  - The issues with converting CBR to modulus values are discussed in later sections.

- Underlying support is also required to obtain the above CBR values (Chapter 11). Figure 13.1 shows the laboratory model of a rigid base must be translated to the field to use the laboratory CBR value
- At the edge of an embankment (lack of edge support), CBR value is not applicable.

13.14 Properties of mechanically stable gradings

- The gradation is the key aspect to obtaining a mechanically stable pavement.
- This is the first step in development of a suitable specifications.

Table 13.14 Properties of mechanically stable gradings for pavements (adapted from Woolorton, 1947).

Application	*% passing 75 micron "fine material"*	*% passing 425 micron medium sand or less*	*% >2 mm gravel size*
Unstable in wet due to high volume change	>50%	>80%	0%
Light traffic	40% to 20%	70% to 40%	0% to 40%
Heavy traffic wearing course	20% to 10%	40% to 20%	40% to 60%
Heavy traffic base course	15% to 10%	20% to 10%	60% to 70%

13.15 Soil stabilisation with additives

- The main types of additives are lime, cement and bitumen.

Table 13.15 Soil stabilisation with additives.

Soil property		*Typical additive*
% Passing 75 micron sieve	*Atterberg*	
>25%	PI < 10%	Bitumen, cement
	PI > 10%	Cement, lime
<25%	PI < 10%	Cement
	PI = 10–30%	Lime, cement, lime + bitumen
	PI > 30%	Cement, lime + cement

- Cement additive typically 5 to 10%, but can vary from 0.5 to 15%. Best suited to clayey sands (SC).
- Lime additives typically 1.5% to 8%. Best suited to silts and clays.
- Bitumen additives typically 1 to 10%. Best suited to clayey gravels (GC).

13.16 Soil stabilisation with cement

- If the subgrade has insufficient strength then stabilisation of the subgrade may be required.
- Adding cement is just one of the means of acquiring additional strength.
- Above 8% cement may be uneconomical, and other methods should be considered.

Table 13.16 Typical cement content for various soil types (Ingles, 1987).

Soil type		*Cement requirement*
Fine crushed rock		0.5%–3%
Well graded and poorly graded gravels	GW, GP	2%–4%
Silty and clayey gravels	GM, GC,	
Well graded sands	SW	
Poorly graded sand, silty sands, clayey sands	SP, SM, SC	4%–6%
Sandy clay, silty clays Low plasticity inorganic clays and silts	ML, CL	6%–8%
Highly Plastic inorganic clays and silts	MH, CH	8%–12%
Organic clays	OL, OH	12%–15% (pre treatment with lime)
Highly organic	Pt	Not suitable

- The table presents a typical range, but a material specific testing programme should be carried out to confirm the most economical cement content.

13.17 Effect of cement soil stabilisation

- The stabilisation of pavement layers is also used to produce higher strengths, and minimise the pavement thickness.
- These may be cement treated base (CTB) or cement treated sub bases (CTSB).

Table 13.17 Soil stabilisation (Lay, 1990; Ingles, 1987).

Stages	*Soil*	*Modified soil*	*Cemented soil*	*Lean Mix*	*Concrete*
Cement content for granular material	0%	<5%		>5%	>15%
Tensile strength		<80 kPa		>80 kPa	
Failure mode	Plastic ---------------------------------------→ Brittle				

- For each 1% cement added, an extra unconfined compressive strength of 500 kPa to 1000 kPa may be achieved.
- Shrinkage concerns for cement >8%.
- Tensile strength ~10% Unconfined compressive strength.

13.18 Soil stabilisation with lime

- Applicable mainly to high plasticity materials.
- The table presents a typical range, but a material specific testing programme should be carried out to confirm the most economical lime content.
- Use the lime demand test first (pH of 12.4 to produce long term reactions), before testing for other material properties. Without this test, there would be uncertainty on the permanent nature of the lime stabilisation.
- Soil modification is used as a construction expedient while soil stabilisation is the permanent process.

Table 13.18 Typical lime content for various soil types (Ingles, 1987).

Soil type		*Lime requirement*
Fine crushed rock		0.5%–1%
Well graded and poorly graded gravels	GW, GP	0.5–2%
Silty and clayey gravels	GM, GC	
Well graded and poorly graded sands	SW, SP	
Silty sands, clayey sands	SM, SC	2%–4%
Sandy clay, silty clays, low plasticity inorganic clays and silts	ML, CL	4%–6%
Highly plastic inorganic silts	MH	
Highly plastic inorganic clays	CH	5%–8%
Highly organic	OL, OH, Pt	Not recommended

- For strength improvements requirements, the UCS or CBR test is used in the literature.
- Test results may show CBR values above 100%. Irrespective of test results a subgrade design CBR of 20% maximum should be used.
- For strength, a target CBR value (at 7 days) of 60% used.
- For strength, a target UCS value (at 28 days) of 1MPa used. 7 Day UCS ~ ½ 28 Day UCS.
- Hydrated lime tends to be used in the laboratory but quicklime is used more extensively in the field
- 3% quicklime is approximately 4% hydrated lime.
- Less dusting and safety requirements are the benefits of hydrated lime, but this is usually offset by the disadvantage that hydrated lime is both more expensive and less concentrated than quicklime.
- Soil modification is used as a construction expedient while soil stabilisation is the permanent process.

13.19 Lime stabilisation rules of thumb

- Rules of thumbs are useful first indicators, but are mainly guides to assist the likely mix for the laboratory testing. These guides should not be used as a standalone.

Table 13.19 Rules of thumb for liming requirements.

Parameter	*Lime addition*	*Comment*
Percentage clay	1% lime for each 10% of clay in sample	Therefore up to 10% lime in 100% clay material, but unusual to use above 8%
Lower limit	2% lime	Below this % is not usually cost effective
Plasticity index	1% lime results in 3% reduction in PI	Unusual to target PI < 10%
Field liming	Add 1% additional lime above the laboratory test requirements	Accounts for unevenness in mixing in the field

13.20 Soil stabilisation with bitumen

- Bitumen is a good waterproofing agent, and preserves the natural dry strength.
- Asphalt, bitumen and tar should be distinguished (Ingles, 1987). These material properties are temperature dependent:

- Asphalt – most water repellent, but most expensive.
- Bitumen – most widely available.

Table 13.20 Typical bitumen content for various soil types (Ingles, 1987).

Soil type		*Bitumen requirement*
Fine crushed rock – open graded		3.5%–6.5%
Fine crushed rock – dense graded		4.5–7.5%
Well graded and poorly graded gravels	GW, GP	
Silty and clayey gravels	GM, GC	
Well graded and poorly graded sands	SW, SP	2%–6%
Silty sands	SM	
Clayey sands	SC	
Sandy clay, silty clays, low plasticity inorganic clays and silts	ML, CL	
Highly plastic inorganic silts	MH	
Highly plastic inorganic clays	CH	4%–7%
Highly organic	OL, OH, Pt	Not recommended

13.21 Pavement strength for gravels

- The pavement strength requirement is based on the type of road.

Table 13.21 Typical pavement strength requirements.

Conditions	*CBR strength*	*Comments*
"Standard" requirements	80% Soaked	On major roads at least 100 mm of pavement layer >80% CBR
Low traffic roads	60% unsoaked	Top 100 mm of base layer
	30%	Sub base
Rural traffic roads/arid to semi-arid regions	>30% unsoaked	Upper sub base
	>15%	Lower sub base

13.22 CBR values for pavements

- The applicable CBR values depend on both the pavement layer and closeness to the applied load.

Table 13.22 CBR values for pavements.

Pavement layer	*Design traffic (ESA Repetitions)*	*Minimum CBR %*
Base	$>10^6$	80
	$<10^6$	60
Upper sub base	$>10^6$	45
	$<10^6$	35
Lower sub base	$>10^6$	35
	$<10^6$	25
Capping	N/A	10

13.23 CBR swell in pavements

- The CBR swell should also be used to assess pavement quality.

Table 13.23 Soaked CBR swell in pavement materials.

Pavement layer	*Pavement type*	*Soaked CBR swell (%)*
Base	Rigid, flexible, CTB	<0.5
Subbase	Rigid, CTSB	<1.0
	Flexible	<1.5
Capping	Rigid overlying	<1.5
	CTB overlying with granular sub base	<2.0
	CTSB overlying	<1.5
	Flexible overlying	<2.5

- For low rainfall areas (<500 mm), soaked CBR swell <1.5% may be acceptable for the base layer.

13.24 Plasticity index properties of pavement materials

- Plasticity Index of the pavement influences its performance.

Table 13.24 Plasticity index for non-standard materials (adapted from Vic Roads 1998).

Pavement type	*Pavement layer*	*Rainfall*	
		<500 mm	*>500 mm*
Unsealed	Base/shoulder	PI < 15%	PI < 10%
	Sub base	PI < 18%	PI < 12%
Sealed	Base/shoulder	PI < 10%	PI < 6%
	Sub base	PI < 12%	PI < 10%

- Pavements for unsealed roads/rural roads/light traffic based on 80% probability level.
- Pavements for sealed roads/moderate to high traffic based on 90% probability level – slighter thicker pavement.

13.25 Typical CBR values of pavement materials

- The modified compaction is typically applied to paving materials. Do not use for subgrade materials.
- The achieved density and resulting CBR is higher than the standard compaction result.
- The modified CBR result for the full range of USC materials is provided for completeness, but non-granular materials would not be applicable to paving materials.

Table 13.25 Typical CBR values for paving materials.

Soil Type	*Description*	*USC symbol*	*CBR % (Modified)*
Gravels	Well graded	GW	40–80
	Poorly graded	GP	30–60
	Silty	GM	20–50
	Clayey	GC	20–40
Sands	Well graded	SW	20–40
	Poorly graded	SP	10–40
	Silty	SM	10–30
	Clayey	SC	5–20
Inorganic silts	Low plasticity	ML	10–15
	High plasticity	MH	<10
Inorganic clays	Low plasticity	CL	10–15
	High plasticity	CH	<10
Organic	With silt/clays of low plasticity	OL	<5
	With silt/clays of high plasticity	OH	<5
Peat	Highly organic silts	Pt	<5

- Actual CBRs depends on the grading, maximum size and percentage fines.
- CBRs of ≥10% for clays should not be used, irrespective of the results or above table which is for modified compaction.

13.26 Typical values of pavement modulus

- Pavements require compaction to achieve its required strength and deformation properties. The level of compaction produces different modulus.
- Existing pavements would have reduced values for asphalt and cemented materials.

Table 13.26 Typical elastic parameters of pavement layers (Austroads, 2004 and 1992).

Pavement layer			*Typical modulus (MPa)*	*Typical Poisson's ratio*
Asphalt at temperature	10°C		11,500	0.4
	25°C		3,500	0.4
	40°C		620	0.4
Unbound granular (Modified/standard Compaction) below thin bituminous surfacings	High quality crushed rock	Over granular material	500/350	0.35
	Base quality gravel		400/300	0.35
	Sub base gravel		300/250	0.35
Cemented material (Standard compaction)	Crushed rock, 2 to 3% cement (lean mix)		7,000	0.2
	Base quality natural gravel 4 to 5% cement		5,000	0.2
	Sub base quality natural gravel 4–5% cement		2,000	0.2

- Degree of anisotropy = Ratio of vertical to horizontal modulus.
- Degree of anisotropy = 1 for asphalt and cemented material.
- Degree of anisotropy = 2 for unbound granular material.

- Flexural modulus applies to pavement layers, while compressive modulus applies to subgrade in pavement design.

13.27 Typical values of existing pavement modulus

- The moduli for existing asphalt and cemented materials are reduced due to cracking.
- Apply cracked value when used with clay subgrades with WPI >2200.

Table 13.27 Typical elastic parameters of pavement layers (Austroads, 2004).

Existing pavement layer		*Cracked modulus (MPa)*
Asphalt at temperature	15°C	1,050
	25°C	880
	40°C	620
Cemented material	Post fatigue phase	500

13.28 Equivalent modulus of sub bases for normal base material

- The equivalent modulus combines the effect of different layer. A minimum support requirement is required.
- The table applies for sub-base materials with a laboratory soaked CBR value of less than 30% with a value of E = 150 MPa.
- These values apply in the back-analysis of an existing pavement system.

Table 13.28 Selecting of maximum modulus of sub-base materials (Austroads, 2004).

Thickness of overlying material	*Suggested vertical modulus (MPa) of top sub-layer of normal base material*					
	Modulus of cover material (MPa)	*1000*	*2000*	*3000*	*4000*	*5000*
40 mm		350	350	350	350	350
75 mm		350	350	340	320	310
100 mm		350	310	290	270	250
125 mm		320	270	240	220	200
150 mm		280	230	190	160	150
175 mm		250	190	150	150	150
200 mm		220	150	150	150	150
225 mm		180	150	150	150	150
≥250 mm		150	150	150	150	150

- Cover material is either asphalt or cemented material or a combination of these materials.

13.29 Equivalent modulus of sub bases for high standard base material

- As above for normal base material.
- The table applies for sub-base materials with a laboratory soaked CBR value greater than 30% with a value of E = 210 MPa used.

Table 13.29 Selecting of maximum modulus of sub-base materials (Austroads, 2004).

Thickness of overlying material	*Suggested vertical modulus (MPa) of top sub-layer of high standard base material*				
Modulus of cover material (MPa)	*1000*	*2000*	*3000*	*4000*	*5000*
40 mm	500	500	500	500	500
75 mm	500	500	480	460	440
100 mm	500	450	410	390	360
125 mm	450	390	350	310	280
150 mm	400	330	280	240	210
175 mm	360	270	210	210	210
200 mm	310	270	210	210	210
225 mm	260	210	210	210	210
≥250 mm	210	210	210	210	210

- Cover material is either asphalt or cemented material or a combination of these materials.

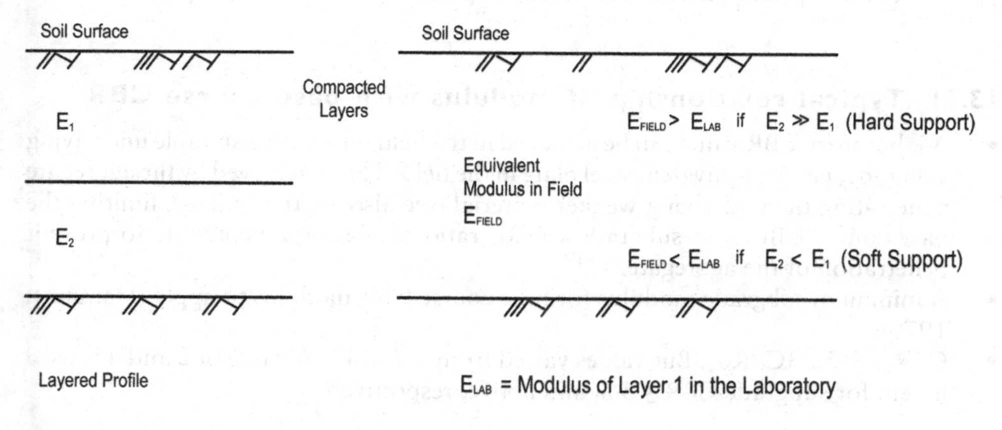

Figure 13.4 Equivalent modulus.

13.30 Typical relationship of modulus with subgrade CBR

- This is the resilient modulus value (dynamic modulus of elasticity), which is significantly higher than the foundation (secant) modulus.
- The CBR test is carried out at a high strain level and low strain rate while subgrades under pavements experience a relatively low strain level and higher stress rates.
- Design modulus = Equivalent modulus, which is dependent on materials above and below.

Table 13.30 CBR/modulus subgrade relationships.

Reference	*Relationship*	*Comments*	*E (MPa) based on CBR (%)*		
			2%	*5%*	*10%*
Heukelom and Klomp (1998)	E ~ 10 CBR (actually 10.35 CBR)	Most common relationship (Range of 20 to 5 for upper to lower bound). CBR < 10%	20	50	N/A
Croney and Croney (1991)	E = 6.6 CBR (from repeat load test data – significant strain)	Zone defined by E = 10 CBR to E = 20 CBR using wave velocity tests – low strain	13	33	66
NAASRA (1950)	$E = 16.2\ CBR^{0.7}$ $E = 22.4\ CBR^{0.5}$	For CBR < 5% For CBR > 5%	26	50	81
Powell, Potter, Mayhew and Nunn (1984)	$E = 17.6\ CBR^{0.64}$	A lower bound relationship (TRRL Study) For CBR < 12%	27	49	77
Angell (1988)	$E = 19\ CBR^{0.68}$	For CBR < 15%	30	57	91

- For weathered rock subgrade E = 2,000 MPa (typically).
- For competent unweathered rock subgrade E = 7,000 MPa (typically).

13.31 Typical relationship of modulus with base course CBR

- A laboratory CBR value can be achieved in the field only with a suitable underlying subgrade, i.e. An equivalent steel plate in the field. This is achieved by the aggregate penetrating the underlying weaker material (see also aggregate loss), limiting the base course CBR_{BC} to subgrade CBR_{SG} ratio or placing a geotextile to prevent penetration of the aggregate.
- A minimum subgrade modulus for base course CBR modulus to apply (Hammitt, 1970).
- $CBR_{BC} = 5.23CBR_{SG}$. But values varied from 1.7 to 17. A value of 2 and 4 is used herein for subgrade CBR ≤ 5% and above, respectively.

Table 13.31 CBR/modulus base relationships.

Reference	*Relationship*	*Comments*	*E* (MPa) based on		
			CBR = 20%	*CBR = 50%*	*CBR = 80%*
AASHTO (1993)	$E = 36CBR^{0.5}$	For CBR > 10%	161	228	322
NAASRA (1950)	$E = 22.4CBR^{0.5}$	For CBR > 5%	100	142	200
Queensland Main Roads (1988)	$E = 21.2CBR^{0.64}$	For CBR > 15% maximum of 550 MPa	144	225	350
Minimum subgrade modulus for base CBR modulus to apply			3.5%	10%	20%

13.32 Aggregate loss to weak subgrades

- Aggregate is lost into the underlying subgrade depending on its strength and the compaction effort being applied to the aggregate to attain its potential (lab) strength.

Table 13.32 Aggregate loss to weak subgrades (FHWA, 1989).

CBR (%) of subgrade	% of design aggregate thickness
<0.5	85–115%
0.5	65–95%
1.0	40–70%
1.5	25–50%
2.0	15–35%
2.5	5–25%
3.0	0–20%

13.33 Elastic modulus of asphalt

- Asphalt strength varies with temperature.
- Weighted mean annual temperature (WMAPT) is used. These temperatures correspond to depth of 50 mm to 75 mm for the asphalt layer.
- Asphalt is a visco-elastic material but at normal operating temperatures, it may be treated as an elastic solid.
- Asphalt response is linear below 1000 microstrain.
- Other variables such as air voids, asphalt content, loading rate, age of asphalt, etc., also affect the modulus values.
- Poisson's ratio of 0.4 typical.

Table 13.33 Asphalt temperature zones and corresponding modulus.

Typical Queensland area	*Temperature range °C*	*Representative temperature °C*	*Asphalt modulus MPa*
Western Queensland, Mt Isa, Cairns, Townsville, Barcaldine	WMAPT > 35	36	970
Roma, Gladstone, Mackay, Gladstone	35 > WMAPT > 32	30	1400
Brisbane, South East Queensland	32 > WMAPT > 29	30	2000
Toowoomba, Warwick, Stanthorpe	29 < WMAPT	28	2500

13.34 Poisson ratio

- Some variability is likely in the vertical, horizontal and cross direction for all materials.

Table 13.34 Poisson ratio of road materials.

Material	*Poisson ratio*
Asphaltic	0.40
Granular	0.35
Cement Treated	0.20
Subgrade soils	0.25 to 0.40
Weathered rock subgrade	0.3
Unweathered bedrock subgrade	0.15

- Variation of Poisson ratio values close to the above values typically has little effect on the analysis.

13.35 Specific gravity

- Most soils have a specific gravity value (G_s) typically in the range 2.6 to 2.8
- This value is critical in the assessment of the degree of saturation which is a derived parameter calculated from the G_s, moisture content and void ratio.
- In pavement specifications (Chapter 17) the degree of saturation (say below 70%) may be used in moisture control, hence the need to accurately assess this value. Typical values for various rock types in Queensland is provided inn table.

Table 13.35 Typical specific gravity values for rock types in Queensland (Main Roads, 1989).

Rock type	*Name*	*Specific gravity*
Igneous		
– Acid	Adamelite	2.62
	Dacite	2.64
	Granite	2.61
	Garanidiorite	2.70
	Rhyolite	2.61
	Tuff	2.56–2.82
– Basic	Basalt	2.80–3.00
	Diorite	2.87
	Gabbro	2.90
– Intermediate	Dolerite	2.83
	Trachyandesite	2.68
	Trachyte	2.62
Metamorphic	Amphibiote	3.04
	Greenstone	2.97
	Hornfels	2.67–3.13
	Metagreywacke	2.72
	Quartzite	2.65
	Slate	2.70
Sedimentary and duricrust	Dolomite	2.74
	Limestone	2.68
	Mudstone	2.63

Chapter 14

Slopes

14.1 Slope measurement

- Slopes are commonly expressed as 1 Vertical: Horizontal slopes (highlighted in table). This physical measurement is easier to construct (measure) in the field, although for analysis and design purpose the other slope measurements may be used.

Table 14.1 Slope measurements.

Descriptor	*Degrees*	*Radians*	*Tangent*	*Percentage*	*1 Vertical: Horizontal*	*Design considerations*
Flat	0	0.000	0.000	0%	∞	Drainage
Moderate	5	0.087	0.087	9%	11.4	Slope design
	10	0.174	0.176	18%	5.7	
Steep	11.3	0.197	0.200	20%	5.0	
	15	0.262	0.268	27%	3.7	
	18.4	0.322	0.333	33%	3.00	
	20	0.349	0.364	36%	2.75	
	25	0.436	0.466	47%	2.14	
Very steep	26.6	0.464	0.500	50%	2.00	
	30	0.524	0.577	58%	1.73	
	33.7	0.588	0.667	67%	1.50	
	35	0.611	0.700	70%	1.43	
	40	0.698	0.839	84%	1.19	
Extremely steep	45	0.785	1.000	100%	1.00	Reinforced design if a soil slope
	50	0.873	1.192	119%	0.84	
	55	0.960	1.428	143%	0.70	
	60	1.047	1.732	173%	0.58	
	63	1.107	2.000	200%	0.50	
	65	1.134	2.145	214%	0.47	
Sub-vertical	70	1.22	2.75	275%	0.36	Wall design if a soil slope
	75	1.31	3.73	373%	0.27	
	76	1.33	4.0	400%	0.25	
	80	1.40	5.7	567%	0.18	
	85	1.48	11.4	1143%	0.09	
Vertical	90	1.57	∞	∞	0.00	

  - Typically soil slopes do not exceed very steep unless some reinforcement or wall is used.

- Rock slopes can be extremely steep to vertical.
- Typically only slightly weathered or fresh natural slopes are sub-vertical to vertical.

14.2 Factors causing slope movements

- The macro factors causing slope movements are outlined below.

Table 14.2 Macro factors causing slope movements.

Macro factor	*Effects*
Tectonics	Increased height that results in an angle change.
Weathering	Chemical and physical processes resulting in disintegration and break down of material. Subsequent removal of the material by water.
Water	Removes material, either in a small-scale surface erosion or major undercutting of cliffs and gullies. Aided by wind and gravity. Water Increases dead weight of material and/or increased internal pressure to dislodge the material.
Gravitational	Downward movements of material due to its dead weight.
Dynamic	Due to natural vibrations such as earthquakes, waves or man-made such as piling and blasting.

- Slope degradation can be divided into mass movement and surficial erosion. The former is covered in this chapter while the latter will be covered in the next chapter.

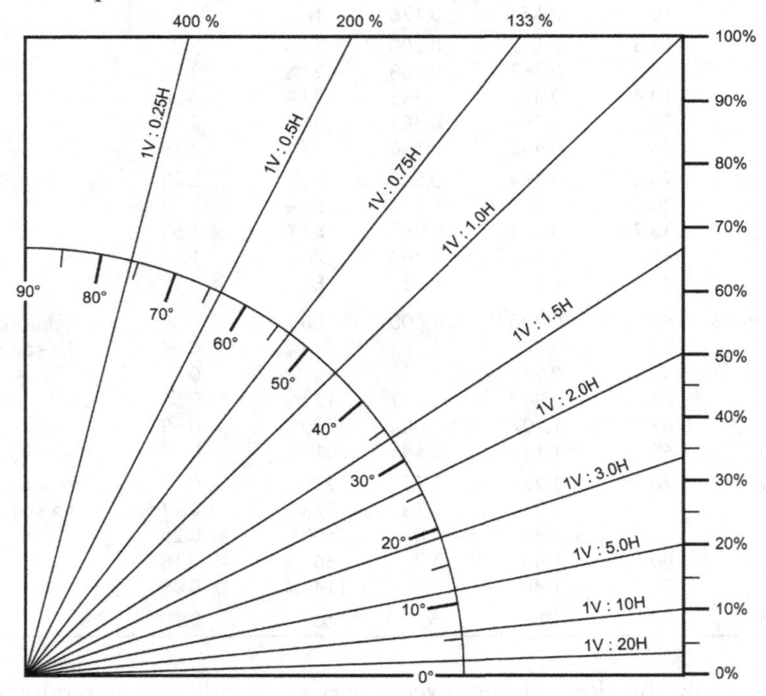

Figure 14.1 Slope definitions.

14.3 Causes of slope failure

- The micro scale effects causing slope movement are covered in the next table.
- Slope failure occurs either due to a decrease in soil strength or an increase in stress.
- Slopes are affected by load, strength, geometry and water conditions.
- The load may be permanent, such its own weight or transient (dynamic from a blast).

Table 14.3 Causes of slope failure (adapted from Duncan and Wright, 2005).

Decrease in soil strength	*Increase in shear stress*
Increased pore pressure (reduced effective stress). Change in water levels (often from change in land use). Uncontrolled discharge of water from drains. High permeability soils have rapid changes – includes coarse grained soils, clays with cracks, fissures and lenses.	Loads at the top of the slope. Placement of fill and construction of buildings on shallow foundation near crown of slope.
Cracking. Tension in the soil at the ground surface. Applies only in soils with tensile strength. Strength is zero in the cracked zone.	Water pressure in cracks at the top of the slope. Results in hydrostatic pressures. If water in cracks for extended periods seepage results, with an increase in pore pressures.
Swelling. Applies to highly plastic and over consolidated clays. Generally a slow process (10 to 20 years). Low confining pressures and long periods of access to water promote swell.	Increase in soil weight. Change in water content due to changes in the water table, infiltration or seepage. Increasing weight of growing trees and wind loading on those trees. Vegetation has a stabilising effect initially (cohesion effect of roots).
Development of slickensides. Applies mainly to highly plastic clays. Can develop as a result of tectonic movement.	Removal of lateral support at toe of slope. Can be man-made or due to erosion at base of slope.
Decomposition of clayey rock fills. Clay shales and claystone may seem like hard rock initially, but when exposed to water may slake and degrade in strength.	Change of slope grade. Steepening of slope either man made (mainly) or by natural processes.
Creep under sustained load. Applies to highly plastic clays. May be caused by cyclic loads such as freeze – thaw or wet – dry variations.	Drop in water level adjacent to slope. Water provides a stabilising effect. Rapid drawdown effect when this occurs rapidly.
Leaching. Change in chemical composition. Salt leaching from marine clays contributes to quick clays, which have negligible strength when disturbed.	Dynamic loading. Usually associated with earthquake loading or blasting. A horizontal or vertical acceleration results. This may also result in a reduction in soil strength.
Strain Softening. Applies to brittle soils.	
Weathering. Applies to rocks and indurated soils.	
Cyclic Loading. Applies to soils with loose structure. Loose sands may liquefy.	

- The analytical model and its interpretation influence the perceived stability.
- Shallow (surficial) failures often occur following high rainfall events. An infinite slope analysis with steady state seepage parallel to the slope applies.

Note that a significant volume of soil mass can be mobilised in surficial failures, and surficial does not necessarily mean a small slide.

- Deep seated failures use both translational and rotational slope stability analysis.
- Water is involved in most of the above factors that cause instability.

14.4 Factors of safety for slopes

- The factor of safety is the ratio of the restoring over the activating condition.
- The condition may be forces or moments being analysed.
- Moment equilibrium is generally used for the analysis of rotational slides. Circular slip surfaces are analysed.
- Force equilibrium is generally used for rotational or translational slides. Circular, plane, wedge or polygonal slip surfaces may be analysed.
- The requirement for different factors of safety depending on the facility and its effect on the environment.

Table 14.4 Factor of safety dependency.

Variable	*Effect on factor of safety*	*Comment*
Strength • Lowest value • Lower quartile • Median	Lower quartile should be typically used. Higher or lower should have corresponding changes on acceptable factor of safety. Need to consider "softened" strength i.e. reduced cohesion, and strain compatibility between layers	Mean values should not be used due to the non-normality of soil and rock strength parameters. Deep failures not typically associated with low cohesion values
Geometry • Height • Slope • Benching • Stratification/ discontinuities	Higher slopes at a given angle would be more unstable than a low height slope. Dip of weakness plane towards slope face influences result	Benching also useful to reduce erosion, provides rock trap area, and as a maintenance platform
Load • Weight • Surcharge • Water conditions	Water is the most significant variable in design. Buoyant unit weight then applies at critical lower stabilizing part of slope, i.e. soil above is heavier than soil below	The weight acts both as an activating and restoring force
Analytical methods • Method of slices • Wedge methods	Different methods (and some software programs) give different outputs for the same data input. Moment equilibrium and force equilibrium methods can sometimes produce different results, especially with externally applied loads. Rigorous methods use both force and moment equilibrium	Probability of failures/displacement criteria should also be considered in critical cases. Factor of safety for 3-dimensional effect ~15% greater than 2-D analyses.

- Choice of factor of safety also depends on quality of available geotechnical information and choice of parameters, i.e. worst credible to probabilistic mean, or conservative best estimate.

- Temporary works may use reduced factors of safety.
- Critical areas or projects would use higher factors of safety.

14.5 Factor of safety for different input assumptions

- The factor of safety is a decision making tool. The number used in isolation can be misinterpreted in terms of "safety".
- The value has meaning only when considered against the background of consequences as well as the underlying input assumptions.
- The table shows how the values may be considered equivalent for a risk decision assessment yet numerically different. The table is illustrative only and should be assessed on a project specific basis.

Table 14.5 Possible "equivalent" minimum factors of safety for long term risk assessment.

Type of investigation	*Limited*		*Detailed*	
Parameter used	Typical	Typical	Lower bound	Lower bound
Strength input	Ultimate (peak)	Critical state	Ultimate (peak)	Critical state
Factor of safety	1.5	1.4	1.4–1.3	1.3–1.2

- For a low or high risk consequence the factors of safety may be reduced or increased by 0.1, respectively.
- Above is for the most likely water condition. Extreme water condition should be checked with short term strength parameters as the governing condition.

14.6 Comparison of factor of safety with probability if failure

- The corresponding probability of failure with factor of safety can be calculated in most modern day slope stability software using material variability parameters discussed in chapter 10.
- One still has to also account for surcharge, geometry, and water variability
- The table below provides the corresponding probability of failure typically associated with various factors of safety.

Table 14.6 Probability of failure with factors of safety (Chowdury et al., 2010).

Factor of safety	*0.8*	*1.0*	*1.2*	*1.4*
Probability of failure	100%	50%	10%	<1%

14.7 Factors of safety for new slopes

- New slopes require a higher factor of safety applied as compared with existing slopes.
- This accounts for possible future (minor) changes, either in load on strength reductions with time due to weathering or strain softening.

Table 14.7 Factors of safety for new slopes (adapted from GEO, 1984).

Economic risk	*Required factor of safety with loss of life for a 10 years return period rainfall*		
	Negligible	*Low*	*High*
Negligible	>1.1	1.2	1.4
Low	1.2	1.3	1.5
High	1.4	1.5	>1.5

14.8 Factors of safety for existing slopes

- Existing slopes generally have a lower factor of safety than for new slopes.
- An existing slope has usually experienced some environmental factors and undergone some equilibration.

Table 14.8 Factors of safety for new slopes (adapted from GEO, 1984).

Risk	*Required factor of safety with loss of life for a 10 years return period rainfall*
Negligible	>1.1
Low	1.2
High	1.3

14.9 Risk to life

- The risk to life includes both the number of people exposed as well as the length of time exposed to the hazard.

Table 14.9 Risk to life (adapted from GEO, 1984).

Situation	*Risk to Life*
Open farmland	Negligible
Country parks, lightly used recreation areas	Negligible
Country roads and low traffic intensity B roads	Negligible
Storage compounds (Non-hazardous goods)	Negligible
Town squares, sitting out areas, playgrounds and car parks	Negligible
High traffic density B roads	Low
Public waiting areas (e.g. railway stations, bus stops)	Low
Occupied buildings (residential, commercial, industrial and educational)	High
All A roads, by-passes and motorways, including associated slip roads, petrol stations and service areas	High
Buildings storing hazardous goods, power stations (all types), nuclear, chemical, and biological complexes	High

14.10 Economic and environmental risk

- Environmental risk can also include political risk, and consequences to the perception of the project.

Table 14.10 Economic and environmental risk (adapted from GEO, 1984).

Situation	*Risk*
Open farmland, country parks, lightly used recreation areas of low amenity value	Negligible
Country roads and low traffic intensity B roads, open air car parks	Negligible
Facilities whose failure would cause only slight pollution	Negligible
Essential services (e.g. gas, electricity, water, whose failure would cause loss of service)	Low
Facilities whose failure would cause significant pollution or severe loss of amenity (cultivated public gardens, with established and mature trees)	Low
High traffic density B roads and all A roads, residential, low rise commercial, industrial and educational properties	Low
Facilities whose failure would cause significant pollution	High
Essential services whose failure would cause loss of service for a prolonged period	High
All A roads, by-passes and motorways, including associated slip roads, petrol stations and service areas	High
Buildings storing hazardous goods, power stations (all types), nuclear, chemical, and biological complexes	High

14.11 Cut slopes

- The stability is dependent on the height of the slope. Table applies only to low to medium height slopes.
- Benches may be required.

Table 14.11 Typical batters of excavated slopes (Hoerner, 1990).

Material	*Slope batters (Vertical:Horizontal)*	
	Permanent	*Temporary*
Massive rock	1.5V: 1H to Vertical	1.5V: 1H to Vertical
Well jointed/bedded rock	1V: 2H to 2V: 1H	1V: 2H to 2V: 1H
Gravel	1V: 2H to 1V: 1H	1V: 2H to 1V: 1H
Sand	1V: 2.5H to 1V: 1.5H	1V: 2.5H to 1V: 1H
Clay	1V: 6H to 1V: 2H	1V: 2H to 2V: 1H

- ○ Water levels often dictate the slope stability.
- ○ Table assumes no surcharge at the top.
- ○ A guide only. Slope stability analysis required.

14.12 Fill slopes

- The strength of underlying materials often dictates the slope stability.

Table 14.12 Typical batters of fill slopes (Hoerner, 1990).

Material	*Slope batters (Vertical : Horizontal)*
Hard rock fill	1V: 1.5H to 1V: 1H
Weak rock fill	1V: 2H to 1V: 1.25H
Gravel	1V: 2H to 1V: 1.25H
Sand	1V: 2.5H to 1V: 1.5H
Clay	1V: 4H to 1V: 1.5H

  - Table assumes no surcharge at the top.
  - A guide only. Depends on risk acceptable, surcharge, water table and ground underlying embankment. Slope stability analysis required.

14.13 Factors of safety for dam walls

- Dam walls can typically have complex geometry with cores and outer zones.

Table 14.13 Factors of safety for dam walls.

Seepage condition	*Storage*	*Required factor of safety*	*Design consideration*
Steady seepage	With maximum storage pool	1.5	Long term condition
Sudden drawdown	From maximum pool	1.1	Short term condition
	From spillway crest	1.3	
End of construction	Reservoir empty	1.3	Short term condition
Earthquake	With maximum storage pool	1.1	Pseudo-static approach. Long term condition

  - A guide only. Depends on risk level.
  - Use of dynamic analysis where F.S. < 1.1. Deformations then govern.

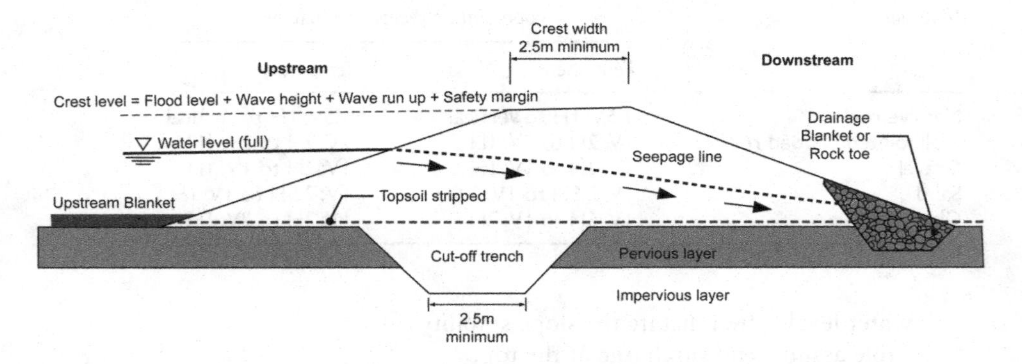

Figure 14.2 Typical small earth dam.

14.14 Typical slopes for low height dam walls

- The size of dams discussed herein is <5 m (low); 5 to 15 m medium; >15 m high.
- In a risk-based design, size is judged on volume of water retained, and its effect on the people and environment. Typically a dam with height less than 5 m is a low risk to the community, although it can affect those locally on the property.

Table 14.14 Typical slopes of low height, homogeneous dam walls (USDI, 1965).

Subject to drawdown	*Soil classification*	*Upstream slope*	*Downstream slope*
No	GW, GP, SW, SP	N/A (Pervious)	N/A (Pervious)
Usual farm design storage	GC, GM, SC, SM	1V: 2.5H	1V: 2.0H
Designs	CL, CH	1V: 3.0H	1V: 2.5H
	CH, MH	1V: 3.5H	1V: 2.5H
Yes	GW, GP, SW, SP	N/A (Pervious)	N/A (Pervious)
Drawdown rates > 150 mm/day	GC, GM, SC, SM	1V: 3.0H	1V: 2.0H
	CL, CH	1V: 3.5H	1V: 2.5H
	CH, MH	1V: 4.0H	1V: 2.5H

- Other dam considerations on seepage below and through dam walls, as well as overtopping needs to be considered.
- Drawdown rates as low as 100 mm/day can be considered rapid in some cases.

14.15 Effect of height on slopes for low height dam walls

- In the design of dam walls, zoned embankments provide the advantage of steeper slopes, and to control drawdown/seepage effects.
- Zoned embankments are recommended for dam heights exceeding 6 m.
- Slope stability analysis required for zoned walls. The slope guidance shown is for homogeneous earth dams.

Table 14.15 Typical slopes of homogeneous dam walls (Nelson, 1985).

Height of wall (m)	*Location*	*Slope*			
		GC	*SC*	*CL*	*CH*
<3	Upstream	←	1V: 2.5 H	→	1V: 3.0 H
	Downstream	←	1V: 2.0 H	→	1V: 2.5 H
3 to 6	Upstream	←	1V: 2.5 H	→	1V: 3.0 H
	Downstream	←	1V: 2.5 H	→	1V: 3.0 H
6 to 10	Upstream	←	1V: 3.0 H	→	1V: 3.5 H
	Downstream	1V: 2.5 H	←	1V: 3.0 H	→

14.16 Design elements of a dam walls

- Some design elements of dam walls are summarised below.
- Dam design and construction for medium to high walls needs detailed considerations of all elements. These are covered in Fells et al. (2005).

- Dam walls experience an unsymmetrical loading, yet many (small to medium) dam walls are constructed as symmetrical. These cross-sections are relevant only for ease of construction, and with an abundant supply of the required material.
- Diaphragm walls are the most material efficient design, where sources of clayey material are limited.

Table 14.16 Design elements of dam walls.

Design element	*Consideration*	*Some dimensions for H < 10 m*	*Comments*
Type	• Homogeneous • Zoned • Diaphragm	• Applicable for < 6 m • Minimum core width = H • Thickness = 1.5 m for H < 10 m	Type cross-section depends on the availability of material.
Seepage cut offs	• Horizontal upstream blanket • Cut-off at base	• 0.5 m minimum thick extending for >5H • Minimum 2.5 m width	Blanket not effective on highly permeable sands or gravels. See section 15.
Crest widths	• Maintenance	• Not less than 3 m	Capping layers at top.
Free board	• Overtopping	• 1 m for small dams (0.5 m for flood flows + 0.5 m wave action)	This is a critical design element for dam walls. Most dams fail by overtopping.
Settlement	• Height dependent	• Allow 5% H for well-constructed dam wall	Allow for this in free board.
Slope protection	• Rip rap	• 300 mm minimum thickness	Angular stones.
Outlet pipes	• Cut-off collars	• Placed every 3 m, typically 1.2 m square for 150 mm diameter pipe	Compaction issues.

  - In a staged dam raising the capping layers are still required in the years between each stage. However it must be removed prior to each lift.

14.17 Stable slopes of levees and canals

- The stability of a slope needs consideration of factors, other than limit equilibrium type analysis. Some other factors are listed in the table below.
- Steeper slopes are possible, than those indicated.

Table 14.17 Typical stable slopes for levees and canals.

Criteria	*Slope*	*Comments*
Ease of construction	1V:2H	For stability of riprap layers
Maintenance	1V:3H	Conveniently traversed with mowing equipment and walked on during construction
Seepage	1V:5H	To prevent damage from seepage with a uniform sandy material
Seepage	1V:6H	To prevent damage from seepage with a uniform clayey material

  - Minimum width for maintenance and feasible for construction with heavy earthmoving equipment = 2.5 m.

14.18 Slopes for revetments

- Revetments are required to protect the slope against erosion, and based on the type of material may govern the slope design.
- Safety aspects may also influence the slope angle, e.g. adjacent to recreational water bodies.

Table 14.18 Slopes for different revetment materials (McConnell, 1998).

Revetment type	*Optimum slope*	*Maximum slope*
Rip–rap	1V:3H	1V:2H to 1V:5H
Rock armour		1V:1.5H
Concrete blocks		1V:2.0H
Concrete mattresses		1V:1.5H
Asphalt – OSA on LSA filter layer	1V:3H	1V:2.0H
Asphalt – OSA on geotextile anchored at top		1V:1.5H
Asphalt – mastic grout		1V:1.5H

- OSA – Open Stone Asphalt is a narrowly graded stone precoated with an asphalt mastic, typically 80% aggregate (20–40 mm) and 20% mastic.
- LSA – Lean sand asphalt typically 96% sand and 4% bitumen 100 pen.
- Mastic grout is a mixture of sand, filler and bitumen, typically 60% sand, 20% filler and 20% bitumen 100 pen.

14.19 Crest levels based on revetment type

- The crest levels are based principally on design wave heights (based on fetch, wind and water depths).
- Other controlling factors are slope and revetment type.
- The required freeboard is then based on consideration of all of the above factors.
- Design wave height factored according to the next 2 tables.

Table 14.19 Design wave height, H_D (McConnell, 1998).

Revetment type	*Crest configuration*	*Design wave height, H_D*
Concrete/masonry	–	$0.75H_s$
Rock fill	Surfaced road	$1.0H_s$
Earth fill with reinforced downstream face	Surfaced road	$1.1H_s$
Earth fill with grass downstream face	Surfaced road	$1.2H_s$
	Grass crest	$1.3H_s$
All embankment types – no still water or wave surcharge carryover permitted		$1.67H_s$

- Significant water depth $= H_s$.

14.20 Crest levels based on revetment slope

- The design wave height is factored according to the run-up factor × H_D.
- The run-up factor is based on the dam slope provided in table below.

Table 14.20 Run-up factor based on slope (adapted from McConnell, 1998).

Dam slope	*Run-up factor*		
	Maximum (smooth slope)	*Intermediate (rough stone or shallow rubble)*	*Minimum (thick permeable rip–rap)*
1V:5H	1.0	0.85	0.65
1V:4H	1.25	1.05	0.8
1V:3H	1.7	1.35	1.05
1V:2.5H	1.95	1.55	1.2
1V:2H	2.2	1.75	1.35

- Different overtopping limit apply based on the access requirements, type of structure and land use immediately behind.

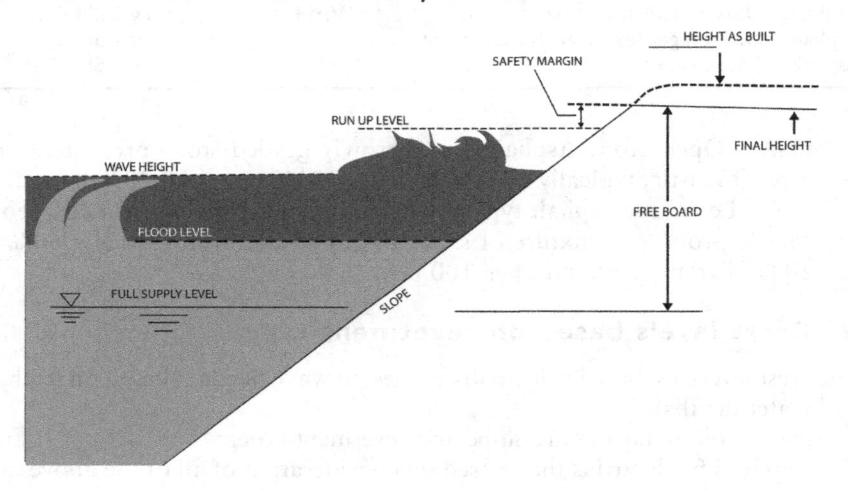

Figure 14.3 Freeboard requirements.

14.21 Stable slopes underwater

- Slope stability analysis alone does not capture the stability of slope under water.
- Slopes fully underwater tend to be stable at much flatter angles than indicated by slope stability analysis.
- This is due to the activity of the water and continuous erosion effects under water.

Table 14.21 Typical slopes under water (ICE, 1995).

Type of material	*Description*	*Slopes in still water*		*Slopes in active water*	
Rock		*Nearly vertical*		*Nearly vertical*	
Clay	Stiff	45°	1V:1H	45°	1V:1H
	Firm	35°	1V:1.4H	30°	1V:1.7H
	Sandy	25°	1V:2.1H	15°	1V:3.7H
Sand	Coarse	20°	1V:2.7H	10°	1V:5.7H
	Fine	15°	1V:3.7H	5°	1V:11.4H
Silt	Mud	10–1°	1V:5.7H to 57H	<5°	1V:11.4H or less

14.22 Side slopes for canals in different materials

- The side slopes in canals depends on the type of natural materials, and the canal depth.
- A canal that is 1.0 m in depth may have material that can have a 1V:1.0H slopes, while at 2.0 m depth a slope of 1V:2.0H may be required.
- The flow velocity in the canal may require revetment protection, and that may govern the slope.

Table 14.22 Typical slopes for earthen canals in different soil materials.

Group symbol	*Material type*	*Minimum side slope*	*Comments*
	Rock	1V:0.25 H	Extent of weathering and joints may affect slope design
	Boulders, cobbles	1V:1.5H	Good erosion resistance Seepage loss
GW, GP SW, SP	Gravels, well or poorly graded Sands, well or poorly graded	1V:2.5H	Good erosion resistance Seepage loss
SC SM	Clayey sands Silty sands	1V: 2.5H	Fine sands have poor erosion resistance
GM GC	Silty gravels Clayey gravels	1V: 1.5H	Medium erosion resistance Medium seepage loss
ML CL OH	Inorganic low plasticity silts Inorganic low plasticity clays Organic low plasticity clays	1V: 1.5H	Poor erosion resistance for low Plasticity index Low seepage loss
MH CH OH	Inorganic high plasticity silts Inorganic high plasticity clays Organic high plasticity clays	1V: 3.0H	Low seepage loss

14.23 Seismic slope stability

- Pseudo-static analysis is performed by applying an acceleration coefficient in the analysis.
- The long term parameters are considered appropriate, however both types of analysis are presented in the table below. There seems to be a divided opinion in the literature in using long term or short-term analysis.

Table 14.23 Seismic slope stability

Consideration	*Long term seismic*	*Short term seismic*
Reasons for	The soil has reached its long-term strength parameters, when the seismic event is likely to occur. Short-term (undrained) parameters are appropriate only during construction	Seismic load, therefore soils (except for some coarse gravels and cobbles) will not drain properly during seismic shaking. The event is short term
Method	• Use effective stress parameters. Softened (Constant volume) values • Apply a horizontal seismic coefficient	• Use undrained shear strength, that has reached its equilibrium, i.e. due to swelling/consolidation • Apply a shear strength reduction factor of 0.8 • Apply a horizontal seismic coefficient
Factor of safety	>1.15 (OBE) >1.0 (MCE)	>1.0 (OBE)
Liquefiable zone	Use $c' = 0$, $\phi' = 0$ for a layer that is liquefiable	

- Horizontal seismic coefficient (k_h) = a_{max}/g.
- Peak ground acceleration (a_{max}) is derived from the Operational Basis Earthquake (OBE) or Maximum Credible Event (MCE).
- OBE derived from probability of occurrence, and usually provided in local codes. However those codes may be 1 in 50 year occurrence and for buildings, which may not be appropriate for some structures e.g. dams.
- MCE derived from consideration of all available fault lengths, near sites, and attenuated acceleration to the site.
- Refer to chapter 23 for a performance based criteria for seismic design.
- Due to the rapid rate of loading (period of 1 sec), conventional strength tests (with time to failure of 10 minutes) may not be appropriate. Typically this rate of loading effect can increase the soil strength by 15% to 20% (Duncan and Wright, 2005). This offsets the above strength reduction factor

14.24 Stable topsoil slopes

- This is a surficial failure common during construction and following rainfall events, when the vegetation has not yet established to stabilise the slopes.
- This surface sliding is common as the topsoil is meant to promote vegetation growth and has been loosely placed (and tamped) on the compacted embankment/slope.

Table 14.24 Topsoil placement considerations.

Consideration	*Slope requirements*	*Comments*
Placing by machine	Slopes > 1 in 5 (19 degrees) required	
Adhering to slope	Slopes > 1 in 3.5 (27 degrees) required	
Grassing and Planting	Slopes > 1V in 2H	Lesser slopes has increasing difficulty to plant and adherence of topsoil
Thickness	Slopes < 1V in 2H: Use 200 mm maximum Slopes 1V in 2H to 1V in 3H: Use 300 mm maximum Slopes > 1V in 3H: Use 400 mm maximum	Greater thickness may be used with geocell or geo mats

- The short-term conditions governs the soil thickness. Greater thickness usually results in gullying and slumping of the topsoil. Once the vegetation has been established the overall slope stability and erosion resistance increases.
- Tree roots have a tensile strength of 10 to 40 kPa and helps reinforce slope with added "cohesion".

14.25 Design of slopes in rock cuttings and embankments

- The slopes for embankments and cuttings are different even for the same type of material.
- Materials of the same rock type but different geological age may perform differently when exposed in a cutting or used as fill.

Table 14.25 Typical slopes in rock cuttings and embankments (adapted from BS 6031 – 1981).

Types of rock/geological age	*Cuttings: safe slopes*	*Embankments: angle of repose*	*Resistance to weathering*
Sedimentary			
• Sandstones: strong, massive Triassic; Carboniferous; Devonian	70° to 90°	38° to 42°	Very resistant
• Sandstones; Weak, bedded Cretaceous	50° to 70°	33° to 37°	Fairy resistant
• Shales Jurassic; Carboniferous	45° to 60°	34° to 38°	Moderately resistant
• Marls Triassic; Cretaceous	55° to 70°	33° to 36°	Softening may occur with time
• Limestones; strong massive Permian; Carboniferous	70° to 90°	38° to 42°	Fairly resistant
• Limestones; weak Jurassic	70° to 90°	33° to 36°	Weathering properties vary considerably
• Chalk Cretaceous	45° to 80°	37° to 42°	Some weathering
Igneous	80° to 90°	37° to 42°	
• Granite, Dolerite, Andesite, Gabbro			Excellent resistant.
• Basalt			Basalts exfoliate after long periods of exposure
Metamorphic	60° to 90°	34° to 38°	
• Gneiss, Quartzite,			Excellent resistant
• Schist, Slate			Weathers considerably

- Angles referred to the horizontal.
- Consider if weaker layer underneath.
- Even in weather resistant rocks, tree roots may open joints causing dislodgement of blocks.

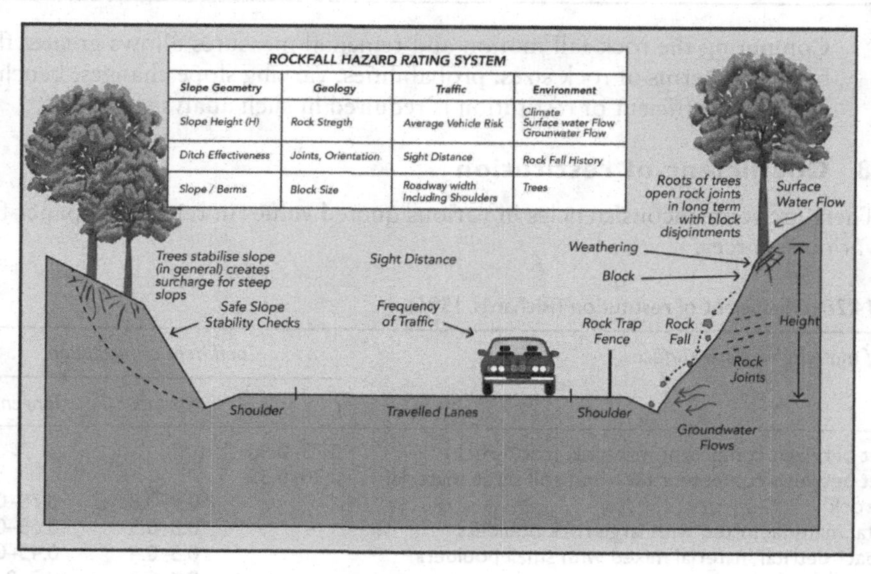

Figure 14.4 Rockfall hazard.

14.26 Factors affecting the stability of rock slopes

- The stability of rock slopes is sensitive to the slope height.
- For a given height the different internal parameters may govern as shown in the table.

Table 14.26 Sensitivity of rock slopes to various factors (after Richards et al., 1978).

Rank	*Slope height*		
	10 m	*100 m*	*1000 m*
1	← ---------- Joint inclination ---------- →		
2	Cohesion	← ---------- Friction angle ---------- →	
3	Unit weight	Cohesion	Water pressure
4	Friction angle	Water pressure	Cohesion
5	Water pressure	← ---------- Unit weight ---------- →	

14.27 Rock falls

- The rock fall motion governs rock trajectory, and design of rock traps (fences and ditches).

Table 14.27 Rockfall motions and effect on slope heights up to 40 m (Ritchie, 1963).

Slopes	*Rock fall motion*	*Effect on trap depth*	*Effect on trap width*
>75°	Falling	1.0 m to 1.5 m	1.0 m (Low H) to 5.5 m (High H)
45 to 75°	Bouncing	Largest depth at a given height 1.0 m to 2.5 m	1.0 m (Low H) to 5.5 m (High H)
<45°	Rolling	1.0 m to 1.5 m	<1.0 m (Low H) to 2.5 m (High H)

  - Computing the rock fall motion and remedial measures allows greater flexibilities, in terms of rock sizes, probabilities, varying slope changes, benching, etc. The coefficient of restitution is required in such analysis

14.28 Coefficient of restitution

- There are some inconsistencies in various quoted values in referenced paper from various sources.

Table 14.28 Coefficient of restitution (Richards, 1991).

Type of material on slope surface	*Coefficient of restitution*		
	r	*Normal r_n*	*Tangential r_t*
Impact between competent materials (rock–rock)	0.75–0.80		
Impact between competent rock and soil scree material	0.20–0.35		
Solid rock		0.9–0.8	0.75–0.65
Detrital material mixed with large rock boulders		0.8–0.5	0.65–0.45
Compact detrital material mixed with small boulders		0.5–0.4	0.45–0.35
Grass covered slopes or meadows		0.4–0.2	0.3–0.2

14.29 Rock cut stabilization measures

- Rock slopes that are considered unstable need stabilization or protective measures needs to be considered.

Table 14.29 Rock slope stabilization considerations.

Consideration	*Solution*	*Methods*	*Comment*
Eliminate problem	Rock removal	• Relocate structure/service/road/rail • Resloping • Trimming and scaling	Relocation is often not possible. Resloping requires additional land
Stabilization	Reinforcement	• Drainage • Berms • Rock bolting and dowels • Tied back walls • Shotcrete facings	Often expensive solutions
Reduce hazard	Protection measures	• Mesh over slope • Rock trap ditches • Fences • Berms • Barriers and impact walls • False tunnels	Controls the rock falls. Usually cheapest solution. Requires some maintenance, e.g. clearing rock behind mesh

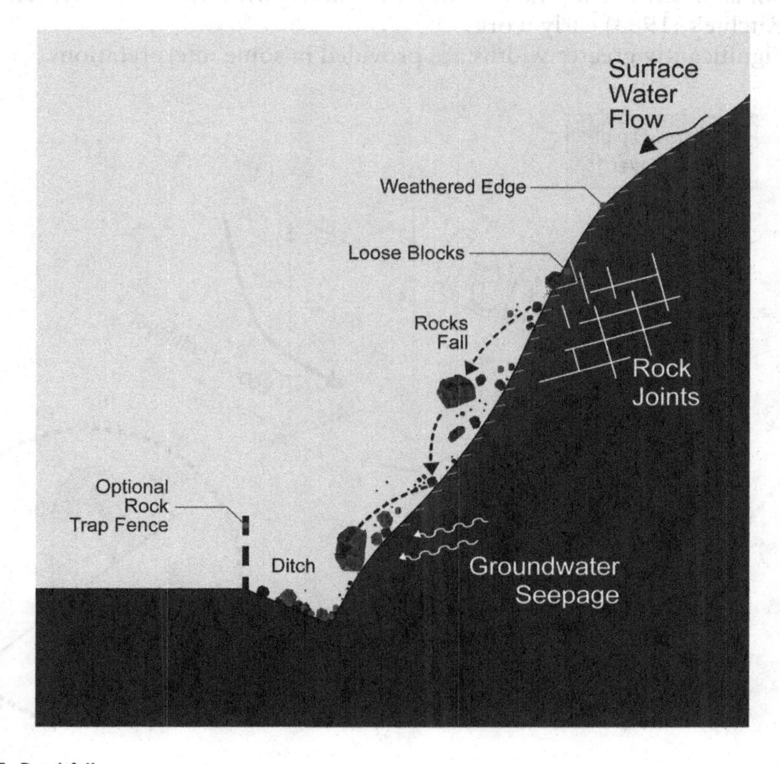

Figure 14.5 Rockfalls.

14.30 Rock trap ditch

- The ditch depth and widths are provided in the table for rock trap measures.
- These can also be used to design fences, e.g. a 1.5 m fence placed 3.0 m from the toe slope provides an equivalent design for a 20 m high slope at 75–55°. Fence must now be designed for impact forces.

Table 14.30 Typical rock trap measures (adapted from graphs from Whiteside, 1986).

Slope height	*Ditch depth * width for slope angles*		
	90–75°	*75–55°*	*55–40°*
5 m	0.75 * 1.0 m	1.0 * 1.0 m	0.75 * 1.5 m
10 m	1.0 * 2.0 m	1.25 * 2.0 m	1.0 * 1.5 m
15 m	1.25 * 3.0 m	1.25 * 2.5 m	1.25 * 2.0 m
20 m	1.25 * 3.5 m	1.5 * 3.0 m	1.25 * 2.5 m
30 m	1.5 * 4.5 m	1.75 * 4.0 m	1.75 * 3.0 m

  - Rock trap benches can be designed from these dimensions, e.g. for a bench of 3 m width plus an suitable factor of safety (additional width, fence, berm) provides an equivalent design for a 20 m high slope at 75–55°.
  - Some inconsistency in the literature here, with various interpretations of Ritchie's (1963) early work.
  - Significantly greater widths are provided in some interpretations.

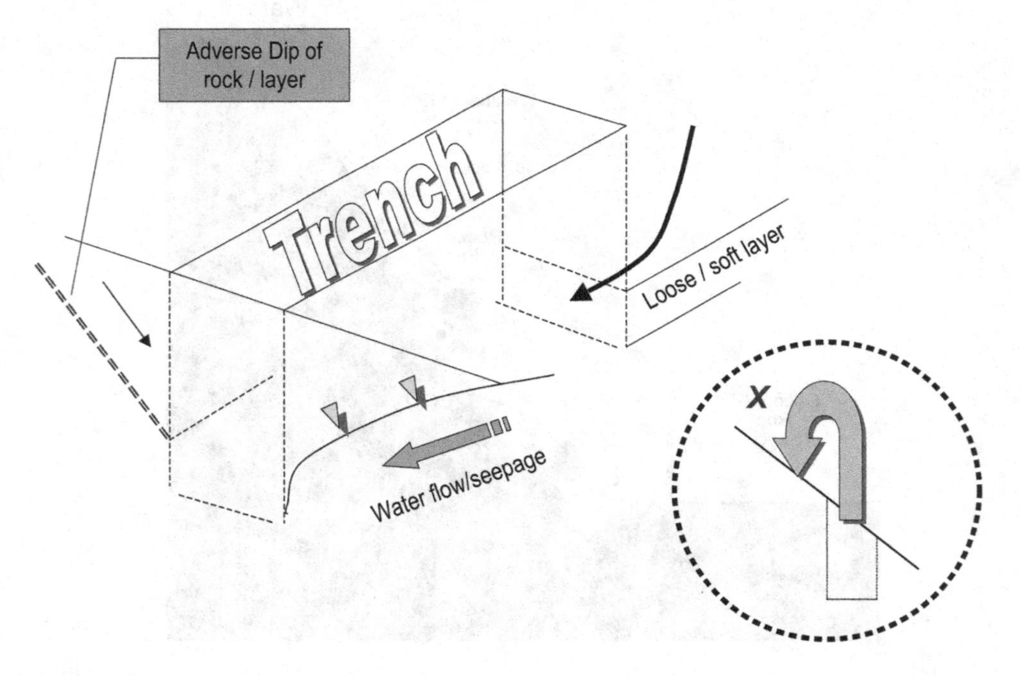

Figure 14.6 Safety in trenching.

14.31 Trenching

- Trenching depth = H.
- Trench width = B.
- Trenching >1.0 m deep typically requires shoring before it is considered safe to enter an excavation.
- When B > 5H, i.e. a wide open cutting, this excavation is now considered an open cutting rather than a trench.

Table 14.31 Safety in trenching.

Risk	*Distance from edge of trench*
High	<(H + B)
Medium	(H + B) to 2(H + B)
Low	>2(H + B)

  - Stockpile/equipment must be placed to minimise risk to the trench, unless trench bracing designed to accommodate the loads.
  - Structures/services at the above distance need to be also considered.
  - Movements when placed at <2(H + B) discussed in later chapters.
  - To minimise risk, corrective action and continuous observations for:
    - Adverse dip of rock/soil layers.
    - Loose/soft layers intersected.
    - Water flow and seepage into trenches.

Chapter 15

Terrain assessment, drainage and erosion

15.1 Terrain evaluation

- Terrain evaluation is particularly useful in linear developments and large projects.
- This involves an extensive desktop study of aerial photos, geology maps, topography, etc., before any need for extensive ground truthing. Phasing of the study is important here. Refer Chapter 1 as various corridor or site options are still under consideration at this stage of the study.

Table 15.1 Terrain evaluation considerations.

Consideration	*Terrain evaluation*	*Comments*
Accuracy of data scale	Geology maps Aerial photos Orthophotos Development plan	The maps are likely to be at different accuracy scales. Using this data in a GIS analysis for example, is likely to produce inconsistencies in accuracy. A trade-off between the largest useable scale and some loss of data accuracy is here made.
Development	Grades size	Construction/access as well as long term.
Geology	Lithology Structure	Rock/soil type. Dip/orientation with respect to proposed slope.
Drainage	Surface Ground Erosion Catchment area	Hydrology considerations. Also affected by vegetation and land cover.
Slope	Transverse batters Longitudinal grades	Affects horizontal resumptions/stability measure required.
Height	Above flood levels Cuttings	Affects vertical alignments, which could mean a horizontal alignment shift if significant cut/fill/stability issues.
Aspect of slope	Orientation	With respect to development as well as true north. southern aspect wetter in southern hemisphere (Greater landslide potential).
Land use	Existing Proposed	Roads, rails, services, and developments. Environmental considerations. Adjacent affects considered here.
Vegetation	Type, intensity	Forested, agricultural, barren.

15.2 Scale effects in interpretation of aerial photos

- The recognition of instability with aerial photographs can only occur at a suitable scale.

Table 15.2 Relative suitability of different scales of aerial photography (Soeters and van Westen, 1996).

Recognition	*Size (m)*	*Scale*		
		1:20,000	*1:10,000*	*1:5,000*
Instability	<20 m	0	0	2
	20–75 m	0→1	1→2	3
	>75 m	1→2	2	3
Activity of unstable area	<20 m	0	0	1
	20–75 m	0	0→1	2
	>75 m	1	1→2	3
Instability elements (cracks, steps, depressions, etc.)	<20 m	0	0	0
	20–75 m	0	0→1	1→2
	>75 m	1	2	3

  - 3 is very suitable while 0 is unsuitable

15.3 Development grades

- The different types of developments require different grades. Typical grades for various developments provided in the table.

Table 15.3 Grades required for development (part from Cooke and Doornkamp, 1996).

Development type	*Grade %*	*Deg. °*	*Vert.:Horiz.*
International airport runways	1	0.6	1V:100H
Main line passenger and freight rail transport	2	1.2	1V:50H
Local aerodrome runways			
To minimize drainage problems for site development			
Acceptable for playgrounds			
Major roads	4	2.3	1V:25H
Agricultural machinery for weeding, seeding	5	2.9	1V:20H
Soil erosion begins to become a problem			
Land development (construction) becomes difficult			
Industrial roads	6	3.4	1V:17H
Upper limit for playgrounds			
Housing roads	8	4.6	1V:12.5H
Acceptable for camp and picnic areas			
Absolute maximum for railways	9	5.1	1V:11.1H
Heavy agricultural machinery	10	5.7	1V:10.0H
Large scale industrial development			

(Continued)

Table 15.3 (Continued)

Development type	*Grade %*	*Deg. °*	*Vert.:Horiz.*
Site development Standard wheel tractor Acceptable for recreational paths and trails Upper limit for camp and picnic areas	15	8.5	1V:6.7H
Housing site development	20	11.3	1V:5.0H
Lot driveways Upper limit for recreational paths and trails Typical limit for rollers to compact	25	14.0	1V : 4.0H
Benching into slopes required	33	18.4	1V : 3.0H
Planting on slopes become difficult without mesh/benches	50	26.6	1V: 2.0H

- Construction equipment has different levels of operating efficiency depending on grade, and riding surface.

15.4 Equivalent gradients for construction equipment

- The rolling resistance is the force that must be overcome to pull a wheel on the ground. This depends on the gradient of the site and the nature of the road.
- Rolling Resistance = Rolling Resistance Factor × Gross Vehicle Weight.

Table 15.4 Rolling resistance and equivalent gradient of wheeled plant (Horner, 1988).

Haul road conditions		*Rolling resistance factor*	
Surface	*Description*	*kg/t*	*An equivalent gradient*
Hard, smooth	Stabilized surfaced roadway, no penetration under load, well maintained	20	2.0%
Firm, smooth	Rolling roadway with dirt or light surfacing, some flexing under load, periodically maintained	32.5	3.0%
With snow	Packed	25	2.5%
	Loose	45	4.5%
Dirt roadway	Rutted, flexing under load, little maintenance, 25 to 50 mm tyre penetration	50	5.0%
Rutted dirt roadway	Rutted, soft under travel, no maintenance, 100 to 150 mm tyre penetration	75	7.5%
Sand/gravel surface	Loose	100	10%
Clay surface	Soft muddy rutted. No maintenance	100–200	10–20%

- This indicates that maintenance of haul road helps to reduce operational cost of plant by a factor of 5 to 10 times if a well maintained surface.
- Safety considerations often govern as construction vehicles may tolerate >100 mm ruts while passenger vehicles using the same access haul road may be unstable.

15.5 Development procedures

- The slope is usually the key factor in consideration of stability. However geology, aspect, drainage etc. also affect the stability of the slopes.

Table 15.5 Development procedures based on slope gradients only.

Vert.:Horiz.	*Deg.* °	*Grade %*	*Slope risk*	*Comments on site development*
>1V:2H	>27	>50	Very high	Not recommended for development
1V:2H to 1V:4H	27 to 14	50 to 25	High	Slope stability assessment report
1V:4H to 1V:8H	14 to 7	25 to 12.5	Moderate	Standard procedures apply
<1V:8H	<7	<12.5	Low	Commercially attractive

15.6 Terrain categories

- Categorisation of the terrain is the first stage in its assessment.

Table 15.6 Terrain categories.

Terrain category	*Slope*			*Common elements*
	%	*Deg.°*	*Vert.:Horizontal*	
Steep hill slopes	>30%	>16.7	1V:3.3H	
High undulating rises	20–30	11.3–16.7	1V:5.0H to 1V:3.3H	Ridges, crests and upper slopes
Moderate undulating rises	10–20	5.7–11.3	1V:10H to 1V:5H	Mid slopes
Gently undulating to level plains	<10%	5.7	1V:10H	Lower and foot slopes

15.7 Landslide classification

- Steeper slopes have a higher potential for landslides, but depends on material type.
- This does not cover rock falls, which was covered in previous chapters.

Table 15.7 Typical landslide dimensions in soils (Skempton and Hutchinson, 1969).

Landslide type	*Depth/length ratio (%)*	*Slope inclination lower limit (Deg. °)*
Debris slides, avalanches	5–10	22–38
Slumps	15–30	8–16
Flows	0.5–3.0	3–20

15.8 Landslide velocity scales

- Rapid landslides cause greater damage and loss of life than slow landslides.

Table 15.8 Landslide velocity scale (Cruden and Varnes, 1996).

Description	*Velocity (mm/s)*	*Typical velocity*	*Probable destructive significance*
Extremely rapid			Catastrophe of major violence; buildings destroyed by impact of displaced material; many deaths, escape unlikely.
	5×10^3	5 m/second	
Very rapid			Some lives lost; velocity too great to permit all persons to escape.
	5×10^1	3 m/minute	
Rapid			Escape evacuation possible; structures, possessions, and equipment destroyed.
	5×10^{-1}	1.8 m/hour	
Moderate			Some temporary and insensitive structures can be temporarily maintained.
	5×10^{-3}	13 m/month	
Slow			Remedial construction can be undertaken during movement; insensitive structures require frequent maintenance work if total movement is not large during a particular acceleration phase.
	5×10^{-5}	1.6 m/year	
Very slow			Some permanent structures undamaged by movement.
	$<5 \times 10^{-7}$	16 mm/year	
Extremely slow			Imperceptible without instruments; construction possible with precautions.

15.9 Slope erodibility

- The slope erodibility is controlled by the grades and type of soil. The latter is provided in later tables.
- The minimum gradients are usually required for drainage purposes, e.g. 1% gradient for drainage – a cleansing velocity, but higher velocities are required to minimise flood conditions on higher ground.
- The greater slope lengths produce greater erosion potential.

Table 15.9 Slope erodibility with grades.

Erosion potential	*Grade %*
High	>10%
Moderate	10–5%
Low	<5%

15.10 Erodibility hierarchy

- Simplified hierarchy based on soil type provided based on USC system. Refer to Figure 15.1.

- Note next table highlights soil topography (LS) and cropping factor (C) is of greater significance than soil type

Table 15.10 Soil erodibility based on soil type.

Erodibility	*Highly erodible* ←----------							→	*Least erodible*
Soil type	ML SM	SC MH	OL OH	CL CH	SW SC	GM GC	GP GW	Boulders & cobbles
Size and cohesion	Fine grained/low cohesion			Fine grained with cohesion	Coarse grained			Very large

15.11 Soil erosion

- Surface erosion is due to water, wind and ice.
- Catchment and precipitation affects erosion hence hydrology need to be considered.
  - Internal erosion (piping) is the process where soil is removed by groundwater and is discussed separately in the sections on seepage.
- The universal soil Loss equation provides the governing elements for the annual soil loss (A).

Table 15.11 Universal soil loss equation.

Symbol	*Description*	*Comments*
A	Average annual soil loss	A = R. K. LS. C. P. (tonnes /hectare)
R	Rainfall factor	R = f{E I_{30}} where E is kinetic energy, and I_{30} is the greatest rainfall in a 30 minute period. Maps of R developed for USA but needs to be calculated for elsewhere (Refer Fookes et al., 2005)
K	Soil erodibility factor	K = f{permeability, soil structure, soil composition}. Can be obtained from nomograph
L	Slope length factor	L and E evaluated as single topographical factor
S	Slope gradient factor	Unlike previous factors, this factor can be modified and has significant effect
C	Cropping management factor (vegetation)	C = f{canopy, ground and vegetative cover}. This factor can also be modified and has significant effect
P	Erosion control practice factor	P = f {slope contouring and terracing}. P = 1.0 for no erosion control factor

  - Developed originally for agricultural purposes. Limited use for direct application to linear road or small scale developments. But the principles outlined in the table would apply.

- o Vegetation reduces erosion but also reinforces soils with its roots and extracts moisture hence increasing slope stability. At steep slopes trees act as a surcharge with an adverse effect on stability.

15.12 Soil dispersivity

- Soil erodibility also affected by its dispersivity properties. Some of these indicative tests are outlined below.
- Similar to geomechanical tests one should not judge by 1 poor test results as significant testing variation can occur. Soil and environmental scientists often use any one poor test result as a rationale for classification as a dispersive site and is inconsistent with a geotechnical type evaluation.
- Site observations, gullying effects and site topography need to be also assessed before implementation.

Table 15.12 Soil dispersion.

Test	*Value/degree of dispersion*			
Exchangeable sodium potential ESP (%)	<7 Non-dispersive	7–10 Intermediate	≥10 Dispersive	≥15 Highly dispersive
Percent dispersion (also called double hydrometer test)	<30 Non-dispersive	30–50 Intermediate	≥50 Dispersive	
Emerson class number	8 non-dispersive	5–7 some dispersion	1–4 treat with caution for dams	
Pinhole dispersion	ND1 completely erosion resistant, ND2, non-dispersive	PD1, PD2 intermediate	D2 – dispersive	D1 – highly dispersive
Emerson aggregate stability (EAT)	7,8 very low	3, 4 – intermediate estimated dispersion (50)	2 – high estimated dispersion (65)	1 – very high, estimated dispersion (70)
Dispersion index	≥30 non-dispersive	2.5–3.0 possibly dispersive	0–2.5 dispersive	

15.13 Erosion thresholds

- Soil, slope and hydrologic variables combine to affect soil erosion. Dietrich et al. (2001) provide the generalised delineation of slope stability fields.
- Some of those findings provided in Table for $T/q = 1259$ m (transmittivity/ total runoff per unit area) surface overland flow (SOF) in a given catchment or watershed i.t.o. contributing area per contour length.
- T/q is hydrologic term and represents ability of subsurface to transmit water flow relative to applied runoff. The ground saturates for small ratios (large q, and small T).

Table 15.13 Erosion stability fields based on slope and area/contour length (here extracted from graphs in (Schor and Gray, 2007).

Slope		*Slope stability fields*	*Contributing per cell area width (m)*	*Comments*
Tangent	Degrees			
0.1	5.7°	Stable unconditional Saturated – unsaturated	1.2×10^2	SOF without erosion > 1.2×10^2 Stable diffusion dominated < 1.2×10^2
0.4	21.8°		4.5×10^2	SOF without erosion > 4.2×10^2 Stable diffusion dominated < 4.2×10^2
0.5	26.5°	Unstable saturated – stable unsaturated	4.7×10^2 4.5×10^2	SOF with/without erosion > 4.5×10^2 Unstable unsaturated Stable diffusion dominated < 4.0×10^2
0.7	35°		6×10^2 4.0×10^2	SOF with/without erosion > 6.0×10^2 Unstable unsaturated Stable diffusion dominated < 4.0×10^2
0.9	42°		7×10^2 1.2×10^2	SOF with/without erosion > 6.5×10^2 Unstable unsaturated Stable diffusion dominated < 1.2×10^2
1.0	45°	Unconditionally unstable Saturated – unsaturated	9×10^2	SOF erosion and landsliding > 9×10^2 Landsliding only < 9×10^2

- Table will vary based on rainfall (T/q) but highlights erosion occurs mainly at steeper slopes (25°) and when saturated.

15.14 Sediment loss from linear vs. concave slopes

- Sediment loss is greatly reduced for a concave as compared to a linear slope.
- Table compares a simulated loss from Hancock (2003) – here from Schor and Gray, 2007).

Table 15.14 Sediment loss from linear vs. concave slopes (here from Schor and Gray, 2007).

Average slope (%)	*Slope length (m)*	*Sediment loss (t/ha/yr)*	
		Linear slope	*Concave slope*
20	200	22	4
25	170	34	6
35	120	69	12
45	90	100	21

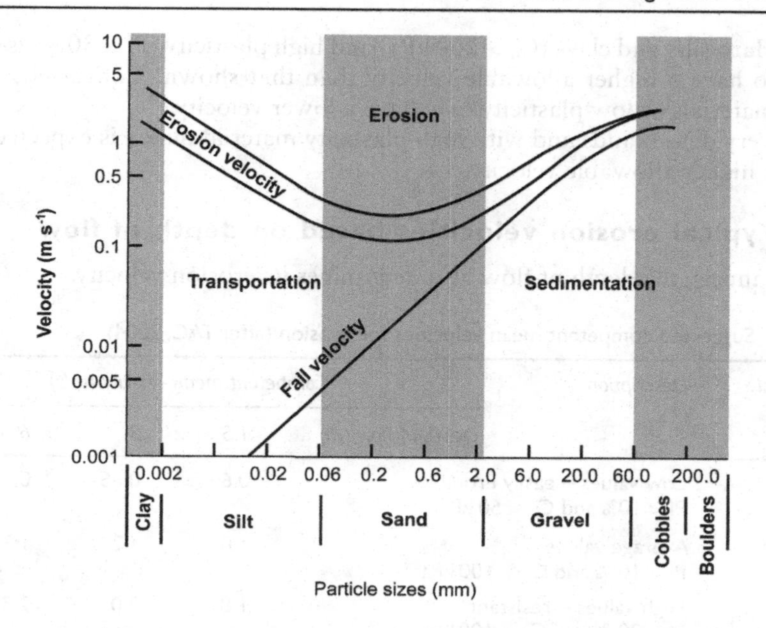

Figure 15.1 Erosion and deposition process (here from Bell, 1998, after Hjulstrom, 1935).

15.15 Typical erosion velocities based on material

- The definition of erosion depends on its application, i.e. whether internal or surface erosion. Surface erosion against rainfall is also different from erosion in channels.
- The ability of a soil to reduce erosion depends on its compactness.
- The soil size (gradation characteristics), plasticity and cohesiveness also affect its erodibility.
- Fine to medium sand and silts are the most erodible, especially if uniformly graded.
- The table is based on Hjulstrom's chart (Figure 15.1) based only on particle size for stream flow velocities. However the state of the soil (compactness) and the relative proportion of materials also influence its allowable velocity.

Table 15.15 Typical erosion velocities.

Soil type	*Grain size*	*Erosion velocity (m/s) Particle size only*
Cobbles, cemented gravels, conglomerate, soft sedimentary rock	>60 mm	3.0
Gravels (coarse)	20 mm to 60 mm	2.0
Gravels (medium)	6 mm to 20 mm	1.0
Gravels (fine)	2 mm to 6 mm	0.5
Sands (coarse)	0.6 mm to 2 mm	0.25
Sands (medium)	0.2 mm to 0.6 mm	0.15
Sands (coarse)	0.06 mm to 0.2 mm	0.25
Silts (coarse to medium)	0.006 mm to 0.06 mm	0.5
Silts (fine)	0.002 mm to 0.006 mm	1.0
Clays	<0.002 mm	3.0

- Hard silts and clays ($C_u > 200$ kPa) and high plasticity (PI > 30%) is expected to have a higher allowable velocity than that shown. Conversely, very soft materials of low plasticity may have a lower velocity.
- Very dense sands and with high plasticity material mixed is expected to have a higher allowable velocity.

15.16 Typical erosion velocities based on depth of flow

- In channels, the depth of flow also determines its erosion velocity.

Table 15.16 Suggested competent mean velocities for erosion (after TAC, 2004).

Bed material	*Description*		*Competent mean velocity (m/s)*			
		Depth of flow (m)	*1.5*	*3*	*6*	*15*
Cohesive	Low values – easily erodible PI < 10% and C_u < 50 kPa		0.6	0.65	0.7	0.8
	Average values PI > 10 % and C_u < 100 kPa		1.0	1.2	1.3	1.5
	High values – resistant PI > 20 % and C_u > 100 kPa		1.8	2.0	2.3	2.6
Granular	Medium sand	0.2–0.6 mm	0.65	1.0	1.4	2.2
	Coarse sand	0.6–2.0 mm	0.75	1.1	1.5	2.2
	Fine gravel	2.0–6 mm	0.9	1.2	1.6	2.3
	Medium gravel	6–20 mm	1.2	1.5	1.8	2.5
	Coarse gravel	20–60 mm	1.7	2.0	2.2	2.9
	Cobbles	60–200 mm	2.5	2.8	3.3	4.0
	Boulders	>200 m	3.3	3.7	4.2	5

15.17 Erosion control

- Erosion control depends on the size and slope of the site. The use of contour drains, silt fences or vegetation buffers are typical control measures.

Table 15.17 Erosion control measures.

Consideration		*Typical erosion control measures – spacing*		
		Vegetation buffers	*Contour drains*	*Silt fences*
Slope	5%	75 m	50 m	25 m
	10%	50 m	40 m	15 m
	15%	25 m	30 m	10 m
Typical details		10 m strips of thick grass vegetation to trap sediment	250 mm ditch to divert flow with soil excavated from the formed ditch placed as compacted earth ridge behind	0.5 m high posts with filter fabric buried 250 mm at the bottom
Application		Adjacent to waterways	Temporary protection at times of inactivity. Diverts water runoff to diversion channels	Temporary sediment barrier for small sites

- Suitably sized vegetation buffers and contour drains may also be used as permanent erosion control features.
- Refer Chapter 16 for added details on silt fences.

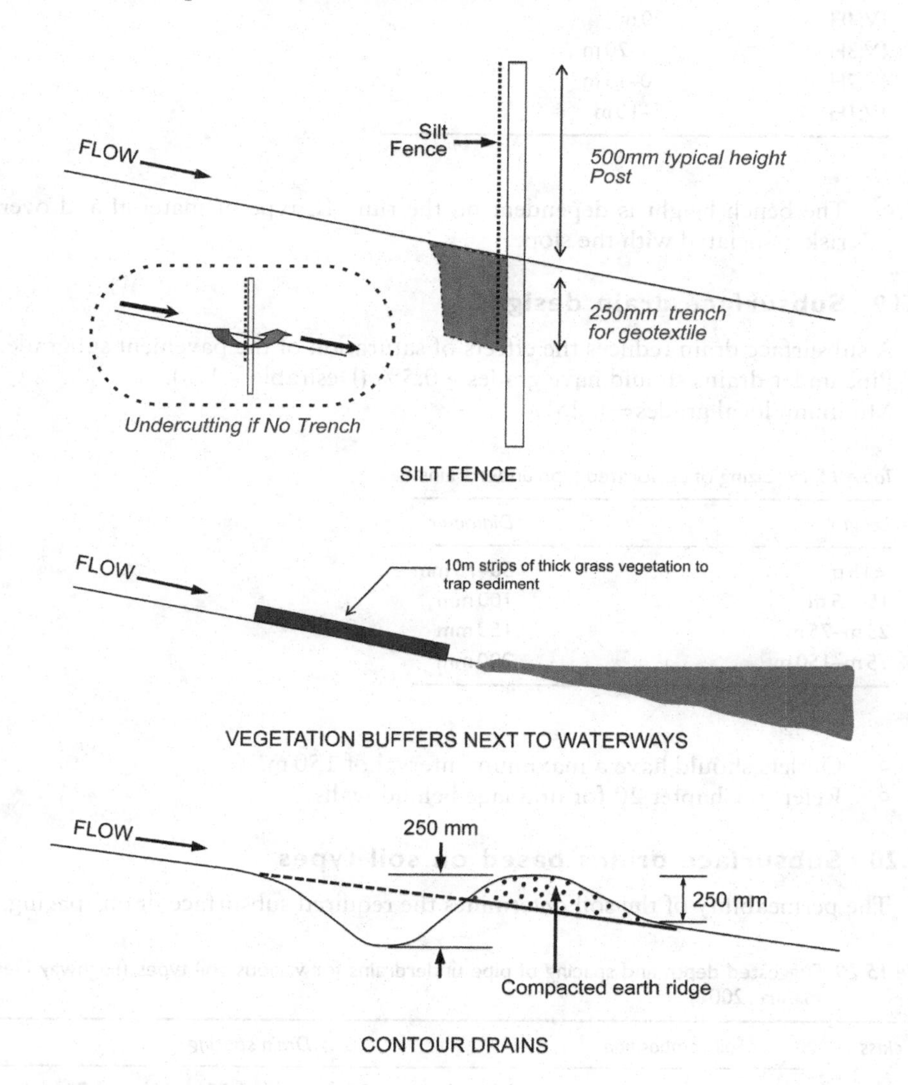

Figure 15.2 Erosion protection.

15.18 Benching of slopes

- Benching of slopes reduces concentrated run off – which reduces erosion. Benching also aids in slope stability.
- Apply a reverse slope of 10–15%, and a minimum depth of 0.3 m.
- The bench width is typically 2–4 m. But this should consider rock fall bench width requirements, and maintenance access requirements.

Table 15.18 Typical benching requirements.

Slope	*Vertical height between benches*
1V:4H	20 m
1V:3H	15–20 m
1V:2H	10–15 m
1V:1H	5–10 m

- The bench height is dependent on the run off, type of material and overall risk associated with the slope.

15.19 Subsurface drain designs

- A subsurface drain reduces the effects of saturation of the pavement subgrade.
- Pipe under drains should have grades ≥ 0.5% (Desirable > 1%).
- Minimum local grades = 0.25%.

Table 15.19 Sizing of perforated pipe underdrains.

Length	*Diameter*
<15 m	50–75 mm
15–25 m	100 mm
25 m–75 m	150 mm
75 m–150 m	200 mm

- Outlets should have a maximum interval of 150 m.
- Refer to Chapter 20 for drainage behind walls.

15.20 Subsurface drains based on soil types

- The permeability of the soil determines the required subsurface drain spacing.

Table 15.20 Suggested depth and spacing of pipe underdrains for various soil types (Highway Design Manual, 2001).

Soil class	*Soil composition*			*Drain spacing*			
	% sand	*% silt*	*% clay*	*1.0 m deep*	*1.25 m deep*	*1.50 m deep*	*1.75 m deep*
Clean sand	80–100	0–20	0–20	35–45	45–60	–	–
Sandy loam	50–80	0–50	0–20	15–30	30–45	–	–
Loam	30–50	30–50	0–20	9–18	12–24	15–30	18–36
Clay loam	20–50	20–50	20–30	6–12	8–15	9–18	12–24
Sandy clay	50–70	0–20	30–50	4–9	6–12	8–15	9–18
Silty clay	0–20	50–70	30–50	3–8	4–9	6–12	8–15
Clay	0–50	0–50	30–100	4 (max)	6 (max)	8 (max)	12 (max)

- Trench widths should be 300 mm minimum.
- Minimum depth below surface level = 500 mm in soils and 250 mm in rock.

15.21 Open channel seepages

- Earthen channels are classified as lined or unlined.

Table 15.21 Seepage rates for unlined channels (Typical data extracted from ANCID, 2001).

Type of material	*Existing seepage rates (litres/m²/day)*
Clays and clay loams	75–150
Gravelly clays, silty and silty loams, fine to medium sand	150–300
Sandy loams, sandy soils with some rock	300–600
Gravelly soils	600–900
Very gravelly	900–1800

- A seepage of 20 litres/m²/day is the USBR benchmark for a water-tight channel with sealed joints.
- Concrete linings are typically 75 mm to 100 m thick.
- Refer Section 17 for typical compacted earth linings.
- Compacted clay linings at the bottom of a channel typically 0.5 m thick can reduce the seepage by 80% to 50% for very gravely soils to fine sand materials, respectively.
- Geosynthetic Clay Liners (GCLs) and Geomembranes can also be used with 250 mm minimum soil cover.

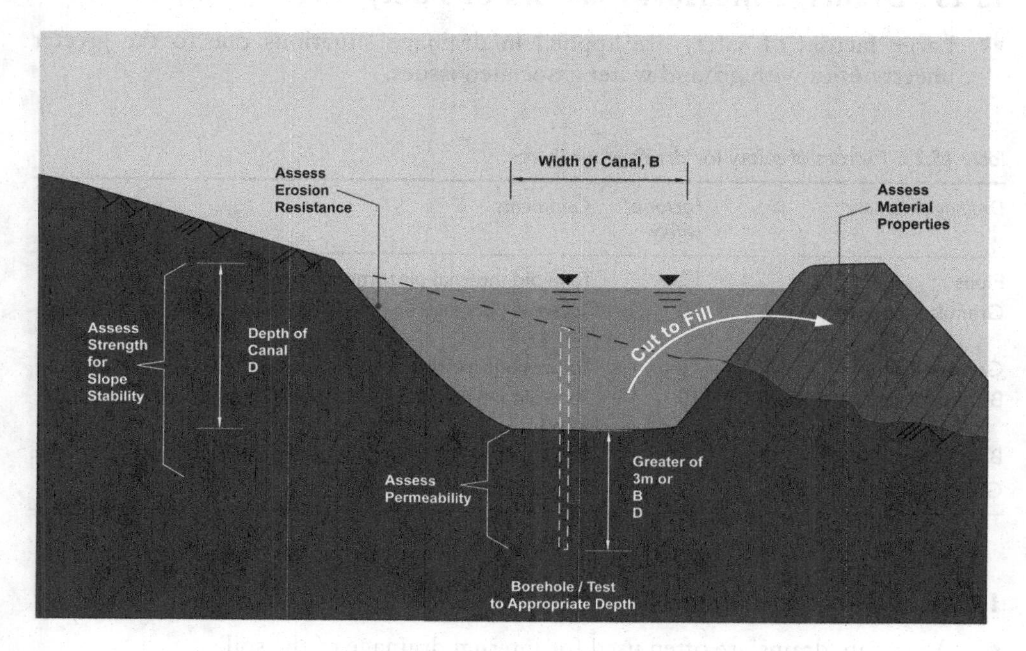

Figure 15.3 Canal issues to be assessed during investigation.

15.22 Comparison between open channel flows and seepages through soils

- Hydraulic gradient of 0.01 in all cases.

Table 15.22 Comparisons between flows in open channels and pipes and seepage through soils and aggregates, Cedergren (1989).

Flow medium	*Effective channel diameter*	*Flow (m^3/s)*	*Area (m^2) for discharge of 50 mm pipe*
Smooth channel	24 m = 2R	12,000	
Smooth pipe	2.4 m = d	20	
	0.30 m = d	0.1	
	50 mm = d	4×10^{-4}	50 mm pipe (0.2 m^2)
25 mm to 40 mm gravel	5 mm	# 4×10^{-4}	0.1
12 mm to 25 mm gravel	2.5 mm	# 1×10^{-4}	0.3
5 mm to 10 mm gravel	0.75 mm	# 2×10^{-5}	2.0
Coarse sand	0.25 mm	# 3×10^{-6}	17
Fine sand, or graded filter aggregate	0.05 mm	# 3×10^{-8}	1.7×10^3
Silt	0.006 mm	# 3×10^{-11}	1.7×10^6
Fat clay	0.001 mm	# 3×10^{-13}	1.8×10^8

- # Per 0.93×10^{-3} square metre area.

15.23 Drainage measures factors of safety

- Large factors of safety are applied in drainage situations due to the greeter uncertainties with ground water associated issues.

Table 15.23 Factors of safety for drainage measures.

Drainage element	*Factor of safety*	*Comments*
Pipes	2	To avoid internal piezometric pressures.
Granular material	10	To avoid permeability reduction due to fines or turbulent flows.
Geotextiles	10	To account for distortion and clogging.
Blanket drain on flat slope	10	To avoid permeability reduction due to fines or turbulent flows.
Blanket drain on steep slope	5	E.g. chimney drains, which uses graded filter or geotextile.
Geocomposite	4	To account for crushing.

15.24 Aggregate drains

- Aggregate drains are often used for internal drainage of the soil.

Table 15.24 Aggregate drains.

Aggregate type	*Advantages*	*Disadvantages*
Open graded gravels – French drain	Good flow capacity	Clogging by piping from surrounding soils
Well graded sands – filter sands	Resists piping. Useful in reduction in pore water pressures	Low flow capacity
Open graded gravels wrapped in Geotextile	Resists piping. Reasonable flow capacity	Depth limitation

15.25 Aggregate drainage properties

- Aggregate drains are sometimes used with or in place of agricultural perforated pipes. The pipes channel the already collected water while the aggregate drains the surrounding soils.
- The equivalent permeability for various size aggregate is provided in the table.
- There is a significant advantage of using large size aggregate in terms of increased permeability (flows) and reduced size.

Table 15.25 Equivalent aggregate cross sections as a 100 mm OD corrugated plastic pipe (Forrester, 2001).

Drainage element	*Size*	*Area (m^2)*	*Comments/Permeability*
Corrugated plastic pipe	100 mm, ID = 85.33 mm	0.0057	Flow Q = 2.7 Litres/sec: Piezometric gradient, i = 1%
20 mm aggregate	1.87 m * 1.87 m	3.5	k = 0.075 m/s
14 mm aggregate	2.45 m * 2.45 m	6	k = 0.045 m/s
10 mm aggregate	3.32 m * 3.32 m	11	k = 0.025 m/s
7 mm aggregate	4.24 m * 4.24 m	18	k = 0.015 m/s
5 mm aggregate	5.83 m * 5.83 m	34	k = 0.008 m/s

- No factors of safety apply.
- I = 1% to minimise turbulent effects in the aggregate.

15.26 Discharge capacity of stone filled drains

- The aggregate size affects the flow capacity. Following seepage analysis, the appropriate stone sizing may be adopted.

Table 15.26 Discharge capacity of 0.9 m * 0.6 m cross-section stone filled drains (Cedergren, 1989).

Size of stone	*Slope*	*Capacity (m^3/s)*
19 mm to 25 mm	0.01	200
	0.001	20
9 mm to 12 mm	0.01	50
	0.001	5
6 mm to 9 mm	0.01	10
	0.001	1

15.27 Slopes for chimney drains

- Chimney drains are used to cut of the horizontal flow paths through an earth dam.

Table 15.27 Slope for chimney drains.

Drainage material	*Slope (1 Vertical:Horizontal)*
Sand	1V:1.75H
Gravel	1V:1.5 H
Sand/gravel	1V:1.75H
Gravel wrapped in geotextile	1V:1.5H

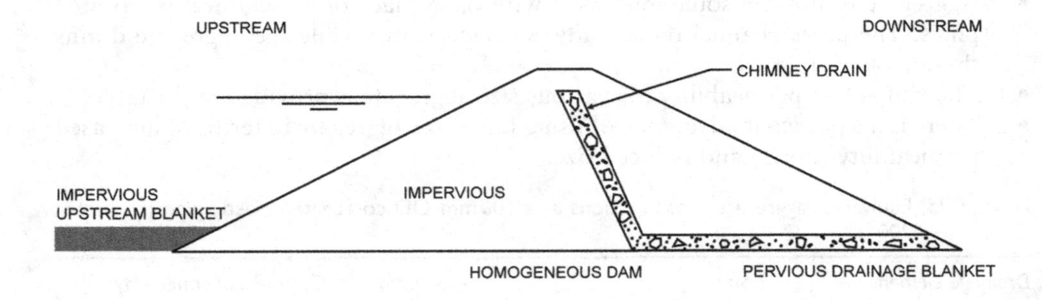

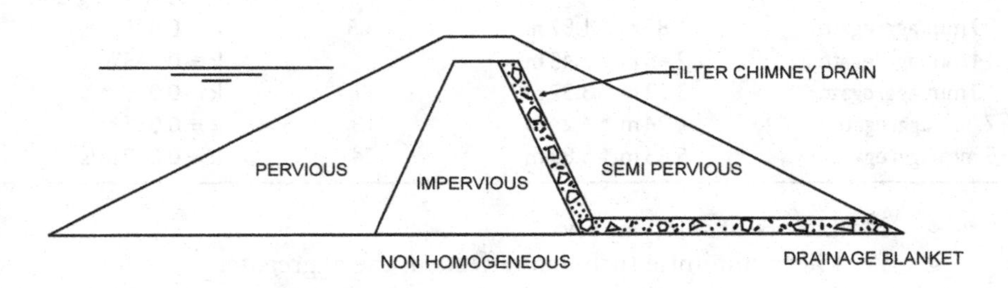

Figure 15.4 Seepage control.

15.28 Drainage blankets

- Drainage blankets are used below roads or earth dams.
- The size should be based on the expected flow and length of the flow path.

Table 15.28 Drainage blanket design requirements below roads.

Criteria	*Thickness of drainage blanket*	*Comment*
No settlement	300 mm minimum compacted	
With settlement	500 mm minimum	or allowance for expected consolidation settlement

15.29 Resistance to piping

- Piping is the internal erosion of the embankment or dam foundation caused by seepage.
- Erosion starts at the downstream toe and works backwards towards the inner reservoir forming internal channels pipes.

Table 15.29 Resistance of a soil to piping.

Resistance controlled by	*Suitability*	*Property*
Plasticity of the soil	Suitable	PI = 15–20%
	Poor	PI < 12%; PI > 30%
Gradation	Suitable	Well graded
	Poor	Uniformly graded
% Stones	Suitable	10% to 20%
	Poor	<10% or >20%
Compaction level	Suitable	Relative compaction = 95%
	Poor	Relative compaction < 90%

15.30 Soil filters

- The permeability of the filter should be greater than the soil it is filtering, while preventing washing out of the fine material.

Table 15.30 Filter design.

Criterion	*Design criteria*	*Comments*
Piping	$D_{15(Filter)} < 5D_{85(soil)}$ Maximum sizing	Filter must be coarser than soil yet small enough to prevent soil from passing through filter – and forming pipe
Permeability	$D_{15(Filter)} > 5D_{15(soil)}$ Minimum sizing	Filter must be significantly more permeable than soil. Filter should contain < 5% Fines
Segregation	Moderately graded $2 < U < 5$	Avoid gap graded material, but with a low uniformity coefficient U
	$D_{50(Filter)} > 25D_{50(soil)}$	For Granular filters below revetments

- Medium and high plasticity clays not prone to erosion, filter criteria can be relaxed.
- Dispersive clays and silts prone to erosion, filter criteria should be more stringent.
- Refer to Chapter 16 for use of geotextiles as a filter.
- Thickness of filter typically $> 20D_{max}$.

15.31 Seepage loss through earth dams

- All dams leak to some extent. Often this is not observable. Design seeks to control that leakage to an acceptable level.
- Guidance on the acceptable seepage level is vague in the literature.

- The following is compiled from the references, but interpolating and extrapolating for other values. This is likely to be a very site and dam specific parameter.

Table 15.31 Guidance on typical seepage losses from earth dams (adapted from Quies, 2002, Kutzner, 1997).

Dam height (m)	*Seepage, litres/day/metre, (Litres/minute /metre)*	
	O.K.	Not O.K.
<5	<25 (0.02)	>100 (0.07)
5–10	<50 (0.03)	>500 (0.3)
10–20	<100 (0.07)	>1000 (0.14)
20–40	<200 (0.14)	>2000 (1.4)
>40	<400 (0.28)	>4000 (2.8)

  - All noted wet spots to be investigated

15.32 Clay blanket thicknesses

- A clay blanket can be used at the base of a canal or immediately inside of a dam wall to increase the seepage path (L), thus reducing the hydraulic gradient ($i = h/l$).
- The actual thickness should be based on permeability of cover material and more permeable materials underlying, head of water and acceptable seepage loss.
- In canals allowance should be made for scour effect.

Table 15.32 Clay blanket thickness for various depths of water (Nelson, 1985).

Water depth	*Thickness of blanket (mm)*
<3.0 m	300
3.0 to 4.0	450
4.0 to 5.0	650
5.0 to 6.0	800
6.0 to 7.0	950
7.0 to 8.0	1150
8.0 to 9.0	1300
9.0 to 10.0	1500

Chapter 16

Geosynthetics

16.1 Type of geosynthetics

- The type of geosynthetics to be used depends on the application.
- The terms geosynthetics and geotextiles are sometimes used interchangeably although geosynthetics is the generic term and geotextile is a type of product.

Table 16.1 Geosynthetic application.

Application	*Typical types*	*Examples*
Reinforcement	Geogrids, geotextiles	Stabilization of steep slopes and walls Foundation of low bearing capacity
Filter	Non-woven geotextiles, Geocomposites	Filters beneath revetments and drainage blankets Separation layer beneath embankment
Drainage	Geonets, geocomposites	Erosion Control on slope faces Drainage layer behind retaining walls
Screen	Geomembranes, Geosynthetic Clay Liner (GCL)	Reservoir containment Landfills

  - Geogrids are usually biaxial and uniaxial types. The latter usually has a higher strength, but in one direction only.
  - Geonets differ from geogrids in terms of its function, and are generally diamond shaped as compared to geogrids, which are planar.
  - Geocomposites combine one or more geosynthetic product to produce a laminated or composite product. GCL is a type of geocomposite.
  - Geomembrane is a continuous membrane of low permeability, and used as a fluid/barrier liner. It has a typical permeability of 10^{-13} to 10^{-15} m/s. This is a theoretical number only as in practice the net permeability is based on water loss through wrinkles seams and holes which typically occur.

16.2 Geosynthetic properties

- The main polymers used in the manufacture of geosynthetics are shown below.
- The basic elements are carbon, hydrogen and sometimes nitrogen and chlorine (PVC). They are produced from coal and oil.

Table 16.2 Basic materials (Van Santvoort, 1995).

Material	*Symbol*	*Unit mass (kg/m^3)*	*Tensile strength at 20°C (N/mm^2)*	*Modulus of elasticity (N/mm^2)*	*Strain at break (%)*
Polyester	PET	1380	800–1200	12000–18000	8–15
Polypropylene	PP	900	400–600	2000–5000	10–40
Polyethylene	PE	920	80–250	200–1200	200–1200
• High density	HDPE	950	350–600	600–6000	10–45
• Low density	LDPE	920	80–250	20–80	20–80
Polyamide	PA	1140	700–900	3000–4000	15–30
Polyvinylchloride	PVC	1250	20–50	10–100	50–150

- PP is the main material used in geotextile manufacture due to its low cost.
- PP is therefore cost effective for non-critical structures and has good chemical and pH resistance.
- For higher loads and for critical structures PP loses its effectiveness due to its poor creep properties under long term and sustained loads. PET is usual in such applications.
- Stiffness assessed as strength/strain where the strain is typically used as 5% to 10% for geogrids
- Allow for 120 year design life by factoring stiffness value by 0.6 typically.

16.3 Geosynthetic functions

- The geosynthetic usually fulfils a main function shown in the table below, but often a minor function as well.
- The table highlights the key properties. Strength, creep, cost and resistance to chemicals are some of the considerations.

Table 16.3 Functional applications.

Material	*Application*					
	Reinforcement/Filter		*Drainage*	*Screen*	*Properties*	
	Geotextile	*Geogrid*	*Geonet*	*Geomembrane*	*High*	*Low*
PET	X	X			Strength, modulus, cost, unit weight	Creep, resistance to alkalis
PP	X	X			Creep, resistance to alkalis	Cost, unit weight, resistance to fuel
PE	X			X	(PE) Strain at failure, creep, resistance to alkalis	(PE) unit weight, strength, modulus, cost
– HDPE		X	X			
– MDPE			X			
– LDPE						
– CSPE				X		
– CPE				X		
PA	X				Resistance to alkalis and detergents	
PVC				X	Strain at failure, unit weight	Strength, modulus

- PET is increasingly being used for geogrids. It has an excellent resistance to chemicals, but low resistance to high pH environments. It is inherently stable to ultra violet light.
- PP and PE have to be stabilised to be resistant against ultra violet light.

16.4 Leakage rates

- Leakage rates are dependent on hydraulic head, type of liner, its thickness and its placement.

Table 16.4 Leakage rate per unit area in litres per hectare per day (lphd) through various liners (Giroud et al., 1994).

Liner type	*Permeability k (m/s)*	*Leakage rate (lphd) with hydraulic head (m)* ~0.01	~0.3
Compacted clay liner (CCL)	$\sim 1 \times 10^{-8}$	9000	15000
0.3 < thickness (D) <0.9 m	$\sim 1 \times 10^{-9}$	900	1500
Geomembrane (GM)	$k_{soil} \sim 1 \times 10^{-2}$	600	3000
GCL thickness (D) = 6 mm	1×10^{-11}	25	450
Composite Liner, GM/CCL	$k_{CCL} \sim 1 \times 10^{-9}$	0.05	1
Composite Liner, GM/GCL	$k_{GCL} \sim 1 \times 10^{-11}$	0.002	0.2

16.5 Static puncture resistance of geotextiles

- An increase in geotextile robustness required for an increase in stone sizes.
- An increased robustness is also required for the weaker subgrades.

Table 16.5 Static puncture resistance requirement (adapted from Lawson, 1994).

Subgrade strength CBR %	*Geotextile CBR puncture resistance (N) for maximum stone size d_{max}* $d_{max} = 100$ mm	$d_{max} = 50$ mm	$d_{max} = 30$ mm
1	2500	2000	1500
2	1800	1500	1200
3	1200	1000	800

- Table applies for geotextiles with CBR puncture extensions ≥40%.

16.6 Construction survivability ratings

- Construction survivability may often govern the geotextile specification.
- The equipment pressure, aggregate and subgrade conditions affects its survivability.

Table 16.6 Construction survivability ratings (AASHTO, 1990)

CBR at installation (soaked)		*<1%*		*1–2%*		*≤3%*	
Equipment ground pressure (kPa)		*≥350*	*<350*	*≥350*	*<350*	*≥350*	*<350*
Cover thickness (compacted)[1]	100 mm[2,3]	NR	NR	H	H	M	M
	150 mm	NR	NR	H	H	M	M
	300 mm	NR	H	M	M	M	M
	450 mm	H	M	M	M	M	M

- H – High; M – Moderate; NR – Not recommended.
- [1]Maximum aggregate thickness not to exceed half the compacted cover thickness.
- [2]For low volume unpaved roads (ADT < 200 vehicles).
- [3]The 100 mm minimum cover is limited to existing road bases and not intended for use in new construction.

16.7 Physical property requirements

- The geotextile strength based on the previous survivability rating provided in table below.
- Drainage and filtration requirements should be also checked.

Table 16.7 Physical property requirements (AASHTO, 1990).

Survivability level	*Grab strength (N)*		*Puncture resistance (N)*		*Tear strength (N)*	
Geotextile elongation →	*<50%*	*≥50%*	*<50%*	*≥50%*	*<50%*	*≥50%*
Moderate	800	500	300	175	300	175
High	1200	800	450	325	450	325

- Values shown are minimum roll average values.
- Values of geotextile elongation do not imply the allowable consolidation properties of the subgrade soil.

16.8 Robustness classification using the G– rating

- G-rating = $(\text{Load} \times \text{Drop Height})^{0.5}$.
- Load (Newtons) on CBR plunger at failure.
- Drop Height (mm) required to make a hole 50 mm in diameter.

Table 16.8 Robustness classification of geotextile – G rating (Waters et al., 1983).

Classification	*G-rating*
Weak	<600
Slightly robust	600–900
Moderately robust	900–1350
Robust	1350–2000
Very robust	2000–3000
Extremely robust	>3000

- This robustness rating is used mainly in Australia. It is used to assess the survivability during construction.

16.9 Geotextile durability for filters, drains and seals

- The construction stresses often determine the durability requirements for the geotextile.
- A non-woven geotextile required in the applications of the table below.

Table 16.9 Geotextile robustness requirements for filters and drains (Austroads, 1990).

Application	*Typical G rating*	*Typical minimum mass (g/m^2)*
Subsoil drains and trenches	900	100
Filter beneath rock filled gabions, mattresses and Drainage blankets	1350	180
Geotextile reinforced chip seals	950	140

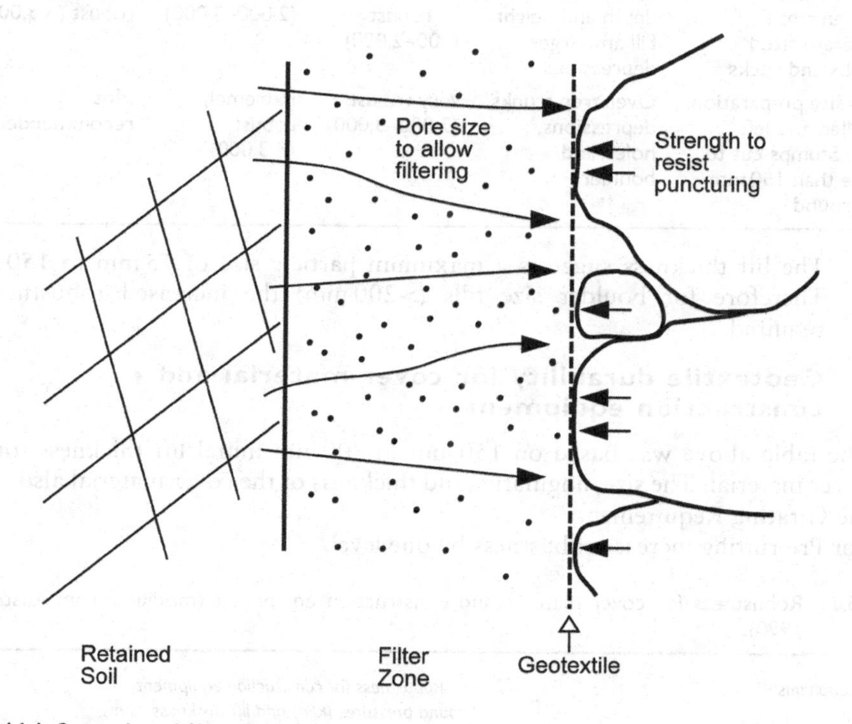

Figure 16.1 Strength and filtering requirements.

16.10 Geotextile durability for ground conditions and construction equipment

- The construction stresses are based on 150 mm to 300 mm initial lift thickness.
- For lift thickness of:
  - 300–450 mm: Reduce robustness requirement by 1 level.
  - 450–600 mm: Reduce robustness requirement by 2 levels.
  - >600 mm: Reduce robustness requirement by 3 levels.
- The design requirements for bearing capacity failure must be separately checked.
- The lift thickness suggests a maximum particle size of 75 mm to 150 mm. Therefore for boulder size fills (>200 mm) the increased robustness is required.

Table 16.10 Robustness required for ground conditions and construction equipment (Austroads, 1990).

Ground conditions		*Robustness for construction equipment ground pressures*		
Natural ground clearance	*Depressions and humps*	*Low (<25 kPa)*	*Medium (25–50 kPa)*	*High (>50 kPa)*
Clear all obstacles except grass, weeds, leaves and fine wood debris	<150 mm in depth and height. Fill any larger depressions	Slightly robust (600–900)	Moderate to robust (900–2,000)	Very robust (2,000–3,000)
Remove obstacles larger than small to moderate sized tree limbs and rocks	<450 mm in depth and height. Fill any larger depressions	Moderate to robust (900–2,000)	Very robust (2,000–3,000)	Extremely robust (>3,000)
Minimal site preparation. Trees felled and left in place. Stumps cut to no more than 150 mm above ground	Over tree trunks, depressions, holes, and boulders	Very robust (2,000–3,000)	Extremely robust (>3,000)	Not recommended

- The lift thickness suggests a maximum particle size of 75 mm to 150 mm. Therefore for boulder size fills (>200 mm) the increased robustness is required.

16.11 Geotextile durability for cover material and construction equipment

- The table above was based on 150 mm to 300 mm initial lift thickness for the cover material. The size, angularity and thickness of the cover material also affect the G-rating Requirement.
- For Pre-rutting increase robustness by one level.

Table 16.11 Robustness for cover material and construction equipment (modified from Austroads, 1990).

Ground conditions		*Robustness for construction equipment Ground pressures (kPa) and lift thickness (mm)*				
Cover material	*Material shape*	*Low (<25 kPa) 150–300 mm*	*Medium (25–50 kPa) 300–450 mm*	*High (>50 kPa) >450 mm*	*Medium (25–50 kPa) 150–300 mm*	*High (>50 kPa) 300–450 mm*
Fine sand to ±50 mm gravel	Rounded to sub angular	Slightly robust (600–900)			Moderately to robust (900–2,000)	
Coarse gravel with diameter up to ½ proposed lift thickness	May be angular	Moderate to robust (900–2,000)			Very robust (2,000–3,000)	
Some to most aggregate >½ proposed lift thickness	Angular and sharp-edged, few fines	Very robust (2,000–3,000)			Extremely robust (>3,000)	

16.12 Robustness geotextile specifications based on strength class

- Road Authorities in Australia use the geotextile specifications based on the strength class required.
- The grab strength, tear and G rating is based on the 80th percentile value.
- Some manufacturers specify the average value in their product specifications; United States specify the minimum average roll value (MARV) which is typically the 97.5 percentile of all test results.
- Elongation ≥30% – non woven geotextile.
- Elongation <30% – woven geotextile.

Table 16.12 Strength class specification (QMR 1999, RTA NSW 2008, Austroads, 2009).

Strength class	*Elongation*	*Grab strength (N)*	*Tensile strength (N)*	*G-rating*
A	≥30%	500	180	900
	<30%	800	300	1350
B	≥30%	700	250	1350
	<30%	1100	400	2000
C	≥30%	900	350	2000
	<30%	1400	500	3000
D	≥30%	1200	450	3000
	<30%	1900	700	4500
E	≥30%	1600	650	4500

16.13 Establishing geotextile strength class

- Geotextile strength class established from material immediately adjacent to geotextile.
- For horizontal placement the size of the aggregate placed above the geotextile is considered with the underling subgrade CBR.
- For vertical placement, the size of the backfill and depth of the trench determines the strength class.

Table 16.13 Strength class based on material adjacent to geotextile (QMR 1999, Austroads, 2009)

D_{85} (mm) of material to be placed over geotextile	*Strength class*			
	Material below geotextile		*Trench depth*	
	CBR ≤ 3	*CBR > 3*	*<2 m*	*<3 m*
≤37.5	C	A	A	B
≤75	C	B	B	C
≤200	D	C	C	D
≤400	E (N/A with elongation <30%)	D		
≤600	Not applicable	E		

16.14 Establishing geotextile strength class adjacent to walls

- Smooth concrete facing walls have a reduced strength class as compared to wall with stones such as gabions where puncture can occur.

Table 16.14 Strength class based on material adjacent to geotextile (QMR 1999).

Type of structure	*Strength class*
Concrete retaining walls	B
Segmental block walls; Reinforced soil concrete panel walls	
Gabion walls; Crib walls;	C Rock filled mattresses

- Other specifications are
  - For drains G rating ≥900: Mass >100 g/m^2
- Filter beneath rock filled gabions, mattresses, and drainage blankets
  - G rating ≥1350: Mass >180 g/m^2

16.15 Pavement reduction with geotextiles

- The pavement depth depends on ESAs and acceptable rut depth.
- Elongation of geotextile = ε.
- Secant modulus of geotextile = k.

Table 16.15 Typical pavement thickness reduction due to geotextile (adapted from Giroud and Noiray, 1981).

In situ CBR (%)	*Maximum pavement reduction for acceptable rut depth*						
	30–75 mm	*250 mm (ε = 10%)*	*250 mm (ε = 7%)*	*250 mm (ε = 5%)*	*250 mm (k = 10 kN/m)*	*250 mm (k = 100 kN/m)*	*250 mm (k = 300 kN/m)*
0.5	175 mm	450 mm	300 mm	100 mm	150 mm	200 mm	300 mm
1	125 mm	250 mm	100 mm	0 mm	125 mm	150 mm	225 mm
2	100 mm	100 mm	0 mm		75 mm	125 mm	100 mm
3	40 mm	30 mm			30 mm	30 mm	30 mm
4	0 mm	0 mm			0 mm	0 mm	0 mm

16.16 Bearing capacity factors using geotextiles

- The geotextiles provide an increase in allowable bearing capacity due to added localised restraint to the subgrade.
- The strength properties of the geotextile often do not govern provided the geotextile survives construction and the number of load cycles is low.
- Subgrade strength C_u = 23 CBR for undisturbed condition.
- Ultimate Bearing Capacity $q_{ult} = N_c C_u$.

Table 16.16 Bearing capacity factors for different ruts and traffic conditions (Richardson, 1997; Steward et al., 1977; Tingle and Webster, 2003).

Geotextile	*Ruts (mm)*	*Traffic (passes of 80 kN axle equivalent)*	*Bearing capacity factor, N_c*
Without	<50	>1000	2.8
	>100	<100	3.3
With geotextile	<50	>1000	5.0
	>100	<100	6.0
With geogrid	<50	>1000	5.8

- ○ During construction 50 to 100 mm rut depth is generally acceptable.
- ○ Dump truck (8 m^3) with tandem axles would have a dual wheel load of 35 kN.
- ○ Motor grader would have a wheel load approximately 20 kN to 40 kN.
- ○ Placement of the geogrid at the subgrade surface does not have a beneficial effect. Grids perform better when placed at the lower third of aggregate.

16.17 Geotextiles for separation and reinforcement

- A geotextile is used as separation and reinforcement depending on the subgrade strength.
- A geotextile separator is of little value over sandy soils.
- A geogrid over a loose sand subgrade reduces the displacement.

Table 16.17 Geotextile function in roadways (Koerner, 1995).

Geotextile function	*Unsoaked CBR value*	*Soaked CBR value*
Separation	≥8	≥3
Separation with some nominal reinforcement	3–8	1–3
Reinforcement and separation	≤3	≤1

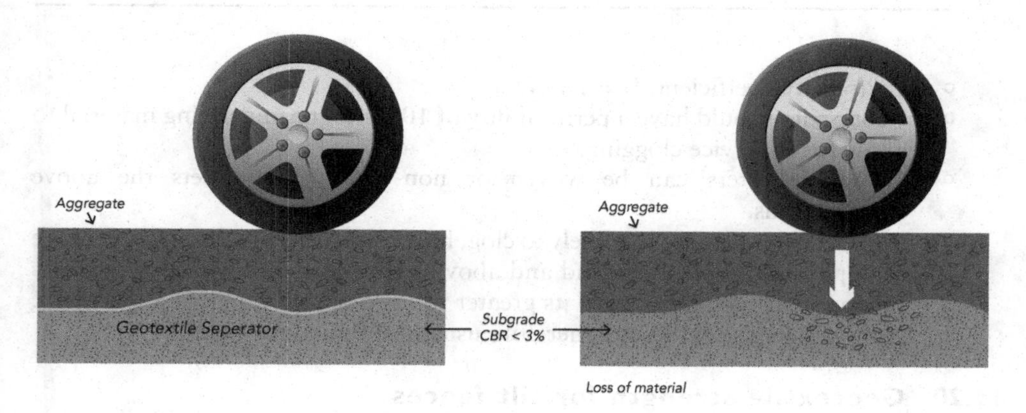

Figure 16.2 Geotextile as a separator.

16.18 Reinforcement location

- Geogrids used when reinforcement function required.
- Separation issue often masked the reinforcement benefit at low subgrade strengths (CBR < 3%).

Table 16.18 Reinforcement location (Berg et al., 2000).

Thickness	*Optimal location of geosynthetic reinforcement*	*Comment*
≥300 mm	Middle of layer	
200–300 mm	Bottom of layer	
<200 mm	Geotextile geogrid composite required	Separation an issue for thin bases

- Placement of <150 mm over a geotextile not recommended.
- Thickness = 150 mm required over subgrades with CBR < 3%.

16.19 Geotextiles as a soil filter

- The geotextile filter pore sizes should be small enough to prevent excessive loss of fines.
- The geotextile filter pore size should be large enough to allow water to filter through.
- The geotextile should be strong enough to resist the stresses induced during construction and from the overlying materials.
- Geotextile permeability is approximately equivalent to clean coarse gravel or uniformly graded coarse aggregate ($>10^{-2}$ m/s).

Table 16.19 Criteria for selection of geotextile as a filter below revetments (McConnell, 1998).

Soil type		Pore size of geotextile O_{90}	
Cohesive		$O_{90} \leq 10D_{50}$	
Non-cohesive	Uniform (U<5), uniform	$O_{90} \leq 2.5D_{50}$	$O_{90} \leq D_{90}$
	Uniform (U<5), well graded	$O_{90} \leq 10D_{50}$	
	Little or no cohesion and 50% by weight of silt	$O_{90} \leq 200\ \mu m$	

- Uniformity coefficient, $U = D_{60}/D_{10}$.
- Geotextiles should have a permeability of 10 times the underlying material to allow for in service clogging.
- Geotextile filters can be woven or non-woven that meets the above specifications.
- Woven geotextiles are less likely to clog, however have a much narrower range of applicability (medium sand and above). However, non-woven geotextiles predominate as filters due to its greater robustness and range of application. Non-woven geotextiles are therefore usually specified for filters.

16.20 Geotextile strength for silt fences

- The geotextile strength required depends on the posts spacing and the height of impoundment (H).

Table 16.20 Geotextile strength for varying post spacing (adapted from Richardson and Middlebooks, 1991).

Post spacing (m)	Tension in silt fence geotextile (kN/m)		
	H = 0.5 m	H = 0.6 m	H = 0.9 m
1	5 kN/m	7 kN/m	12 kN/m
1.5	N/A	10 kN/m	18 kN/m
2	N/A	12 kN/m	25 kN/m
2.5	N/A	N/A	30 kN/m

- The ultimate strength of a typical non-reinforced silt fence geotextile is 8–15 kN/m.
- For unreinforced geotextiles, impoundment height is limited to 0.6 m and post spacing to 2 m.
- For greater heights, use of plastic grid/mesh reinforcement to prevent burst failure of geotextile.

16.21 Typical geotextile strengths

- The geotextile strength depends on the application, with the greatest strength required below embankments founded on compressible clays.

Table 16.21 Typical geotextile reinforcement strengths (adapted from Hausman, 1990).

Application	*Description*	*Fabric wide strength, kN/m*	*Fabric modulus, kN/m*
Retaining structures	Low height	10–15	35–50
	Moderate height	15–20	40–50
	High	20–30	60–175
Slope stabilization	Close spacing	10–20	25–50
	Moderate spacing	15–25	35–70
	Wide spacing	25–50	40–175
Unpaved roads	CBR $\leq$ 4%	10–20	50–90
	CBR $\leq$ 2%	15–25	90–175
	CBR $\leq$ 1%	35–50	175–525
Foundations (Increase in bearing capacity)	Nominal	25–70	175–350
	Moderate	40–90	350–875
	Large	70–175	875–1750
Embankments over soft soils	$C_u > 10$ kPa	100–200	875–1750
	$C_u > 5$ kPa	175–250	1750–3500
	$C_u > 2$ kPa	250–500	3500–7000

16.22 Geotextile overlap

- The geotextile overlap depends on the loading and the ground conditions.
- A 500 mm minimum overlap required in repairing damaged areas.

Table 16.22 Geotextile overlap based on load type and in situ CBR value (adapted from Koerner, 1995).

CBR value	*Required overlap distance for traffic loading*		
	Light duty – access roads	*Medium duty – typical loads*	*Heavy duty – earth moving equipment*
≤0.5%	800 mm	1000 mm or sewn	1000 mm or sewn
0.5–1.0%	700 mm	900 mm	1000 mm or sewn
1.0–2.0%	600 mm	750 mm	900 mm
2.0–3.0%	500 mm	600 mm	700 mm
3.0–4.0%	400 mm	450 mm	550 mm
4.0–5.0%	300 mm	350 mm	400 mm
>5.0%	250 mm minimum	250 mm minimum	250 mm minimum
All roll ends	800 mm or sewn	100 mm or sewn	100 mm or sewn

16.23 Modulus improvements with Geosynthetic inclusions

- The literature is vague on the composite modulus improvements using various geosynthetic inclusions as most improvements have been derived from field trials on reduced rutting or repetitions and focused on strength improvements
- This table provides the relative modulus improvements in a uniform loose gravel with 250 mm loose gravel under

Table 16.23 Relative modulus improvement due to geosynthetic inclusions (Lacey et al, 2013).

Geosynthetic type		*Relative modulus improvements*	*Comments*
None		1.0	Reference
Geotextile	Low grade	1.25	No cover and low stress (<60 kPa). At higher stresses or increased cover the higher grade had no improved performance
	High grade	1.4	
Geogrid	Biaxial	2.0	Cover ≥100 mm
	Triaxial	2.25	
Geocell		2.5	Cover ≥100 mm

  - Increased cover of gravel over geosynthetic generally increased the composite modulus above the values shown

Chapter 17

Fill specifications

17.1 Specification development

- Specifications typically use the grain size as one of the key indicators of likely performance.
- The application determines the properties required. For example, greater fines content would be required for an earthworks water retention system, while low fines would be required for a road base pavement.
- Applying a specification provides a better confidence in the properties of the fill. Specifications use low cost indicator tests while the primary considerations are strength, modulus or permeability.
- Importing a better quality fill can provide a better consistency than using a stabilised local fill. However, the latter may be more economical and reduced environmental "cost" and this has to be factored into the design performance.

Table 17.1 Desirable material properties.

Requirement	*Typical application*	*Desirable material property*			
		Gravel %	*Gravel size*	*Gradation*	*Fines*
High strength	Pavement	Increase	Increase	Well graded	Reduce
Low permeability	Liner	Reduce	Reduce	Well graded	Increase
High permeability	Drainage layer	Increase	Increase	Uniformly/poorly graded	Reduce
Durability	Breakwater	Increase	Increase	–	Reduce

17.2 Pavement material aggregate quality requirements

- Pavement materials are typically granular with low fines content.
- Larger nominal sizing has the greatest strength, but an excessive size creates pavement rideabilty and compaction issues.
- The optimum strength is obtained with a well graded envelope.
- Some fines content is useful in obtaining a well graded envelope but an excessive amount reduces the overall strength.

Table 17.2 Developing a specification for pavement materials.

Nominal sizing	*Material property*	*Aggregate quality required*			
		High (Base)	*Medium (sub base)*	*Low (capping)*	*Poor*
40 mm	% Gravel	>20%	>20%	>20%	<20%
	% Fines	<10%	<15%	<20%	>20%
30 mm	% Gravel	>25%	>25%	>20%	<20%
	% Fines	<15%	<20%	<25%	>25%
20 mm	% Gravel	>30%	>30%	>20%	<20%
	% Fines	<20%	<25%	<30%	>30%

- Natural river gravels may have about 10% more fines than the crushed rock requirements shown in the table, but 10% to 20% more gravel content.
- "Ideal" grading for maximum density based on
  - $p = (d/d_{max})^n \times 100$
  - p = % passing particle size (d, mm)
  - d_{max} = maximum particle size (mm)
  - n = 0.5 (Fuller's curve)

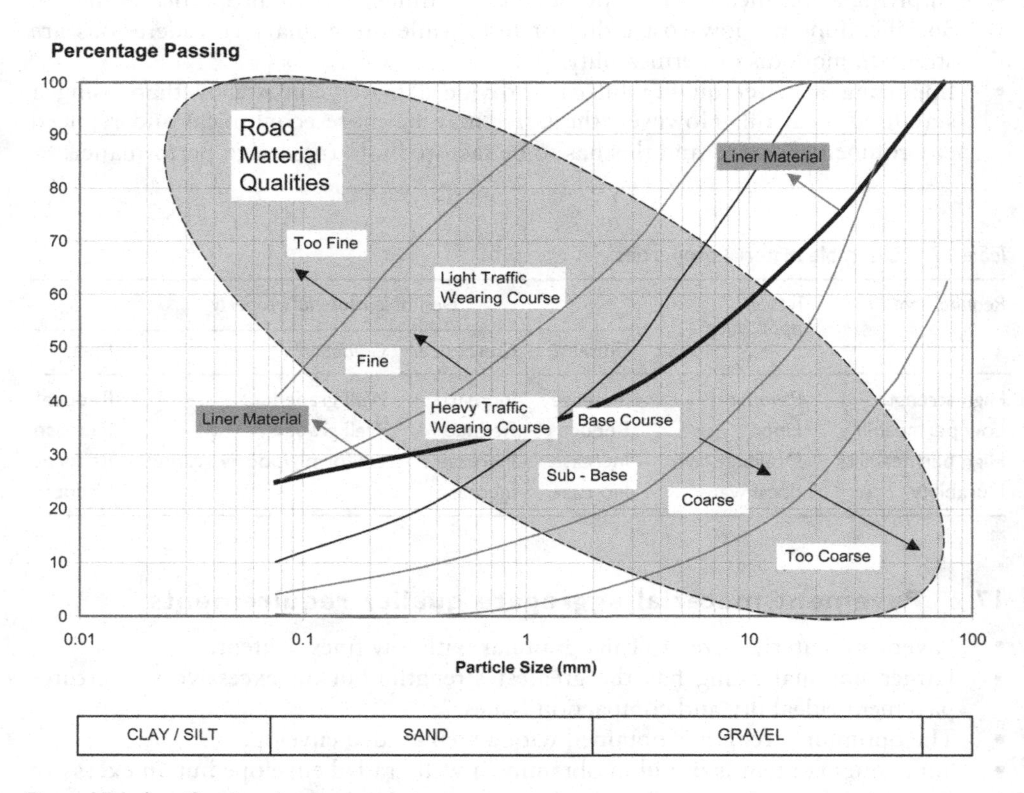

Figure 17.1 Specification development.

17.3 Backfill requirements

- Backfill shall be free from organic or deleterious materials.
- A reinforced soil structure should have a limit on the large sizes to avoid damage to the reinforcing material. Water should be drained from the system, with a limitation on the percentage fines.
- A reinforced soil slope can tolerate greater fines and larger sizes. This design is to limit water intruding into the sloping face, although not the ideal drainage material for reducing groundwater effects.

Table 17.3 Backfill requirements (Holtz et al. 1995).

Property	*Specification requirement*	
	Reinforced soil structure	*Reinforced soil slope*
Sieve size	*Percent passing*	
100 mm	100	100
20 mm	100	100–75
4.75 mm	100–20	100–20
0.425 mm	60–0	60–0
0.075 mm	15–0	50–0
Plasticity Index	PI < 12%	PI < 22%

17.4 Typical grading of granular drainage material

- Granular drainage materials should be uniformly graded and be more permeable than the surrounding soil, as well as prevent washing of fines from the material being drained.

Table 17.4 Grading of filter material (Department of Transport, 1991).

Sieve size	*Percentage by mass passing*
63 mm	100%
37.5 mm	85–100
20 mm	0–25
10 mm	0–5

  - When used as a drainage layer below sloping faces such as revetments or chimney drains, angular material should be used.
  - Refer table 8.8 which shows 7% fines may reduce permeability by a factor of 100.
  - "Clean" sands would have less than 12% fines for general compaction applications, but for drainage related applications than less than 7% fines is required.

17.5 Pipe bedding materials

- A well-graded envelope provides the optimum strength and support for the pipes. However, this requires compaction to be adequate. Pipes in trenches may not have a large operating area and obtaining a high compaction is usually difficult.
- A reduced level of compaction is therefore usually specified and with a single size granular material which would be self-compacting.
- The larger size provides a better pipe support, but is unsuitable for small size pipes.

Table 17.5 Granular materials for pipe beddings.

Pipe size	*Maximum particle size*
<100 mm	10 mm
100–200 mm	15 mm
200–300 mm	20 mm
300–500 mm	30 mm
>500 mm	40 mm

- ○ Pipes are usually damaged during construction and proper cover needs to be achieved, before large equipment is allowed to cross over.
- ○ Typically 300 mm minimum cover, but 750 mm when subjected to heavy construction equipment loads. This varies with strength of pipe itself.
- ○ Proper compaction at the haunches of pipes is difficult to achieve and measure. This relies on suitable bedding to reduce the haunch. Therefore self-compacting material is required and do not compact any material place in middle third below the pipe to allow "bedding".

17.6 Compacted earth linings

- The key design considerations for earth linings are adequate stability and impermeability.
- The low permeability criteria require the use of materials with >30% clay fines.
- Density of 95% of Standard Maximum Dry Density typically used.
- Control Tests of at least 1 per 1000 m^3 placed would be required.

Table 17.6 Typical compacted earth lining requirements.

Depth of water (m)	*Canal design*		
	Side Slope (1V:H)	*Side thickness (m)*	*Bottom thickness (m)*
≤0.5 m	1V:1.5 H	0.75 m	0.25 m
1.5 m	1V:1.75 H	1.50 m	0.50 m
3.0 m	1V:2.0 H	2.50 m	0.75 m

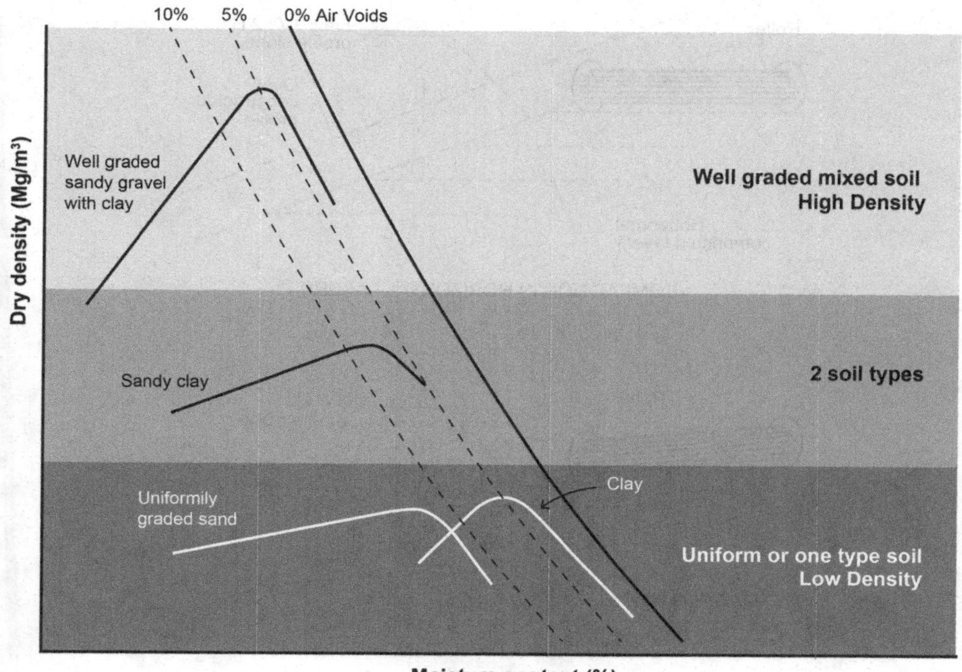

Figure 17.2 Grading influence on compaction curve.

17.7 Constructing layers on a slope

- Inadequate compaction may result at the edges or near sloping faces. Large equipment is unable to compact on steep slopes. Layers are placed either horizontally or on a minor slope. Benching may be required to control the water run-off, and hence erosion.
- Proper compaction requires moisture content of soil near to its plastic limit.
- The thickness of placed layers is typically 0.40 m (compacted) for a 10 tonne roller, but depends on the type of material being placed.
- The thickness of placed layers is typically 0.20 m (compacted) for 3 tonne roller.

Table 17.7 Constructing layers on a slope.

Method	*Place and compact material in horizontal layers*	*Place layers on a 1V:4H slope*
Advantage	Fast construction process	For limited width areas
Disadvantage	Edge not properly compacted	Side profile variability
Remedy	Over construct by • 0.5 m for light weight rollers • 1.0 m for heavy rollers And trim back to final design profile	Regular check on side profile

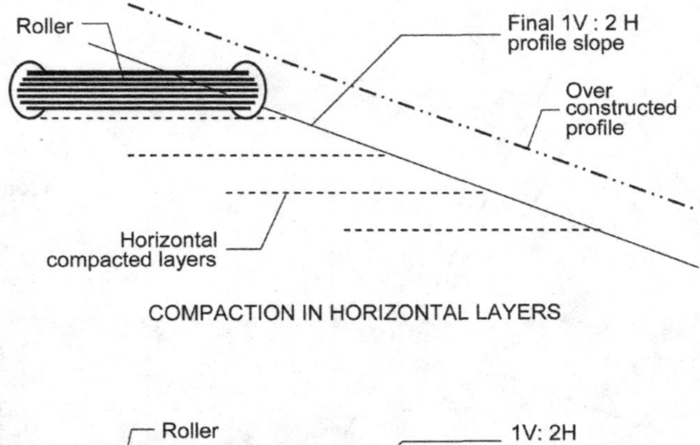

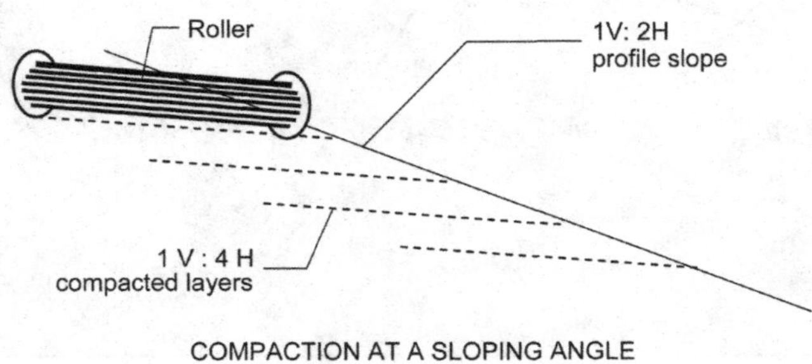

Figure 17.3 Placement and compaction of materials.

17.8 Durability of pavements

- The pavement material is usually obtained from crushed aggregate.
- The wearing and base courses would have a higher durability requirement than the sub base.

Table 17.8 Durability requirements for a pavement.

Parameter	*Wearing course*	*Base course*	*Sub base*	
			Upper	*Lower*
Water absorption	<2%	<3%	<4 %	<5%
Aggregate crushing value	<25%	<30%	<35%	<40%
Los Angeles abrasion	<30%	<35%	<40%	<45%
Sodium sulphate soundness	<10%	<15%	<20%	<25% loss
Flakiness index	<35	<40	<40	<45
Ten percent fines (wet)	>150 kN	>100 kN	>75 kN	>50 kN
Wet/dry strength variation	<30%	<40%	<50%	<50%

17.9 Dams specifications

- The dam core material should be impermeable – have a significant fines proportion.
- The core should also be able to resist internal erosion.

Table 17.9 Dam core material classification to minimise internal erosion.

Consideration	*Reduce erosion*		*Erosion resistance*	
Criteria	Rate of erosion decreases with increasing plasticity index (PI)	Higher compacted density reduces rate of erosion	Addition or inclusion of stone chips improves erosion resistance	Maximum stone size to allow compaction
Measure ideal	PI = 15% to 20%	Dry Density (DD) $\geq$98% (Standard Proctor)	Stones = 10% to 20%	Stone size = 2 mm to 60 mm
Fair	PI $\geq$ 12%	DD $\geq$ 95%	Stones $\geq$ 5% Stones $\leq$ 25%	Stones $\leq$ 100 mm
Poor	PI < 12%	DD < 95%	Stones < 5%	Stones > 100 mm
Very poor	PI < 10%	DD < 90%	Stones > 25%	Stones > 120 mm

- Dam cores should have a material with a minimum clay content of 20%, and preferably 30%.
- High PI clays (CH) are more difficult to place and compact.
- While the presence of some stones reduces erosion potential, a significant quantity of stones will increase the water flow, which is undesirable.

17.10 Frequency of testing

- The frequency of testing is based on the size of the area and project, uniformity of material and overall importance of the layer being tested.

Table 17.10 Guidelines to frequency of testing.

Test	*Field density*	*Grading and plasticity index*
Frequency for large scale operations	For selected material imported to site – Not less than a) 1 test per 1000 m^3, and b) 4 tests per visit c) 1 test per 250 mm layer per material type per 4000 m^2 For on site material imported – less than Not a) 1 test per 500 m^3, and b) 3 tests per visit c) 1 test per 250 mm layer per material type per 2000 m^2	1 test per 2000 m^3 at selected source before transporting to site 1 test per 1000 m^3 for using locally available material on site
Frequency for medium scale operations, e.g. residential lots	Not less than a) 1 test per 250 m^3, and b) 2 tests per visit, and c) 1 test per 250 mm layer per material type per 1000 m^2	1 test per 500 m^3 at selected source before transporting to site 1 test per 250 m^3 for using locally available material on site
Frequency for small scale operations using small or hand operated equipment e.g. backfilling, confined operations, trenches	Not less than a) 1 test per 2 layers per 50 m^2, and b) 1 test per 2 layers per 50 linear m	1 test per 100 m^3, at selected source before transporting to site 1 test per 50 m^3, for using locally available material on site

- Statistical quality control schemes apply for 4 or more tests samples.
- For less than 4 samples per lot use "no value shall be less than".

17.11 Rock revetments

- Rock revetments can be selected rock armour, rip rap or stone pitching.

Table 17.11 Rock revetments (McConnell, 1998).

Revetment type	*Specification*	*Porosity*	*Thickness*
Rip–rap	$D_{85}/D_{15} \sim 2$ to 2.5	35 to 40%	2 to 3 stones/rock sizes thick
Rock armour	$D_{85}/D_{15} \sim 1.25$ to 1.75	30 to 35%	2 rock sizes thick

17.12 Durability

- The degradable materials decompose when exposed to air, as they take on water.
- Sedimentary rocks are the most common rock types, which degrade rapidly, such as shales and mudstones.
- Foliated metamorphic rocks such as slate and phyllites are also degradable.

Table 17.12 Indicators of rock durability.

Test	*Strong and durable*	*Weak and non durable – Soil like*
	Rock like behaviour in long term	*Soil like behaviour in the long term*
Point load index	≥2 MPa	<1 MPa
Free swell	≤3%	>5%
Slake durability test	≥90	<60
Jar slake test	≥6	<2
Los Angeles abrasion	<25%	>40%
Weathering	Fresh to slightly weathered	Extremely weathered
RQD	>50%	<25%

- Several of the above indicators should be in place before classed as a likely non-durable material.
- Strong and durable rocks can be placed at large lift thickness (>500 mm for medium size equipment).
- Weak and durable require small lift thickness (<250 mm for medium size equipment) i.e. such rock materials are treated as a "soil" compaction.

17.13 Durability of breakwater

- The durability should be assessed on the material function.
- Primary armours have a higher durability requirement than a secondary armour.

Table 17.13 Durability requirements for a breakwater.

Parameter	*Stone core*	*Stone armour*		*Comments*
		Secondary	*Primary*	
Rock weathering	DW	DW/SW	SW/FR	Field assessment for suitability
RQD	>50%	>75%	>90%	
Joint spacing	>0.2 m	>0.6 m	>2.0 m	
Water absorption	<5%	<2%	<1%	Control testing
Aggregate crushing value	>25%	>20%	>15%	
Uniaxial compressive strength	>10 MPa	>20 MPa	>30 MPa	
Los Angeles abrasion	<40%	<30%	<20%	
Magnesium sulphate soundness	<15%	<10%	<5% loss	
Nominal rock sizing	>100 kg	>500 kg	>1000 kg	

17.14 Compaction requirements

- The placement density and moisture content depends on the material type and its climatic environment.
- OMC and MDD are not targets. The discussion on whether standard or modified compaction is often a mute point. The OMC and MDD are reference points, i.e. analogous to a unit of measurement. Refer Chapter 12.
- Materials with WPI > 2200 are sensitive to climate, and can wet up or dry back, if compacted at OMC and MDD. This results in a change of density and moisture content with an accompanying volume changes.
- OMC is the moisture content recommended for construction expediency, but can change with time.
- EMC is what the soil will be after construction i.e. several years hence.
- The EMC ~ OMC for rainfall environments of 600 mm to 1000 mm e.g. in California where Proctor was based, or UK conditions. In arid, and tropical (high rainfall environments) placement at the OMC may lead to significant volume changes.

Table 17.14 Acceptance zones for compaction.

Desired property	*Typical application*	*Density (wrt MDD)*	*Moisture content*
Shear strength – high	Pavement	High (near MDD)	Low (≤OMC)
permeability – low	Dams, canals	MDD, but governed by placement moisture content	High (≥OMC)
General fill – typical	WPI ≤ 1200	MDD	OMC
Shrinkage – Low	General embankment fill in dry environments; WPI ≥ 3200	Below MDD but >90% MDD	At EMC – typically 80% OMC
Swelling – Low	General embankment fill in wet environments; WPI ≥ 3200	Below MDD but >90% MDD	At EMC – typically 120% OMC

- EMC – Equilibrium Moisture Content.
- WPI – Weighted Plasticity Index.

17.15 Earthworks control

- Earthworks are controlled mainly by end-result specifications, i.e. measuring the relative compaction.
- Other measures may also be used as shown in the table.

Table 17.15 Earthworks control measures.

Method	*Measurement*	*Typical value*	*Comment*
Relative Compaction (RC), also known as Density Ratio (DR)	Insitu density and Maximum dry density	Trenches: RC 90% Subgrade RC > 95% Pavements RC > 98%	This can be an expensive process due to the large number of tests required
Relative Density (RD)	Insitu density where <12% fines	RD > 35% – médium dense RD > 65% – dense	Applies to "clean" sands – refer Table 12.16
Method specification	Equipment + lift thickness + No. of passes	250 mm 5 No. passes	Useful in rocky material
Degree of Saturation (DOS)	Density, Moisture content and specific gravity	Base DOS < 70% Sub-base DOS < 80% Subgrade DOS ~ 95%	Near OMC
Modulus	Direct, e.g. plate load test	Base E > 400 MPa Sub-base E > 200 MPa Rocky subgrade E > 100 MPa	Useful in rocky material

17.16 Typical compaction requirements

- The minimum compaction requirement depends on the type of layer, thickness, operating area, proximity to services/structures and equipment used.

Table 17.16 Typical compaction requirements.

Type of construction	*Element*		*% Standard compaction*	*Placement moisture content*
Roads and rail	Heavily loaded pavement	Base	>100%	Dry of OMC, DOS < 70%
	Lightly loaded pavement	Subbase	>98%	Dry of OMC, DOS < 80%
	Subgrade	WPI < 2200	>95%	OMC
	General embankment fill	WPI < 2200	>90%	OMC
	Subgrade	WPI > 2200	92% to 98%	EMC
	General embankment fill ≤ 3 m	WPI > 2200	90% to 96%	EMC
	General embankment fill > 3 m		>90%	OMC
Structure	Subgrade	WPI < 2200	>98%	≤OMC
	General fill	WPI < 3200	92% to 98%	EMC to OMC

(Contined)

Table 17.16 (Continued)

Type of construction	Element	% Standard compaction	Placement moisture content
Walls	Backfill, in trenches	90% to 95%	OMC to dry of OMC
Dams	Small	94% to 100%	OMC to wet of OMC
	Large	>97%	OMC to wet of OMC
Landfills	Capping	88% to 94%	EMC
	Liners	94% to 100%	OMC to wet of OMC
Canals	Clay	90% to 95%	OMC to wet of OMC
Landscaping	All soils – under compact for tree survival	80% to 90%	Dry or wet of OMC

- DOS – Degree of Saturation.
- If placement at EMC not practical then equilibration period, stabilisation or zonation of material required.
- EMC can be wet of OMC for climates with rainfall >1000 mm, but dry of OMC for rainfalls <500 mm.
- Relative density instead of relative compaction applies to granular soils with <12% fines.
- Depending on tree type, 15% to 30% may not survive if landscape areas are compacted >90%.
- There is often a mis-match between the specification and quality control values specified and the design values tested, i.e. without intervention the lab defaults to a target of 100% OMC and 100% MDD to test its compaction characteristics and CBR values.

17.17 Typical compacted modulus values

- Density is not directly used in design. The design of a building pad for example is based on the strength value. Density is used as the control parameter because it is an indicator – an index parameter that is a relatively low cost test that can be easily applied in practice. The use of direct strength or stiffness control tests is typically more expensive.
- Modulus can be measured directly with plate loading testing or falling weight deflectometers or through intelligent compaction.
- These in situ strength and modulus measurements are more relevant to the designer than the various density compaction controls.

Table 17.17 Typical compaction modulus values (Bomag, 2005).

Type of Soil	Standard Proctor density (%)	Deformation modulus (MN/m^2)	E_{V2}/E_{V1}
GW, well graded gravel	≥100	≥100	≤2.3
	≥98	≥80	≤2.5
	≥97	≥70	≤2.6
Uniform gravel, uniform, well graded sand (SW)	≥100	≥80	≤2.3
	≥98	≥70	≤2.5
	≥97	≥60	≤2.6

- Plate Loading tests: E_{V1} – First Loading, E_{V2} – Second loading.
- Refer Table 13.30 for CBR – modulus relationships.
- Refer Table 4.16 for LWFD modulus values.

17.18 Compaction layer thickness

- The compaction layer thickness depends on the material type and equipment being used. The operating space for equipment also needs consideration.
- There is a "compact to 250 mm thickness" fixation in many specifications. This assumes only light equipment is available and clay material.
- The 200 mm to 300 mm thickness typical in specification is due mainly to testing and quality control limitation in testing. Depending on the material, modern equipment can achieve compaction specifications with thickness significantly greater with a resulting reduction in time and cost of project. The quality control (i.e. achieved compaction) must still be checked.

Table 17.18 Compaction layer thickness.

Equipment size	*Material type*			
	Rock fill	*Sand & gravel*	*Silt*	*Clay*
Heavy (>10 tonne)	1500 mm	1000 mm	500 mm	300 mm
Light (<1.5 tonne)	400 mm	300 mm	250 mm	200 mm

- Note for light equipment the typical specification is too thick a layer while for non-clayey material with heavy equipment, compaction occurs to greater depths.
- Above assumes appropriate plant, e.g. sheepsfoot roller for clays and grid rollers for rock
- Light equipment typically required behind walls, over or adjacent to services, and in trenches.
- Impact rollers may be used (if no buildings or services nearby) with compaction depths to 3.0 m depending on material type and water level.
- Heavy equipment is not appropriate in urban areas due to vibration induced damage.

17.19 Achievable compaction

- The compaction achievable depends on the subgrade support below.
- Lab CBR values and/or specified compactions may not be achieved without the required subgrade support. A lab CBR is not necessarily a field CBR value. A steel plate model in the lab means an equivalent stiff reaction from the material under in the field to be applicable.
- Hammittt (1970) shows $CBR_{BC}/CBR_{SG} = 5.23$. But his data suggests a lower bound of ~2 for CBR < 5% and 4 for CBR ≥ 5%. This suggests a CBR 50% (lab value) compacted granular material will perform as a CBR 26% to CBR 20% material (field value) if placed on a CBR 5% subgrade.

- Typical achievable compactions with respect to layer thicknesses are provided for a firm clay (CBR = 3% to 5%).
- Lower strength subgrade materials would require an increased thickness specified.

Table 17.19 Achievable compaction for a granular material placed over a firm clay support (from Brandl, 2001).

Relative compaction (Standard Proctor)	*Thickness required to achieve density*	
	Minimum	*Typical*
90%	100 mm	150 mm
92%	175 mm	225 mm
95%	250 mm	350 mm
97%	325 mm	425 mm
100%	425 mm	525 mm

  - The significant depth of material for the support applies to granular and rocky material with suitable compaction equipment.
  - Reduced thickness would require the use of a geotextile and/or capping layer to prevent punching and loss of the material being compacted into the soft support.
  - The above table suggests the inability to achieve high compaction on weak subgrades without sufficient bridging thickness. Such material punches into the weak subgrade to stiffen that underlying material before achieving its compaction. This is a significant "loss" of expensive high quality material to the underlying subgrade.

17.20 Acceptable levels of ground vibration

- The acceptable level of ground vibration for a particular type of structure is usually based on the peak particle velocity.
- The specification of the equipment is required in relation to the environment and surrounds. For example, use of heavy compaction equipment over services or behind walls.
- The table provides the minimum distance a particular weight of vibrating roller should be operating from adjacent buildings to minimise the vibration level. This has an effect on the lift thickness for compaction.

Table 17.20 Minimum recommended distance from vibrating rollers (Tynan, 1973).

Roller class		*Weight range*	*Minimum distance to nearest building*
I	Very light	Maintenance rollers < 1.25 t	Not restricted for normal road use, 3 m
II	Light	1–2 t	Not restricted for normal road use, 5 m
III	Light–medium	2–4 t	5–10 m
IV	Medium–heavy	4–6 t	Not advised for city and suburban street, 10–20 m
V	Heavy	7–11 t	Not advised for built up areas, 20–40 m
VI	Very heavy	>12 t	Restricted to rural areas away from structures

Chapter 18

Rock mass classification systems

18.1 The rock mass rating systems

- Rock mass rating systems are used to classify rock and subsequently use this classification in the design of ground support systems. A few such ratings are provided below.
- Methods developed from the need to provide on site assessment empirical design of ground support based on the exposed ground conditions.
- Relationships exist between the various methods.

Table 18.1 Rock mass rating systems.

Rock mass rating system	*Key features*	*Comments*	*Reference*
Terzaghi's rock classification	7 No. classifications of in situ rock for predicting tunnel support from intact, stratified, moderately jointed, blocky and seamy, crushed, squeezing and swelling. Method did not account for similar classes could having different properties	One of the first rock mass classifications	Terzaghi, 1946
Rock Structure Rating (RSR)	Quantitative method that uses Parameter A – geological structure Parameter B – joint pattern and direction of drive Parameter C – joint condition and groundwater	Specifically related to tunnels	Wickham et al., 1972
Rock Mass Rating (RMR) or Geomechanics classification	Quantitative method that uses • Strength of intact rock • Drill Core Quality (RQD) • Spacing of discontinuities • Condition of discontinuities • Groundwater • Orientation of discontinuities	Based on the RMR classification one can determine: Average stand up time, cohesion and friction angle of the rock mass	Bieniawski, 1973 and 1989
Q System or Norwegian Geotechnical Institute (NGI) method	Quantitative method that uses • Rock quality designation • Joint set number • Joint roughness number • Joint alteration number • Joint water factor • Stress reduction factor	The log scale used provides insensitivity of the solutions to any individual parameter, and emphasizes the combined effects. Extensive correlations	Barton et al., 1974

(*Continued*)

Table 18.1 (Continued)

Geological Strength Index (GSI)	Main structure description adopted from Terzaghi's classification and combined with a joint condition factor • Intact or massive • Blocky • Very blocky • Blocky/folded • Crushed/disintegrated • Laminated/sheared	Covers both hard and weak rocks. Uses rationale basis to assess rock cohesion and friction for numerical analysis.	Hoek and Brown (1997)

- Only the 3 main classification systems in use are discussed further. These are the GSI, Q and RMR systems.

18.2 Rock Mass Rating System – RMR

- The classes provided in the table below are the final output. The derivation and design implications of that rating are provided in the subsequent tables.
- This RMR class provides the basis for strength assessment and support requirements.

Table 18.2 Rock mass classes (Bieniawski, 1989).

RMR class no.	*Description*	*Rating*
I	Very good rock	100–81
II	Good rock	80–61
III	Fair rock	60–41
IV	Poor rock	40–21
V	Very poor rock	<20

- The rating is the sum of the parameters given in the following tables
  - Strength & RQD
  - Discontinuities
  - Groundwater
  - Discontinuity orientations

18.3 RMR system – strength and RQD

- The strength is assessed in terms of both the UCS and Point Load index strengths – I_s (50). A conversion of 25 is assumed, however this relationship can vary significantly for near surface and soft rock. Refer Chapter 6.
- The RQD use the standard classification of poor (<25%) to excellent (>90%).

Table 18.3 Effect of strength and RQD (Bieniawski, 1989).

Parameter		*Range of values*						
Strength of intact rock	I_s (50), MPa	>10	4–10	2–4	1–2	For this low range UCS preferred		
	UCS, MPa	>250	100–250	50–100	25–50	5–25	1–5	<1
	Rating	15	12	7	4	2	1	0
Drill Core Quality	RQD, %	90–100	75–90	50–75	25–50	<25		
	Rating	20	17	13	8	3		

18.4 RMR system – discontinuities

- The discontinuity rating shows it to be the most important parameter in evaluating the rock rating.
- Persistence is difficult to judge from borehole data, and needs to be reassessed during construction.

Table 18.4 Effect of discontinuities (Bieniawski, 1989).

Parameter		*Range of values*				
Discontinuity	Spacing	>2 m	0.6–2 m	200–600 mm	60–200 mm	<60 mm
	Rating	20	15	10	8	5
Discontinuity condition Classification	Surfaces	Very rough 6	Rough 5	Slightly rough 3	Smooth 1	Slickenslided 0
	Persistence	<1 m 6	1–3 m 4	3–10 m 2	10–20 m 1	>20 m 0
	Separation	None 6	<0.1 5	0.1–1 mm 4	1–5 mm 1	>5 mm 0
	Infilling (Gouge)	None 6	Hard filling <5 mm 4	Hard filling >5 mm 2	Soft filling <5 mm thick 2	Soft filling >5 mm 0
	Weathering	FR 6	SW 5	MW 3	HW 1	XW 0
	Total rating	30	25	20	10	0

18.5 RMR – groundwater

- The groundwater flow would be dependent on the discontinuity (e.g. persistence and separation).

Table 18.5 Effect of groundwater (Bieniawski, 1989).

Parameter		*Range of values*				
Groundwater	Inflow per 10 m tunnel length (m)	None	<10	10–25	25–125	>125
	Joint water pressure/ Major principal axis	0	<0.1	0.1–0.2	0.2–0.5	>0.5
	General conditions	Completely dry	Damp	Wet	Dripping	Flowing
	Rating	15	10	7	4	0

18.6 RMR – adjustment for discontinuity orientations

- The discontinuity arrangement effect is based on the type of construction.

Table 18.6 Rating adjustment for discontinuity orientations (Bieniawski, 1989).

Parameter	*Range of values*	*Very favourable*	*favourable*	*fair*	*unfavourable*	*Very unfavourable*
Strike and dip of discontinuities	Tunnels and mines	0	−2	−5	−10	−12
	Strike wrt tunnel axis (Drive with/ against Dip°)	≠with (45–90)	≠with (20–45)	≠against (45–90) ///(20–45) irrespective of strike (0–20)	≠against (20–45)	///(45–90)
	Foundations	0	−2	−7	−15	−25
	Dams/Dip°	(0–10)	(30–60)	Downstream (10–30)	Upstream (10–30)	(60–90)
	Slopes	0	−5	−25	−50	−60

Strike perpendicular (≠); Strike parallel (///)

18.7 RMR – strength parameters

- The classes and its associated strength parameters are provided in the table below.
- Refer chapter 11 for modulus relationships.

Table 18.7 Meaning of rock mass classes (Bieniawski, 1989).

RMR class No.	*Rock mass strength*	
	Cohesion of rock mass, kPa	*Friction angle (deg)*
I	>400	>45
II	300–400	35–45
III	200–300	25–35
IV	100–200	15–25
V	<100	<15

18.8 RMR – application to tunnels, cuts and foundations

- The classes and its meaning are provided in the table below.

Table 18.8 Meaning of rock mass classes (Bieniawski, 1993 and Waltham, 1994).

RMR class No.	*Average stand up time*	*Allowable bearing pressure (kPa)*	*Safe cut slopes in rock*
I	20 yr for 15 m span	6000–4400	>70
II	1 yr for 10 m span	4400–2800	65
III	1 wk for 5 m span	2800–1350	55
IV	10 h for 2.5 m span	1350–450	45
V	30 min for 1 m span	450–300	<40

- ○ The allowable bearing pressure by this method seems low even for a low quality rock.
- ○ Settlement of 12 mm expected may explain low bearing pressure.

18.9 RMR – excavation and support of tunnels

- The classes and its application to tunnel design are provided in the table below.

Table 18.9 Guidelines for excavation and support of 10 m span rock tunnels using RMR Classes (after Bieniawski, 1989).

RMR class no.	*Excavation*	*Support*				
		Rock bolts		*Shotcrete*		*Steel sets*
		Location	*Length × Spacing*	*Location*	*Thickness*	
I	Full face. 3 advance	Generally no support required except spot bolting				
II	Full face. 1–1.5 m advance. Complete support 20 m from face	Locally. In Crown with occasional wire mesh	3 m × 2.5 m	Crown where required	50 mm	None

(Continued)

Table 18.9 (Continued)

RMR class no.	*Excavation*	*Support*				
		Rock bolts		*Shotcrete*		*Steel sets*
		Location	*Length × Spacing*	*Location*	*Thickness*	
III	Top heading and bench. 1.5–3 m advance in top heading. Commence support after each blast. Complete support 10 m from face	Systematic bolts with wire mesh in crown	4 m × 1.5–2 m	Crown Sides	50–100 mm 30 mm	None
IV	Top Heading and bench 1.0–1.5 m advance in top heading. Install support concurrently with excavation, 10 m from face	Systematic bolts with wire mesh in crown and walls	4–5 m × 1–1.5 m	Crown Sides	100–150 mm 100 mm	Light to medium ribs spaced 1.5 m where required
V	Multiple drifts 0.5–1.5 m advance in top heading. Install support concurrently with excavation. Shotcrete as soon as possible after blasting	Systematic bolts with wire mesh in crown and walls. Bolt invert	5–6 m × 1–1.5 m	Crown Sides Face	150–200 mm 150 mm 50 mm	Medium to heavy ribs spaced 0.75 m with steel lagging and forepoling if required. Close invert

- 20 mm diameter fully grouted rock bolts assumed.

18.10 Norwegian Q system

- The rock mass quality – Q values is based on a formula with the relationship shown in the table.
- The Q values are then used to predict rock support design.

Table 18.10 Norwegian Q system (Barton et al., 1974).

Parameter	*Symbol*	*Description*
Rock mass quality		$Q = (RQD/J_n) \times (J_r/J_a) \times (J_w/SRF)$
Rock quality designation	RQD	(RQD/J_n) = Relative block size: Useful for distinguishing massive, rock bursts prone rock
Joint set number	J_n	
Joint roughness number	J_r	(J_r/J_a) = Relative frictional strength (of the least favourable joint set or filled discontinuity)
Joint alteration number	J_a	
Joint water factor	J_w	(J_w/SRF) = Relative effects of water, faulting, strength/stress ratio, squeezing or swelling (an "active" stress term)
Stress reduction factor	SRF	

- Qc = Q × UCS/100.
- Unconfined Compressive Strength = UCS.
- The tables that follow are based principally on the 1974 work but with a few later updates as proposed by Barton.

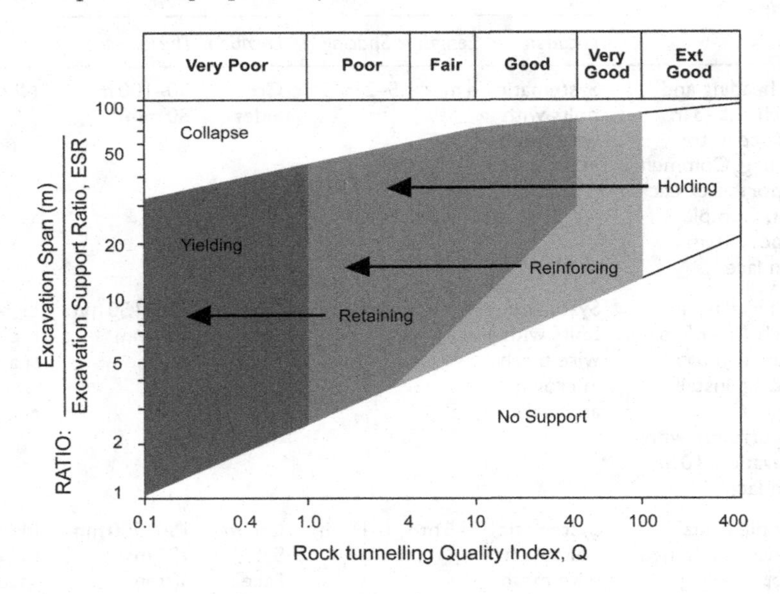

Figure 18.1 Support function (Kaiser et al., 2000).

18.11 Relative block size

- The relative block size is based on the RQD and the joint set number.

Table 18.11 Relative block size (Barton et al., 1974).

Parameter/symbol	*Description*		*Number value*
	Quality	*RQD value*	
Rock	Very poor	0%–10%	10
Quality	Very poor	10%–25%	15, 20, 25
Designation	Poor	25%–50%	30, 35, 40, 45, 50
RQD	Fair	50%–75%	55, 60, 65, 70, 75
	Good	75%–90%	80, 85, 90
	Excellent	90%–100%	95, 100
	Joint set number	*Joint randomness*	
Joint sets	No or few joints	Massive	0.5–1.0
number	One		2
J_n	One	+random	3
	Two		4
	Two	+random	6
	Three		9
	Three	+random	12
	Four or more	+random, heavily jointed earth-like	15
	Crushed rock		20

- o Number value based on RQD ≥10.
- o RQD in intervals of 5.
- o RQD can be measured directly or obtained from volumetric joint count.
- o For tunnel intersections use 3.0 × J_n.
- o For portals use 2.0 × J_n.

18.12 RQD from volumetric joint count

- The RQD may also be assessed by the volumetric joint count.

Table 18.12 Volumetric joint rock (adapted from Barton, 2006).

Block sizes	*Volumetric joint count (J_v) no./m³*		*RQD*	*RQD quality*
	Range	*Likely*		
Massive	≤1			
Large	1–3	≤4	100%	Excellent
Medium	3–10	4–8	90%–100%	Excellent
		8–12	75%–90%	Good
Small	10–30	12–20	50%–75%	Fair
		20—27	25%–50%	Poor
Very small	>30	27–32	10%–25%	Very poor
		32–35	0%–10%	

18.13 Relative frictional strength

- The ratio of the joint roughness number and the alteration number represents inter-block shear strength.

Table 18.13a Relative frictional strength from joint roughness (Barton et al., 1974).

Parameter/ symbol	*Description*			*Value*
	Rock wall contact	*Micro-surface*	*Macro-surface*	
Joint Roughness Number J_r	Rock – wall contact and contact before 10 cm shear	Any	Discontinuous	4.0
		Rough or irregular	Undulating	3.0
		Smooth,	Undulating	2.0
		Slickenslided	Undulating	1.5
		Rough or irregular	Planar	1.5
		Smooth,	Planar	1.0
		Slickenslided	Planar	0.5
	None when sheared	Zone contains minerals or crushed zone thick enough to prevent rock–wall contact		1.0
		Sandy, gravelly or crushed zone thick enough to prevent wall contact		–

Table 18.13b Relative frictional strength from joint alteration (Barton et al., 1974).

Parameter/ symbol	*Description*				*Value*
	Rock wall contact	*Particles Filling*	*Fillings Type*	ϕ_r	
Joint alteration number J_a	No mineral fillings, only coatings	Tightly healed, hard, Non-softening, impermeable	*Quartz*	>35°	0.75
		Unaltered joint walls, None	*Surfacing staining only*	25–35°	1.0
		Slightly altered joint walls, Non-softening mineral coatings	*Sandy particles, clay-free disintegrated rock*	25–30°	2.0
		Non-softening	*Silty or sandy – clay coatings, small clay fraction*	20–25°	3.0
		Softening	*Low friction clay mineral coatings i.e. Kaolinite, mica*	8–16°	4.0
	Thin mineral fillings. Rock wall contact before 10 cm shear	Strongly over-consolidated non-softening fillings	Sandy particles, clay – free disintegrated rock	25–30°	4.0
		Medium or low over-consolidation, softening	Clay mineral (continuous, but <5 mm thickness)	16–24°	6.0
		Depends on access to water and % of swelling clay size particles	Clay mineral fillings (continuous, but <5 mm thickness)	12–16°	8.0
			Swelling – clay fillings i.e. montmorillonite (continuous, but <5 mm thickness)	6–12°	8–12
	No rock wall contact when sheared (thick mineral fillings)	Zones or bands	Disintegrated or crushed rock and clay	6–24°	6, 8 or 8–12
		Zones or bands, small clay fraction (non-softening)	Silty or sandy clays	–	5.0
			Thick continuous zones or bands of clay	6–24°	10,13 or 13–20

18.14 Active stress – relative effects of water, faulting, strength/stress ratio

- The active stress is the ratio of the joint water reduction factor and the stress reduction factor.
- The joint water reduction factor accounts for the degree of water seepage.

Table 18.14 Joint water reduction factor (Barton et al., 1974).

Flow	*Joint flow*	*Approx. water pressure (kPa)*	*J_w value*
Dry excavations or minor inflow	i.e. <5 L/min locally	<100	1.0
Medium inflow or pressure	Occasional outwash of joint fillings	100–250	0.66
Large inflow or high pressure in competent rock	With unfilled joints	250–1000	0.5
Large inflow or high pressure	Considerable outwash of joint fillings		0.33
Exceptionally high inflow	Or water pressure at blasting, decaying with time		0.2–0.1
	Or water pressure continuing without noticeable delay	>1000	0.1–0.05

18.15 Stress reduction factor

- The stress reduction factor is a measure of:
  - The loosening load whee excavations occur in shear zones and clay bearing rock,
  - Squeezing loads in plastic incompetent rock, and
  - Rock stresses in competent rock.

Table 18.15 Stress reduction factor (Barton et al., 1974 with updates).

Rock type	*Zone characteristics*				*SRF value*
	Weakness zones		*Material in zone*	*Depth*	
Weakness zones intersecting excavations which may cause loosening of rock mass when tunnel is excavated	Multiple occurrences, very loose surrounding rock		Clay / Chemically disintegrated rock	Any	10
	Single			≤50 m	5
	Single			>50 m	2.5
	Multiple shear zones, loose surrounding rock		No Clay	Any	7.5
	Single Shear zones			≤50 m	5.0
	Single Shear zones			>50 m	2.5
	Loose, open joints, heavily jointed			Any	5.0
	Stress		UCS/σ_1	σ_ϕ/σ_c	
Competent rock, rock stress problems	Low	Near surface, open joints	>200	<0.01	2.5
	Medium	Favourable stress condition	200–10	0.01–0.3	1
	High	Very tight structure. Usually favourable to stability, may be unfavourable for wall stability	10–5	0.3–0.4	0.5–2
		Moderate slabbing after >1 hour in massive rock	5–3	0.5–0.65	5–50
		Slabbing and rock bursts after a few minutes in massive rock	3–2	0.65–1	50–200
		Heavy rock burst (strain burst) and immediate dynamic deformations in massive rock	<2	>1	200–400

(Continued)

Table 18.15 (Continued)

Rock type	*Zone characteristics*			*SRF value*
	Weakness zones	*Material in zone*	*Depth*	
Squeezing rock, plastic flow of incompetent rock under the influence of high rock pressure	Mild squeezing rock pressure Heavy squeezing rock pressure		1–5 > 5	5–10 10–20
Swelling rock, chemical swelling activity depending on pressure of water	Mild swelling rock pressure Heavy swelling rock pressure			5–10 10–15

- Major and minor principal stresses σ_1 and σ_3.

18.16 Selecting safety level using the Q system

- The excavation support ratio (ESR) relates the intended use of the excavation to the degree of support system required for the stability of the excavation.

Table 18.16 Recommended ESR for selecting safety level (Barton et al., 1974 with subsequent modifications).

Type of excavation	*ESR*
Temporary mine openings	2–5
Permanent mine openings, water tunnels for hydropower, pilot tunnels	1.6–2.0
Storage caverns, water treatment plants, minor road and railway tunnels, access tunnels	1.2–1.3
Power stations, major road and railway tunnels, portals, intersections	0.9–1.1
Underground nuclear power stations, railway stations, sport and public facilities, factories	0.5–0.8

- For temporary supports ESR should be increased by 1.5 times and Q by 5.

18.17 Support requirements using the Q system

- The stability and support requirements are based on the equivalent dimension (D_e) of the excavation.
- D_e = Excavation span, diameter or height/ESR.

Table 18.17 Support and no support requirements based on Equivalent Dimension relationship to the Q value (adapted from Barton et al., 1974).

Q value	*Equivalent dimension (D_e)*	*Comments*
0.001	0.17	Support is required above the D_e value shown. No support is required below that value. The detailed graph provides design guidance on bolts spacing and length, and concrete thickness requirements
0.01	0.4	
0.1	0.9	
1	2.2	
10	5.2	
100	14	
1000	30	

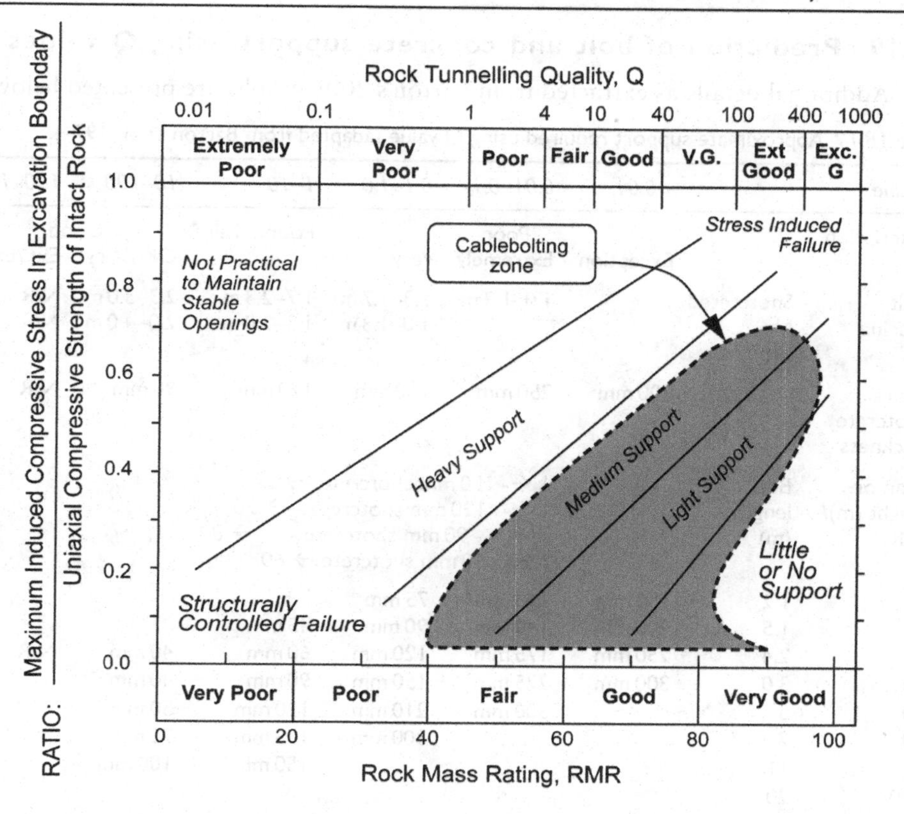

Figure 18.2 Cable bolt support (Hutchinson and Diederichs, 1996).

18.18 Prediction of support requirements using Q values

- Additional details as extracted from Barton's 2006 graphs are presented below.

Table 18.18 Approximate support required using Q value (adapted from Barton et al., 1974).

Q value	*<0.01*	*0.01–0.1*	*0.1–1.0*	*1–10*	*10–100*	*100–1000*
Description	Poor Exceptionally	Poor Extremely	Poor Very	Poor/Fair	Good/Extreme Very good	Good/ Exceptional
Equivalent span/height	0.15	0.25–0.8	0.8–2	2–5	5–12	No rock support 12–30
4–100				4 <-------- Spot bolting --------> 100		
1.5–70			0.15 <---- Systematic bolting ----> 50			
0.3–60		0.3 <------- Bolts and shotcrete -------> 60				
0.15–50	0.15 <------- Bolts and fibercrete -------> 50					
3–40	3 <-- Cast concrete lining --> 40					

18.19 Prediction of bolt and concrete support using Q values

- Additional details as extracted from Barton's 2006 graphs are presented below.

Table 18.19 Approximate support required using Q value (adapted from Barton et al., 1974).

Q Value		*<0.01*	*0.01–0.1*	*0.1–1.0*	*1–10*	*10–100*	*100–1000*
Description			Poor		Poor Fair	Good	
		Exception	Extremely	Very		OK /Very	Ext./Exc.
Bolt spacing	Shotcreted		1.0–1.3 m	1.3–1.7 m	1.7–2.3 m	2.3–3.0 m	N/R
	No shotcrete			1.0–1.3 m	1.3–2.0 m	2.0–4.0 m	N/R
Typical shotcrete thickness		300 mm	250 mm	150 mm	120 mm	90 mm	N/R
Span or height (m)/ ESR	Bolt length (m)		1 ←- 150 mm shotcrete -→ 50 1 ←-- 120 mm shotcrete --→ 70 1 ←---- 90 mm shotcrete -----→ 80 1.5 ← 50 mm shotcrete → 60				
1	1.2	150 mm	110 mm	75 mm			
2	1.5	200 mm	140 mm	90 mm	45 mm		
5	2.4	250 mm	175 mm	120 mm	60 mm	40 mm	N/R
10	3.0	300 mm	225 mm	150 mm	90 mm	40 mm	
20	5		300 mm	210 mm	120 mm	50 mm	
30	7			300 mm	135 mm	75 mm	
50	11				150 mm	100 mm	
100	20						
Steel ribs		0.5 m	0.5–1.0 m	1.0–2.5 m	2.5–5 m	N/R	

- Barton et al.'s research was primarily for tunnel support requirements. Since that time many relationships to other parameters have been developed. Many practitioners have suggested this is beyond its initial scope. However as in many engineering relationships it does provide useful initial guidance to other parameters.
- Some of these relationships are presented below.

18.20 Prediction of velocity using Q values

- The prediction of the P-wave velocity based on the Q value is shown in the Table below.
- This is for hard rock, near the surface.

Table 18.20 P-wave velocity estimate using Q value (adapted from Barton, 2006).

Rock mass quality, Q value	*<0.01*	*0.01–0.1*	*0.1–1.0*	*1–10*	*10–100*	*100–1000*
P-wave velocity V_p (km/s)	<1.5	1.5–2.5	2.5–3.5	3.5–4.5	4.5–5.5	5.5–6.5
RQD %	<5%	5–10%	10–40%	40–80%	80–95%	>95%
Fractures/metre	>27		27–14	14–7	7–3	<3

18.21 Prediction of Lugeon using Q values

- The Lugeon values provide an indication of the rock permeability.
- Chapter 8 related the Lugeon value to the rock jointing characteristics – a key parameter in the Q value assessment.

Table 18.21 Average Lugeon estimate using Q_c value (adapted from Barton, 2006).

$Q_c = Q \times UCS/100$	<0.001	0.01–0.1	0.1–1.0	1–10	10–100	100–1000
Description		Poor		Poor/Fair	Good	
	Exception. Major fault	Extremely Minor fault	Very	Hard – porous	OK/Very Hard – Jointed	Ext./Exc. Hard – massive
Typical Lugeon value	1000–100	100–10	10–1	1–0.1	0.1–0.01	0.01–0.001
Lugeon value at depth						
1000 m	0.01–0.1	~0.01		0.01–0.001	0.01–0.001	0.01–0.001
500 m	0.1–1.0	0.01–0.1		0.1–0.01	0.01–0.001	
100 m	1.0–10	0.1–1.0		0.1–0.01	0.01–0.001	
50 m	10–100	1.0–10		1.0–0.1	0.1–0.01	
25 m	100–1000	10–100		~1.0	0.1–0.01	

18.22 Prediction of advancement of tunnel using Q values

- The tunnel advancement is proportional to the rock quality.
- The Q value has therefore been used by Barton to estimate the average tunnel advancement.
- The TBM rates decline more strongly with increasing tunnel length.

Table 18.22 Average tunnel advancement estimate using Q value (adapted from Barton, 2006).

Rock mass quality, Q value	<0.01	0.01–0.1	0.1–1.0	1–10	10–100	100–1000
Description		Poor		Poor/Fair	Good	
	Exception.	Extremely	Very		OK/Very	Ext./ Exc.
	Delays due to support required				Lack of joints	
Tunnel boring machine	≤10	10–40	40–200	200–140	140–80	80–40 m/wk
Drill and blast	≤10	10–25	25–50	50–120	120 m/week	

18.23 Relative cost for tunnelling using Q values

- The lower quality rock would require greater tunnel support and hence costs.
- The Q value has therefore been used by Barton to estimate the relative tunnelling cost.

Table 18.23 Relative cost estimate using Q value (adapted from Barton, 2006).

Rock Mass Quality, Q Value	<0.01	0.01–0.1	0.1–1.0	1–10	10–100	100–1000
	Delays due to support required				Lack of joints	
Relative cost	>1100%	1100–400%	400–200%	200–100%	100%	
Relative time	>900%	900–500%	500–150%	150–100%	100%	

18.24 Prediction of cohesive and frictional strength using Q values

- Barton used the Q value to estimate the rock strength based on the relationships shown in the Table below.

Table 18.24 Average Cohesive and Frictional strength using Q value (adapted from Barton, 2006).

Strength component	*Relationship*	*Relevance*
Cohesive (CC) – MPa	$CC = (RQD/J_n) \times (1/SRF) \times (UCS/100)$	Component of rock mass requiring concrete, shotcrete or mesh support.
Frictional (FC) – degrees	$FC = \tan^{-1}(J_r/J_a) \times (J_w)$	Component of rock mass requiring bolting.

  - The Hoek-Brown failure criterion discussed in later tables is a more reliable indicator to derive these parameters.

18.25 Prediction of strength and material parameters using Q values

- The interrelationship between the Q values and the various parameters provide the following values.

Table 18.25 Typical strength values using Q value (adapted from Barton, 2006).

RQD	*Q*	*UCS (MPa)*	*Qc*	*Cohesive strength (CC) (MPa)*	*Frictional strength (FC)°*	*Vp (km/s)*	*E mass (GPa)*
100	100	100	100	50	63	5.5	46
90	10	100	10	10	45	4.5	22
60	2.5	55	1.2	2.5	26	3.6	10.7
30	0.13	33	0.04	0.26	9	2.1	3.5
10	0.008	10	0.0008	0.01	5	0.4	0.9

18.26 Prediction of deformation and closure using Q values

- Barton used the Q value to estimate the rock deformation based on the relationships shown in the Table below.

Table 18.26 Typical deformation and closure using Q value (adapted from Barton, 2006).

Movement	*Relationship*
Deformation, Δ (mm)	$\Delta = \text{Span (m)}/Q$
Vertical deformation, Δ_v	$\Delta_v = \text{Span (m)}/(100\ Q) \times \sqrt{(\sigma_v/UCS)}$
Horizontal deformation, Δ_h	$\Delta_h = \text{Height (m)}/(100\ Q) \times \sqrt{(\sigma_h/UCS)}$
At rest pressure, K_o	$K_o = \text{Span (m)}/\text{Height (m)}^2 \times (\Delta_h/\Delta_v)^2$

18.27 Prediction of support pressure and unsupported span using Q values

- The support as recommended by Barton et al. (1974) was based on the following pressures and spans.

Table 18.27 Approximate support pressure and spans using Q value (adapted from Barton, 2006).

Rock mass quality, Q value	*<0.01*	*0.01–0.1*	*0.1–1.0*	*1–10*	*10–100*	*100–1000*
Support pressure (kg/sq cm)	5–30	3–15	1–7	0.5–3	0.1–2	0.01–0.2
Unsupported span (m)	≤0.5 m	0.5–1.0 m	1.0–2 m	2–4 m	4–12 m	>12 m

18.28 Geological strength index – structure description

- The Geological strength index (GSI) was introduced by Hoek et al. (1995) to allow for the rock mass strength of different geological settings. The GSI can be related to rock mass rating systems such as the RMR or Q systems.
- From the description of structure and surface conditions of the rock mass estimate the average GSI value.
- The structure descriptions are presented in the table below.

Table 18.28 Structure descriptions for estimating GSI for jointed rocks (adapted from Hoek and Brown, 1997).

Classification	*Description*	*Joint spacing*	
Intact or massive	Intact rock specimen or massive in situ rock with few widely spaced discontinuities	>1000 mm	← Decreasing interlocking of rock pieces
Very blocky	Very well interlocked undisturbed rock mass consisting of cubical blocks formed by three orthogonal discontinuity sets	300–1000 mm	
Blocky	Interlocked, partially distributed rock mass with multifaceted angular blocks formed by four or more discontinuity sets	100–300 mm	
Blocky/folded	Folded and faulted with many intersecting discontinuities forming angular blocks	30–100 mm	
Crushed/disintegrated	Poorly interlocked, heavily broken rock mass with a mixture of angular and rounded blocks	<30 mm	
Laminated/sheared	Lack of blockiness due to close spacing of weak schistosity or shear planes	<10 mm	

18.29 Geological strength index – discontinuity description

- The discontinuity descriptions are presented in the table below.

Table 18.29 Discontinuity surface descriptions for estimating of GSI (adapted from Hoek and Brown, 1997).

Classification	*Very good*	*Good*	*Fair*	*Poor*	*Very poor*
	Decreasing surface quality →				
Description of roughness	Very rough, and fresh	Rough, surface	Smooth	Slickensided	Slickensided
Weathering of surface	Fresh unweathered	Slightly weathered, iron stained	Moderately weathered or altered surface	Highly weathered, or compact coatings with fillings of angular fragments	Highly weathered with soft clay coatings or fillings

18.30 Geological strength index – estimating value

- Using descriptors in previous tables estimate GSI.
- A range is more realistic that precise values.

Table 18.30 Estimate of GSI for jointed rocks (adapted from Hoek and Brown, 1997, Cai et al., 2004).

Structure Classification		*Block Volume cm³*	*Discontinuity at surface conditions*				
			Very Good	*Good*	*Fair*	*Poor*	*Very Poor*
Intact or massive	↓ Decreasing interlocking of rock pieces	10^6–10^5	100–77	90–66	80–55	N/A	N/A
Very blocky		10^5–10^3	86–65	77–55	66–45	55–35	43–25
Blocky			74–55	65–45	55–37	45–28	35–18
Blocky/folded		20–1000	63–47	55–38	45–29	37–20	28–12
Crushed/disintegrated		1–20	54–38	47–32	38–23	29–16	20–3
Laminated/sheared		<1	N/A	N/A	32–18	23–8	16–1

- The Hoek-Brown failure criterion can be used to directly assess specific shear strength situations based on the relationship major (σ_1) and minor (σ_3) principal stresses, and other material characteristics as shown in Figure 9.2. (Hoek et al., 1997).
- $\sigma'_1 = \sigma'_3 + \sigma'_{ci}\ (m_b\sigma'_3/\sigma'_{ci} + s)^a$.
- $a = 0.5$ for hard rock.
- $s = 0$ for poor quality rock mass with no tensile strength or "cohesion" or specimen that fall apart without confinement.

18.31 Relationship of rock constant m

- The m_b value is related to the intact constant as shown in the table below for both RMR and GSI.

Table 18.31 Estimate of rock constant m_b value (Hoek et al., 1997).

Rock mass	*Material constant m_b*	*s*	*a*
Disturbed	Exp {(RMR – 100)/14)}m_i	Exp{(RMR – 100)/6)}	0.5
Undisturbed or interlocking	Exp{(RMR – 100)/28)}m_i	Exp{(RMR – 100)/9)}	0.5
GSI > 25	Exp{(GSI – 100)/28)}m_i	Exp{(GSI – 100)/9)}	0.5
GSI < 25	Exp{(GSI – 100)/28)}m_i	0	0.65 – GSI/200

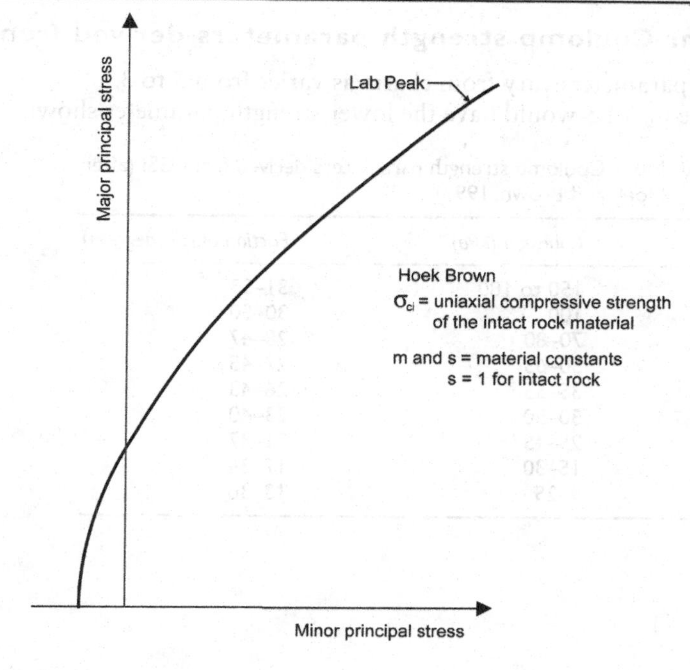

Figure 18.3 Hoek-Brown criteria.

18.32 Geological strength index – values of parameter m for a range of rock types

- The values provided by Hoek and Brown (1997) for the parameter m are for intact rock specimen test normal to bedding or foliation.

Table 18.32 Values of parameter m_i for a range of intact rock types (Hoek and Brown, 1997)

Rock type	*Class*	*Group*	*Texture*			
			Coarse	*Medium*	*Fine*	*Very fine*
Sedimentary	Clastic		Conglomerate 22	Sandstone 19	Siltstone 9	Claystone 4
	Non clastic	Organic		---- Greywacke -- 18 --- Chalk --- 7 --- Coal --- 8–21		
		Carbonate	Breccia 20	Sparitic Limestone 10	Micritic Limestone 8	
		Chemical		Gypstone 16	Anhydrite 13	
Metamorphic	Non-foliated		Marble 9	Hornfels 19	Quartzite 24	
	Slightly foliated		Migmatite 30	Amphibolite 25–31	Mylonites 6	
	Foliated		Gneiss 33	Schists 4–8	Phyllites 10	Slate 9
Igneous	Light		Granite 33 Granodiorite 30 Diorite 28		Rhyolite 16 Dacite 17 Andesite 19	Obsidian 19
	Dark		Gabbro 27 Norite 22	Dolerite 19	Basalt 17	
	Extrusive pyroclastic type		Agglomerate 20	Breccia 18	Tuff 15	

18.33 Mohr-Coulomb strength parameters derived from GSI

- Strength parameters vary from charts as varies from 5 to 35.
- The value m_b of 5 would have the lower strength parameter shown.

Table 18.33 Mohr Coulomb strength parameters derived from GSI (after Hoek and Brown, 1997).

GSI	*Cohesion (kPa)*	*Friction angle (degrees)*
90	150 to 180	31–53
80	100	30–50
70	70–80	28–47
60	50–65	27–45
50	35–55	26–43
40	30–50	23–40
30	25–45	21–37
20	15–30	17–34
10	9–25	13–30

Chapter 19

Earth pressures

19.1 Earth pressures

- Retaining walls experience lateral pressures from:
  - The earth pressures on the wall.
  - Water pressures.
  - Surcharges above the wall.
  - Dynamic loading.
  - Compaction induced.
- Pressure distribution varies with type of wall

Table 19.1 Earth pressures.

Type	*Movement*	*Earth pressure coefficient*	*Stresses*	*Comment*
Active	Soil → Wall	$K_a < K_o$	$\sigma'_h < \sigma'_v$	$K_a = 1/K_p$
At rest	None	K_o	σ'_h, σ'_v	Fixed and unyielding
Passive	Wall → Soil	$K_p > K_o$	$\sigma'_h > \sigma'_v$	Large strains required to mobilise passive resistance

- Horizontal earth pressure $= \sigma'_h$.
- Vertical earth pressure $= \sigma'_v$.
- $K_o = \sigma'_h / \sigma'_v$.
- Water pressures can have a significant effect on the design of the walls.
- Sloping ground exceeding 10° behind or in front of the wall can increase or decrease the active and passive earth pressures, respectively by a factor of 3.
- Refer to appendix for uniform surcharges behind wall.

19.2 Limit state modes

- Walls should be checked for various failure modes based on its earth pressures.
- Global stability should always be checked separate from earth pressure considerations.

Table 19.2 Limit states (AS4678, 2002).

Limit state	*Mode*	*Earth pressure coefficient*
Ultimate	U1	Sliding within or at the base of structure
	U2	Rotation of structure
	U3	Rupture of components
	U4	Pull out of reinforcement elements or anchors
	U5	Global failure
	U6	Bearing failure
Serviceability	S1	Rotation of structure
	S2	Translation or bulging
	S3	Settlement of structure

19.3 Earth pressure distributions

- The earth pressure depends primarily on the soil type, and soil movement.
- The shape of the pressure distribution depends on the surcharge, type of wall, restraint and its movement.

Table 19.3 Types of earth pressure distribution.

Type of wall	*No. of props*	*Example*	*Pressure distribution*	*Comments*
Braced	Multi > 2	Open strutted trench	Trapezoidal/ Rectangular	Fully restrained system H > 5 m
Semi flexible	Two	Soldier pile with two anchors	Trapezoidal/ Rectangular/ Triangular	Partially restrained system H < 5 m
Flexible system – no bracing or gravity system	One None	Soldier pile with one anchor Sheet piling, Gravity wall	Triangular	Shape changes depends on type of wall movement
Any wall with uniform surcharge load at top of wall	Any	Concrete platform at top of wall with 20 kPa traffic	Rectangular	Added to triangular or other pressure distribution
Any with load offset at top of wall	Any	Point load – pad footing Line load – narrow strip footing Strip load – strip footing	Irregular with maximum near top half of wall	Based on the theory of elasticity. This is added to the other loads
During wall construction	Any	Compaction induced pressure distribution	Passive line at the top with vertical drop to the active line	Applies when a heavy static or dynamic construction load is within ½ height of wall

  - Triangular distributions while used for the analysis of any non-braced wall strictly apply only to walls with no movement (at rest condition) and free to rotate about the base.
  - When rotation occurs about the top and/or sliding (translating) occurs, then the shape of the triangular distribution changes with arching near the top.
  - This effect is accounted for by applying a higher factor of safety to overturning as the triangular model calculated means the force is at one-third from

the base, but can be 0.38 to 0.45 H from the base if arching occurs (refer Figure 19.2).

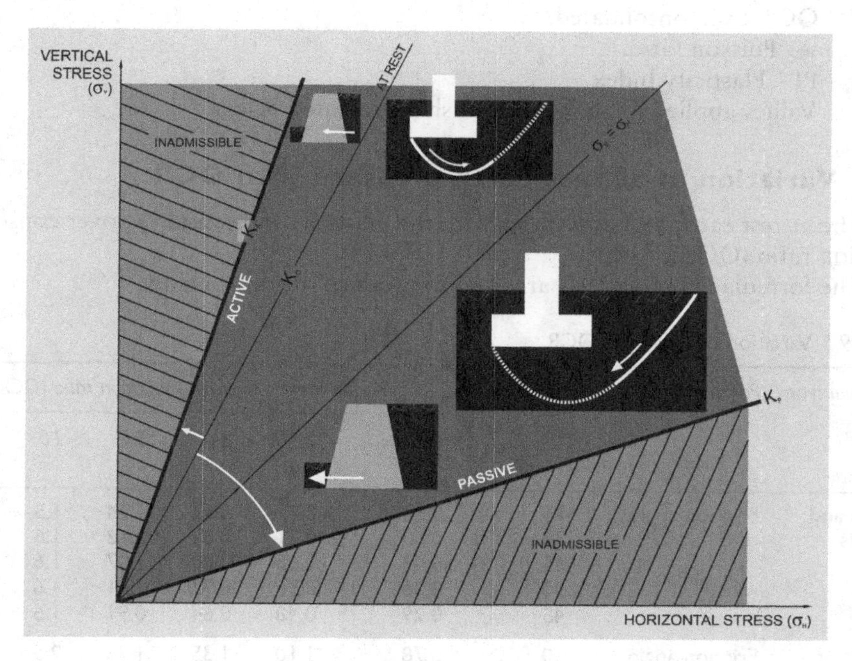

Figure 19.1 Vertical and horizontal stresses.

19.4 Coefficients of earth pressure at rest

- The coefficient of at rest earth pressure (K_o) is based on negligible wall movement.
- For lightly overconsolidated clays $K_o \sim 1.0$.
- For highly overconsolidated (OC) and swelling clays $K_o >> 1$.
- As plastic clays may have high swelling pressures, this material should be avoided where possible.
- The OC formula shown for granular soils and clays produce the same at rest value values for $\phi = 30°$. Below this friction value the clay $K_{o(OC)}$ value is higher, especially for low friction angles.

Table 19.4 Relationships for at rest earth pressure coefficients (part from Brooker and Ireland, 1965).

Soil type	*Relationship*
Normally consolidated	$K_{o\,(NC)} = 1 - \sin\phi$ (Granular soils)
	$K_{o(NC)} = 0.95 - \sin\phi$ (Clays)
	$K_{o(NC)} = 0.4 + 0.007\ PI$ (PI = 0–40%)
	$K_{o(NC)} = 0.64 + 0.001\ PI$ (PI = 40–80%)
Sloping backfill @ angle i	$K_{o(NC)} = (1 - \sin\phi)/(1 + \sin i)$
Over-consolidated	$K_{o(OC)} = (1 - \sin\phi)OCR^{\sin\phi}$ (Granular soils)
	$K_{o(OC)} = (1 - \sin\phi)OCR^{1/2}$ (Clays)
Elastic	$K_o = \nu/(1 - \nu)$

- ϕ – angle of wall friction.
- NC – normally consolidated.
- OC – overconsolidated.
- ν – Poisson ratio.
- PI – Plasticity Index.
- Values applied in above relationship presented below.

19.5 Variation of at rest earth pressure with OCR

- The at-rest earth pressure varies with the plasticity index and the over consolidation ratio (OCR).
- The formulae in table 19.3 are used to produce the table below.

Table 19.5 Variation of (K_o) with OCR.

Material type	*Parameter*	*Value*	*K_o for varying over consolidation ratio (OCR)*					
			OCR = 1 (N.C.)	*2*	*3*	*5*	*10*	*20*
Sands and gravels	Friction angle	25	0.58	0.77	0.92	1.14	1.5	2.0
		30	0.50	0.71	0.87	1.12	1.6	2.2
		35	0.43	0.63	0.80	1.07	1.6	2.4
		40	0.36	0.56	0.72	1.01	1.6	2.4
		45	0.29	0.48	0.64	0.91	1.5	2.4
Clays	Friction angle	10	0.78	1.10	1.35	1.74	2.5	3.5
		15	0.69	0.98	1.20	1.55	2.2	3.1
		20	0.61	0.86	1.05	1.36	1.9	2.7
		25	0.53	0.75	0.91	1.18	1.7	2.4
		30	0.45	0.64	0.78	1.01	1.4	2.0
Clays	Plasticity index with likely friction angle ( )	0 (33)*	0.40	0.57	0.69	0.89	1.3	1.8
		10 (29)	0.47	0.67	0.81	1.05	1.5	2.1
		20 (24)	0.54	0.76	0.94	1.21	1.7	2.4
		30 (20)	0.61	0.86	1.06	1.36	1.9	2.7
		40 (16)	0.68	0.96	1.18	1.52	2.1	3.0
		50 (15)	0.69	0.98	1.20	1.54	2.2	3.1
		60 (14.5)	0.70	0.99	1.21	1.57	2.2	3.1
		70 (14)	0.71	1.00	1.23	1.59	2.3	3.2
		80 (13)	0.72	1.02	1.25	1.61	2.3	3.2

- The table illustrates that the at rest pressure coefficient value can change significantly with change of OCR.
- *Approximate "Equivalent" friction angle from cross calibration of elastic and friction angle formula to obtain K_o. Note the difference in friction angle using this method as compared to that presented in Chapter 5.

19.6 Variation of at rest earth pressure with OCR using the elastic at rest coefficient

- The at rest earth pressure for overconsolidated soils varies from $K_o\ OCR^{\sin\phi}$ to $K_o\ OCR^{½}$ for granular to cohesive sol respectively.

- These formulae are applied below using the K_o derived from elastic parameters, then subsequently using the formulae but an "equivalent" friction angle for the case of sands, gravels and rocks.
- Both formulae are used in the tabulation below to highlight an inconsistency at low Poisson ratio/high friction angle materials, and if OCR used.

Table 19.6 Variation of (K_o) with OCR.

Material type	*Poisson ratio (friction)*	*Formulae used for OCR*	*K_o for varying overconsolidation ratio (OCR)* OCR = 1 (N.C.)	2	3	5	10	20
Rocks	0.1 (63)*	$K_{o(OC)}$	0.11	~~0.21~~	~~0.30~~	~~0.46~~	~~0.86~~	~~1.59~~
Rock/gravels	0.2 (49)	$= K_{o(NC)}\ OCR^{\sin \phi}$	0.25	~~0.42~~	~~0.57~~	~~0.84~~	~~1.41~~	~~2.37~~
Gravel/sand	0.3 (35)		0.43	~~0.64~~	~~0.80~~	~~1.07~~	~~1.60~~	~~2.37~~
Sands	0.4 (20)		0.67	~~0.84~~	~~0.96~~	~~1.14~~	~~1.44~~	~~1.81~~
Rocks	0.1 (63)*	$K_{o(OC)}$	0.11	0.16	0.19	0.25	0.35	0.50
Rock/gravels	0.2 (49)	$= K_{o(NC)}\ OCR^{½}$	0.25	0.35	0.43	0.56	0.79	1.12
Gravel/sand	0.3 (35)		0.43	0.61	0.74	0.96	1.36	1.92
Sands	0.4 (20)		0.67	0.94	1.16	1.49	2.11	2.98
Clay – PI < 12%	0.3 (35)*	$K_{o(OC)}$	0.43	0.61	0.74	0.96	1.36	1.92
Clay – PI = 12–22%	0.4 (20)	$= K_{o(NC)}\ OCR^{½}$	0.67	0.94	1.16	1.49	2.11	2.98
Clays – PI > 32%	0.45 (8)		0.82	1.16	1.42	1.83	2.59	3.67
Undrained clay	0.5 (0)		1.00	1.4	1.7	2.2	3.2	4.5

- The strike out has been used to remove the discrepancy.
- *Approximate "equivalent" friction angle.

19.7 Movements associated with earth pressures

- The active earth pressures (K_a) develop when the soil pushes the wall.
- The passive earth pressures (K_p) develop when the wall pushes into the soil.
- Wall movement is required to develop these active and passive states, and depends on the type and state of the soil.

Table 19.7 Wall movements required to develop the active and passive pressures (GEO, 1993).

Soil	*State of stress*	*Type of movement*	*Necessary displacement*	
Sand	Active	Parallel to wall	0.001 H	0.1% H
		Rotation about base		
	Passive	Parallel to wall	0.05 H	5% H
		Rotation about base	>0.10 H	>10% H
Clay	Active	Parallel to wall	0.004 H	0.4% H
		Rotation about base		
	Passive	–	–	

- Due to the relative difference in displacements required for the active and passive states for the one wall the passive force should be suitable factored or downgraded to maintain movement compatibility.
- Above is for rigid walls, other wall types have other displacement criteria. Refer Chapter 23.
- Soil nail walls deform at the top.
- Reinforced soil walls deform at the base.

19.8 Active and passive earth pressures

- Active and passive earth pressures are based on some movement occurring.
- Rankine (1857) and Coulomb (1776) developed the earth pressure theories for rigid retaining walls with updates by Caquot and Kerisel (1948). While the Rankine and Coulomb theroies are appropriate for the structures of their time, most modern retaining walls are not rigid retaining walls. Rankine presented a simpler and easier to apply triangular model.
- Assumptions and relationship provided below.

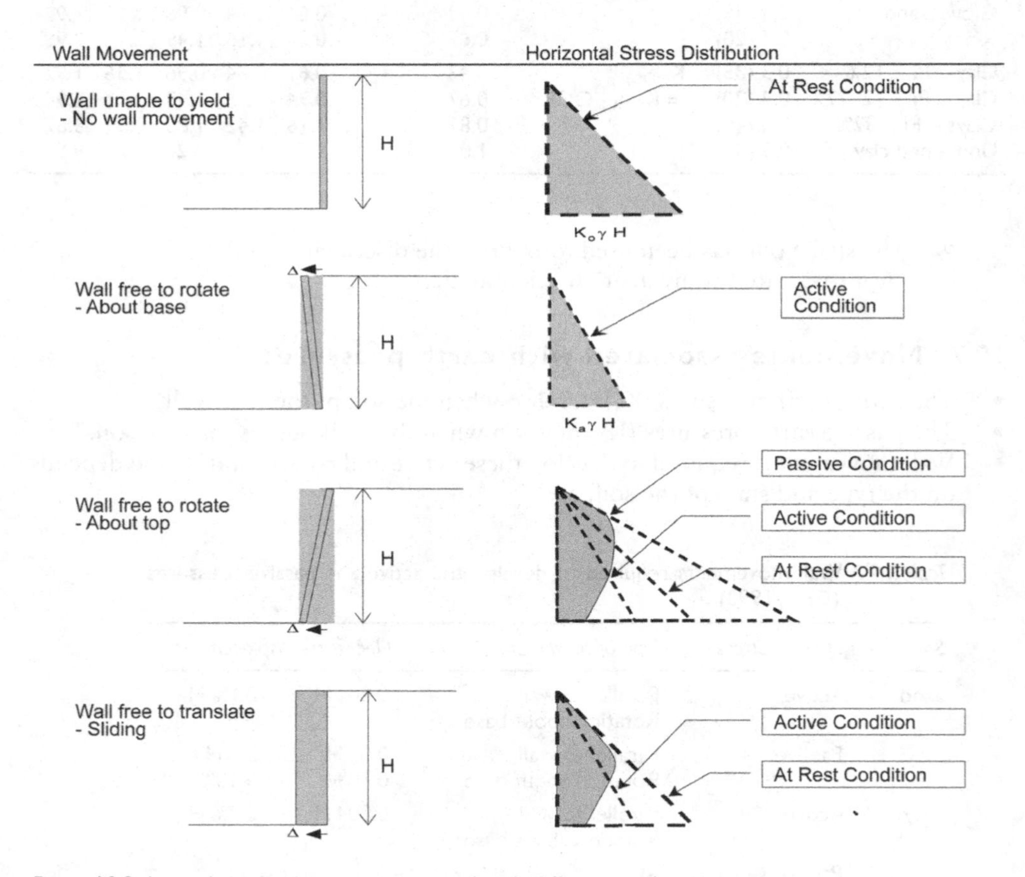

Figure 19.2 Lateral earth pressures associated with different wall movements.

Table 19.8 Earth pressure theories.

Theory	*Rankine*	*Coulomb*	*Caquot and Kerisel*
Based on	Equilibrium of an element	Wedge of soil	
Failure surface	Planar	Planar	Log spiral
Wall friction δ	$\delta = i$: $i = 0$ when ground surface is horizontal	δ	
Pressure distribution	Increases linearly with depth	Provides limiting forces on the wall, but no explicit equivalent pressure distribution	
Resultant active force	At horizontal. At i when ground surface is sloping	δ to normal to back of wall δ to horizontal (wall with a vertical back).	
Active pressure	Rankine similar to Coulomb and Caquot only at $\delta = 0$. As $\delta/\phi \rightarrow 1$ then 10% higher at $\phi < 35°$, but approximately similar at higher ϕ values		
Resultant passive force	At horizontal. At i when ground surface is sloping	δ to horizontal. At $\phi > 5°$ passive force and pressure overestimated. Too high for $\delta > 0.5\phi$	δ to horizontal
Passive pressure	Similar only at $\delta = 0$: Varies significantly for $\phi > 30°$		

- i = slope of backfill surface.
- Passive pressures based on Coulomb theory can overestimate passive resistance.
- Basic Rankine pressures are based on active pressure $K_a = (1 - \sin\phi)/(1 + \sin\phi)$.
- Rankine passive pressure $(K_p) = 1/K_a$.
- Coulomb theory includes wall friction angle, and slope of backfill.
- Active pressure increases considerably for a sloping backfill $i > 10°$.
- Passive pressure decreases considerably for a sloping backfill $i > 10°$.

19.9 Distribution of earth pressure

- The wall pressure depends on the wall movement. For a rigid wall on a competent foundation the movement is reduced considerably.
- The Rankine earth pressure distribution is based on a triangular pressure distribution with the resultant force acting at 1/3 up from the base. This point of application can vary in some cases and first observed by Terzaghi in the 1930s. Therefore calculations should allow for this possibility by either shifting the point of application or factoring the overturning moments accordingly.

Table 19.9 Distribution of earth pressure.

Type of wall foundation material	*Backfill*	*Point of application of resultant force*
Wall founded on soil	Horizontal, $i = 0°$	0.33 H above base
	Sloping at i upwards	0.38 H above base
Wall founded on rock	Horizontal, $i = 0°$	0.38 H above base
	Sloping at i upwards	0.45 H above base

- The triangular earth pressure distribution is not applicable for multi-propped/strutted walls with little movement along its full height.
- Use of FS = 2.0 for overturning and 1.5 for sliding accounted for this possibility of the point of application not always in accordance with an assumed triangular distribution. Limit state procedures factoring strength only do not currently account for the above condition explicitly.

19.10 Application of at rest and active conditions

- Active or at-rest conditions may depend on type of wall.
- While the concept of no wall movement suggests that the at-rest condition should apply, the application is not as self-evident. The cases below illustrate when the higher at rest earth pressure condition applies instead of the active case, and also when the active would always apply.

Table 19.10 Wall types when the active and at-rest condition applies.

Earth pressure condition	*Movement*	*Wall type*
Active	Wall movement occurs due to its inherent flexibility	Sheet piles Gabions
At rest	No/negligible wall movement	Cantilever with stiff basal stems; Rigid counterfort walls Founded on rigid bases, e.g. founded on strong rock or on piles Culvert wing walls; Bridge abutments Basement walls Tanks Tunnels

- Tied back walls may be considered rigid or non-rigid depending on the deflections. If the wall movement calculations (based on section modulus) show little to no deflections then the at-rest condition should apply.
- Walls over designed (with high factors of safety) and based on the active earth pressure condition, may not deflect. The at-rest condition must then be checked for stability.
- Some designers use a value average between the K_o and K_a conditions where uncertainty on the earth pressure condition exists.

19.11 Application of passive pressure

- The passive pressure can provide a significant resisting force based on Rankine and Coulomb theories. However this pressure should be applied with consideration shown in the table below.

Table 19.11 Approaches to consider in application of the passive state.

Issue	*Approach*	*Typical details*	*Comments*
Wall movement incompatibility between the active and passive state	Reduction factor applied to the passive pressure	Reduction factor of 1/3	Approximately ½ of the passive stress would apply for ¼ of the strain.
Desiccation cracks ion front of wall	Passive resistance starts below the depth of the crackled zone	0.5 m cracked zone minimum (typical alpine temperate and coastal areas) to 3.0 m in arid regions	Cracked zone as a proportion of Active zone (H_a) varies from ~1/3 of in temperate areas ~ ½ H_a in wet coastal areas ~ ¾ H_a in arid regions
Non triangular distribution for rotation about the top and sliding	Passive embedment ≥10% H	Wall is unlikely to move in sliding or about the base. Therefore a triangular active condition now applies with rotation about the base	The passive pressure is approximately 10 times the active pressure. Hence 10% H. Similar factors of safety (or partial factors) may then be used for both sliding and overturning. Refer Table 19.8 & Fig. 19.2
Excavation or erosion in front of wall	Reduce passive resistance to that depth	No passive resistance for the top 0.5 m typically used	A heel below the middle or back third of wall can use the full passive resistance

19.12 Use of wall friction

- Coulomb theory considers the effect of wall friction, which reduces the pressure in the active state and increases the passive resistance.
- Application of wall friction to the design should have the following due considerations.

Table 19.12 Use of wall friction.

Consideration	*Value of wall friction, δ*	*Comment*
Active state	0.67ϕ maximum	0.5ϕ for small movements
Passive state	0.5ϕ maximum	0.33ϕ for small movements
Smooth walls	δ = 0 δ = 0.33ϕ	Precast concrete units with smooth finish Precast concrete units with rough finish
Rough walls	δ = 0.67ϕ to 1.0ϕ	Piled walls, lower values where drilling fluids used for installation
Vibration	δ = 0	Adjacent to machinery, railways, vehicular traffic causing vibration
Anchored walls	δ = 0	Negligible movement to mobilise wall friction
Wall has tendency to settle	δ = 0	Uncertainty on the effects of wall friction
Wall supported on foundation slab	δ = 0	Example, cantilever reinforced concrete wall, where virtually no movement of soil relative to back of wall

- The magnitude of δ may not often significantly affect the value of the active force. However the direction is affected and can significantly affect the size of the wall bases.
- Avoid Coulomb values for $\delta > 0.5\phi$.
- Caquout and Kerisel charts preferred over Coulomb or Rankine formula. Driscoll (1979) shows the significant errors (over 50%) for the passive side using either Rankine (underestimation) or Coulomb (overestimation). Errors less than 10% for the active state.

19.13 Values of active earth pressures

- The log spiral surface approximates the active and passive failure surfaces rather than the straight line.
- The value of the active earth pressure coefficient (K_a) is dependent on the soil, friction angle and the slope behind the wall.

Table 19.13 Active earth pressure coefficients (after Caquot and Kerisel, 1948).

Angle of friction		*Active earth pressure coefficient for various slope (i) behind wall*		
Soil (ϕ)	*Wall (δ)*	*$i = 0°$*	*$i = 15°$*	*$i = 20°$*
20	0	0.49	0.65	0.99
	$2/3\phi$	0.45	0.59	0.91
	$\phi = 20°$	0.44	0.58	0.89
25	0	0.41	0.51	0.58
	$2/3\phi$	0.36	0.46	0.56
	$\phi = 25°$	0.35	0.40	0.50
30	0	0.33	0.41	0.46
	$2/3\phi$	0.29	0.35	0.39
	$\phi = 30°$	0.28	0.33	0.37
35	0	0.27	0.32	0.35
	$2/3\phi$	0.23	0.28	0.30
	$\phi = 35°$	0.22	0.27	0.28
40	0	0.22	0.25	0.30
	$2/3\phi$	0.18	0.22	0.23
	$\phi = 40°$	0.17	0.19	0.21

- $i = 0°$ is usually considered valid for $i < 10°$.
- An increase in the active coefficient of 1.5 to 3 times the value with a flat slope is evident.
- If the ground dips downwards, a decrease in K_a occurs. This effect is more pronounced for the K_p value.

19.14 Values of passive earth pressures

- A slope dipping away from the wall affects the passive earth pressure values.

Table 19.14 Passive earth pressure coefficients (after Caquot and Kerisel, 1948).

Angle of friction		*Passive earth pressure coefficient for various slope (i) behind wall*				
Soil (ϕ)	*Wall (δ)*	*i = −20°*	*i = −15°*	*i = 0°*	*i = +15°*	*i = +20°*
20	0	?	1.1	2.0	2.7	3.1
	1/3φ	?	1.2	2.3	3.3	3.6
	1/2φ	?	1.4	2.6	3.7	4.0
25	0	?	1.4	2.5	3.7	4.2
	1/3φ	1.2	1.7	3.0	4.2	5.0
	1/2φ	1.4	1.8	3.5	5.0	6.1
30	0	?	1.7	3.0	4.5	5.1
	1/3φ	1.5	2.2	4.0	6.1	9
	1/2φ	1.7	2.4	4.8	7.0	10
35	0	1.5	2.0	3.7	5.5	10
	1/3φ	2.1	2.9	5.4	8.8	16
	1/2φ	2.2	3.1	6.9	10	12
40	0	1.8	2.3	4.6	7.2	9
	1/3φ	2.8	3.8	7.5	12	17
	1/2φ	3.3	4.3	10	17	21

- $i = 0°$ is usually considered valid for $i < 10°$.
- An increase in the active coefficient of 1.5 to 3 times the value with a flat slope is evident.
- Conversely the values can half for 15° dipping slope.
- ? is shown when the interpolated values are outside the graph range provided.

19.15 Compaction induced pressures

- Earth pressures due to compaction induce a passive horizontal pressure at the surface (structure/equipment pushed into soil) for a depth z_c. This decreases with depth to the active pressure.

Table 19.15 Compaction induced horizontal pressures.

Compaction equipment	*Weight*	*Critical depth (z_c)*	*Horizontal pressure (σ_{hc})*
Vibratory plate	120 kg	0.3 m	12 kPa
	400 kg	0.5 m	16 kPa
Vibratory roller	1.4 t	0.4 m	13 kPa
	3.2 t	0.6 m	20 kPa
Smooth wheel roller	10 t	0.6 m	20 kPa

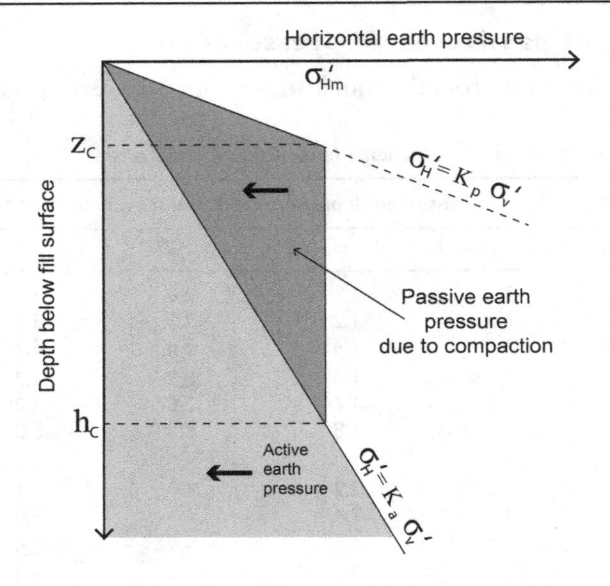

Figure 19.3 Compaction induced stresses on wall.

19.16 Live loads from excavators and lifting equipment

- Excavators and lifting equipment close to a wall impose horizontal stresses on the wall.

Table 19.16 Live loads from equipment adjacent to wall (German society for geotechnics, 2003).

Gross weight (t)	*Minimum distance (m) for dispensing with specific loadings*	*Width of strip load (m)*	*Additional strip load (kPa + 10 kPa)*	
			Adjacent to wall	*0.60 m from wall*
10	1.50	1.50	50	20
30	2.50	2.00	110	40
50	3.50	2.50	140	50
70	4.50	3.00	150	60

  - Unbounded distributed load of 10 kPa applies when minimum distance criteria met.

Chapter 20

Retaining walls

20.1 Wall types

- The classification of earth retention systems can be used to determine the type of analysis.
- Hybrid systems from those tabulated are also available.

Table 20.1 Classification for earth retention systems (adapted from O'Rouke and Jones, 1990).

Stabilization system	*Type*	*Examples*
External	In-situ (Embedded)	Sheet piles; Soldier piles Cast – in situ (slurry walls, secant and contiguous piles) Soil – Cement Precast concrete; Timber
	Gravity	Masonry; Concrete Cantilever; Counterfort Gabion; Crib; Bin Cellular cofferdam
Internal	In-Situ	Soil nailing; Soil dowelling Reticulated micro piles
	Reinforced	Metallic strip; Wire mesh Geotextile; Geogrid Organic inclusions

  - The external walls may be braced/tied back or free standing walls.

20.2 Gravity walls

- Gravity or concrete walls tend to be economical for wall heights <3 m.
- Reinforced soil walls are generally economical for walls >3 m.

Table 20.2 Typical gravity wall designs.

Gravity wall type	*Top width*	*Base width*	*Heights*	*Other design elements*
Gravity masonry	300 mm (minimum)	0.4H to 0.7H	Common for H = 2–3 m Uneconomic for H = 4 m Rare for H = 7 m	0.1H to 0.2H base thickness; 1 Horizontal to 50 Vertical face batter
Reinforced concrete	300 mm (minimum)	0.4H to 0.7H	Suitable for H < 7 m Rare for H = 10 m Counterforts for H > 5 m Counterfort spacing 2/3H but > 2.5 m	0.1H base thickness; 1 Horizontal to 50 Vertical face batter; Counterforts 200 mm minimum thickness
Crib wall	0.5H to 1.0H	0.5H to 1.0H	Suitable for H<5 m	1 Horizontal to 6 Vertical face batter
Gabion wall	0.5 m (minimum)	0.4H to 0.6H	Suitable for H < 10 m	1 Horizontal to 8 Vertical face batter; Unit weight based on porosity n = 0.3 to 0.4
Reinforced soil structure	2.5 m minimum	0.7H minimum 0.8H to 1.1H for sloping fill or external loads	Economical to significant heights Refer table 20.7 for embedment	Minimum horizontal bench width of 1.2 m in front of walls with slopes

- A face batter is recommended for all major walls in an active state. Movement forward is required for the active state. The face batter compensates for this effect.
- Gravity walls may not be economical in deep cut situations due to additional costs associated with temporary support.
- Reinforced soils structures (RSS) are also called mechanically stabilized earth (MSE) walls.
- Precast facing panels may cost as much as the wall structure itself.

20.3 Effect of slope behind walls

- The slope (α) behind the wall can have a significant effect on the wall pressures.
- The slope of the wall itself can also affect the design.
- The embedment (d) and slope (β) in front of wall has a significant effect on the passive wall pressures.

a. Embedded Walls

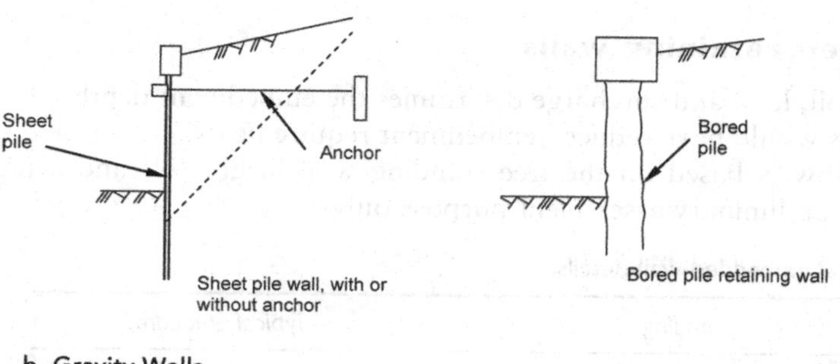

b. Gravity Walls

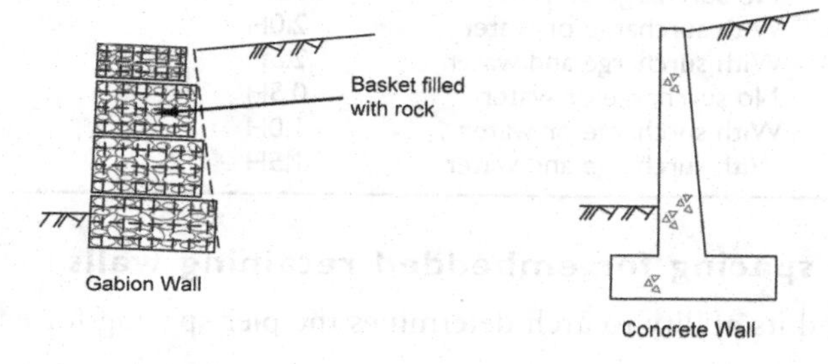

c. Internal Walls

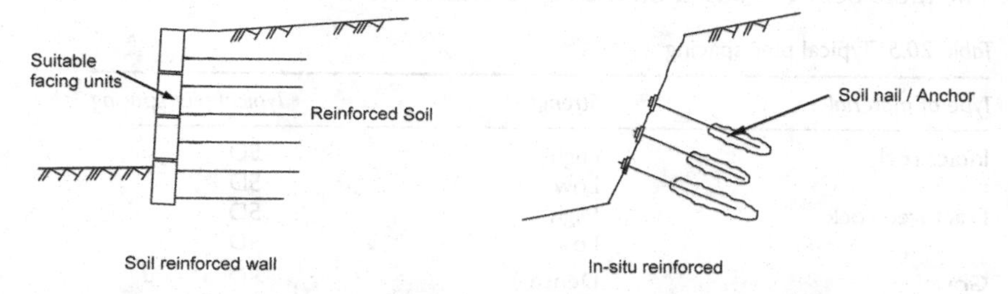

Figure 20.1 Type of walls.

Table 20.3 Typical minimum wall dimension for various sloping conditions.

Sloping area	*Effect on wall dimensions for various slopes*		
α = slope behind the wall Vertical wall $\beta = 0°$	$\alpha < 10°$ $B \geq 0.5$ H	$\alpha \geq 10°$ $B \geq 0.6$ H	$\alpha \geq 25°$ $B \geq 0.7$ H
α = slope behind the wall Wall with slope 6V: 1H $\beta = 0°$	$\alpha < 10°$ $B \geq 0.4$ H	$\alpha \geq 10°$ $B \geq 0.5$ H	$\alpha \geq 25°$ $B \geq 0.6$ H
$\alpha = 0°$ Vertical wall β = slope in front of wall	$\beta < 10°$ $B \geq 0.5$ H d = 10% H or 0.5 m whichever is the greater	$\beta \geq 10°$ $B \geq 0.6$ H (10% H or 0.5 m whichever is the greater) + 300 mm	$\beta \geq 25°$ $B \geq 0.7$ H (10% H or 0.5 m whichever is the greater) + 600 mm

- Refer to Appendix A for effect of live load on sloping backfills.

20.4 Embedded retaining walls

- The type of soil, load and surcharge determines the embedment depth.
- Propped walls would have reduced embedment requirements.
- The table below is based on the free standing wall height (H) and a nominal surcharge for preliminary assessment purpose only.

Table 20.4 Typical embedded wall details.

Type of wall	*Loading*	*Typical embedment depth*
Free cantilever	No surcharge or water	1.5H
	With surcharge or water	2.0H
	With surcharge and water	2.5H
Propped	No surcharge or water	0.5H
	With surcharge or water	1.0H
	With surcharge and water	1.5H

20.5 Typical pier spacing for embedded retaining walls

- The type of soil and its ability to arch determines the pier spacing for embedded retaining walls.
- The table below is based on the pier Diameter (D).

Table 20.5 Typical pier spacing.

Type of material	*Strength*	*Typical pier spacing*
Intact rock	High	>5D
	Low	5D
Fractured rock	High	5D
	Low	4D
Gravel	Dense	3D
	Loose	2.5D
Sand	Dense	2.5D
	Loose	2.0D
Silts	Very stiff	2.0D
	Firm	1.5D
Clays	Very stiff	2.0D
	Firm	1.5D

- Sands and gravels assume some minor clay content.
- Without some clay content and where a high water table exist, the pier spacing would need to be reduced.

20.6 Wall drainage

- All walls should have a drainage system.
- Even walls above the groundwater table must be designed with some water pressure. For a dry site a water pressure of ¼ wall height should be used.

Table 20.6 Typical wall drainage measures.

Wall height	*Drainage measure*	*Typical design detail for rainfall environment*	
		<1000 mm	*>1000 mm*
<1 m	• Weep holes at 250 mm from base of wall or as low as practical • Geotextile wrapped 50 mm perforated pipe at base of wall with outlet	• 50 mm weep holes at 3.0 m spacing, or • 200 mm drainage gravel behind wall	• 50 mm weep holes at 2.0 m spacing, or • 300 mm drainage gravel behind wall
1–2 m	• Weep holes and geotextile wrapped 75 mm perforated pipe at base of wall with outlet	• 50 mm weep holes at 3.0 m spacing, and • 300 mm drainage gravel behind wall	• 50 mm weep holes at 2.0 m spacing, and • 300 mm drainage gravel behind wall
2–5 m	• Weep holes and Geotextile wrapped 100 mm perforated pipe at base of wall with outlet. • Internal drainage system to be considered	• 75 mm weep holes at 3.0 m horizontal and vertical spacing (staggered), and • 200 mm drainage gravel behind wall • Filter drainage material inclined with a minimum thickness of 300 mm	• 75 mm weep holes at 2.0 m horizontal and vertical spacing (staggered), and • 300 mm drainage gravel behind wall • Filter drainage material inclined with a minimum thickness of 300 mm
>5 m	• Weep holes and geotextile wrapped 150 mm perforated pipe at base of wall with outlet. • Internal drainage system necessary • Horizontal drains wrapped in filter to be considered	• 75 mm weep holes at 2.0 m horizontal and vertical spacing (staggered), and • 300 mm drainage gravel behind wall • Typically 5 m long * 75 mm with spacing of 5 m vertically and 5 m horizontally	• 75 mm weep holes at 1.5 m horizontal and vertical spacing (staggered) • 300 mm drainage gravel behind wall • 5 m long * 100 mm with spacing of 3 m vertically and 5 m horizontally

- Drainage layers at rear of gabions and crib walls (free draining type walls) are not theoretically required. The 200 mm minimum thickness of the drainage layer behind these and the low height/low rainfall walls shown above is governed by the compaction requirement more than the drainage requirement.
- Compaction against the back of walls must be avoided, hence the use of a self-compacting "drainage layer" is used behind all walls, without the need to compact against the wall.
- A geotextile filter at the back of the wall drainage gravel (if used) is required to prevent migration of fines
- For intensity rainfall >1500 mm and/or large catchments (sloping area behind wall) more drainage systems than shown may be required.
- For wall lengths >100 m, then 200 mm and 150 mm perforated pipes are typically required for walls ≥5 m, and <5 m respectively. Refer Chapter 15 for added details.
- Clean out lengths every 50 m metres.

20.7 Minimum wall embedment depths for reinforced soil structures

- A minimum embedment of 0.5 m should be provided to allow for shrinkage and swelling potential of foundation soils, global stability and seismic activity.
- Embedment deepening is required to allow for scour or future trenching. Typically 0.5 m or 10% of H, whichever is greater. Reduced embedment may occur where a high level competent rock is at the surface.
- The table provides the minimum embedment depth at the front of the wall.

Table 20.7 Minimum embedment for reinforced soil structures (Holtz et al., 1995).

Slope in front of wall	*Minimum embedment (m)*
Horizontal	
• Walls	H/20
• Abutments	H/10
1V:3H	H/10
1V:2H	H/7
2V:3H	H/5

- ○ For a slope in front of wall a horizontal distance of 1 m minimum, shall be provided to the front of the wall and deepen as required.

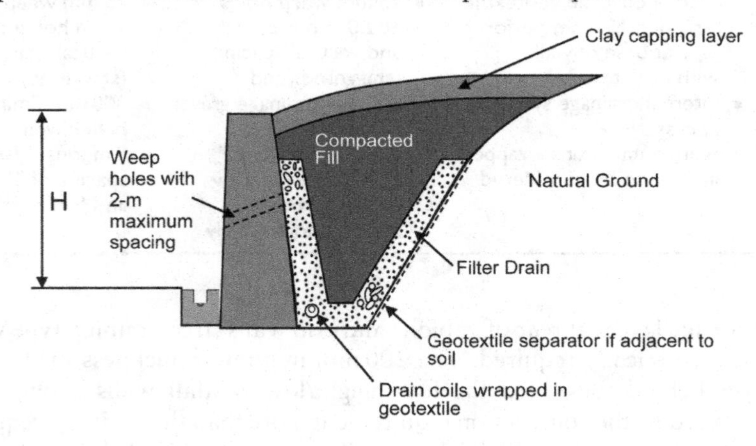

Figure 20.2 Drainage of walls.

20.8 Reinforced soil wall design parameters

- Reinforced soil walls (RSW) are constrained at the top resulting in an increased earth pressure.
- The earth pressure tends towards the at rest condition at the surface top, and decreases linearly to the active condition at 6 m depth (refer Figure 20.3).
- The earth pressure at the top depends on the soil reinforcement. Rigid inclusions move less, with a resulting higher earth pressure.
- The table also shows the soil – reinforcement interface friction angle, based on the friction angle (ϕ) of the soil.

Table 20.8 Variation of earth pressure with depth of wall (TRB, 1995)

Earth pressure coefficient with depth	*Type of reinforcement with friction angle*			
	Geotextile $\frac{2}{3}\phi$	*Geogrid* ϕ	*Metal strip* $\frac{3}{4}\phi$	*Wire mesh* ϕ
0 m (surface)	K_a	$1.5K_a$	$2.0K_a$	$3.0K_a$
≥6 m	K_a	K_a	K_a	$1.5K_a$

- ◦ The geogrids and geotextiles would have to consider the effects of creep and resistance to chemical attack with suitable reduction factors applied to the strength.
- ◦ The metallic reinforcement thickness needs to take into account the effects of corrosion.

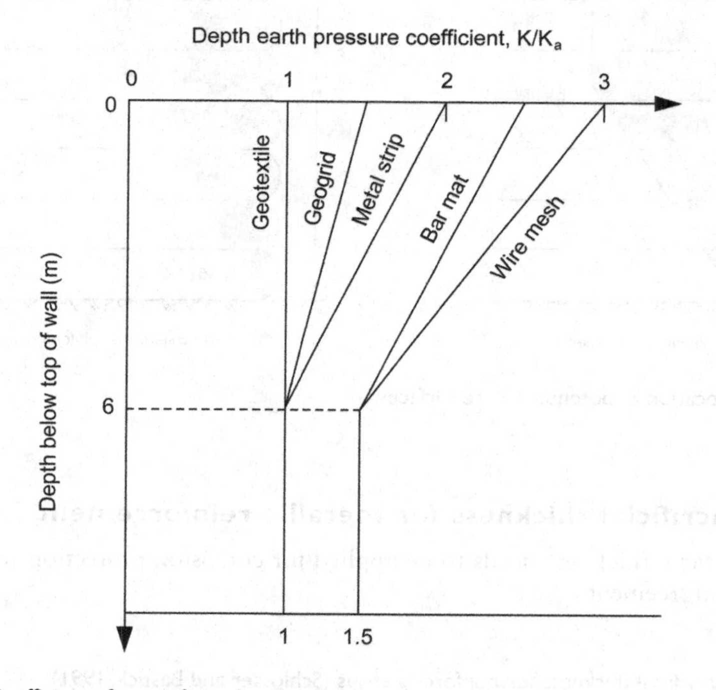

Figure 20.3 Coefficients for reinforced soil walls.

20.9 Location of potential failure surfaces for reinforced soil walls

- The location of the potential failure surface depends on the type of movement.
- Inextensible reinforcement has less movement with an active zone close to the wall face.
- Extensible reinforcement has greater capacity for movement with the typical Rankine active zone.

Table 20.9 Location of potential failure surfaces for RSW.

Type of reinforcement	*Failure surface from base H = Height of wall*	*Distance from wall to failure surface at top*	*Example*
Inextensible	$\tan^{-1}\{0.3\,H/(H/2)\} = \tan^{-1} 0.6$ extending to 0.5H from base	0.3 H	Wire mesh, metal strip Soil nails
Extensible	$(45° + \phi/2)$ extending to surface	$H \tan(45° - \phi/2)$	Geotextile, Geogrids

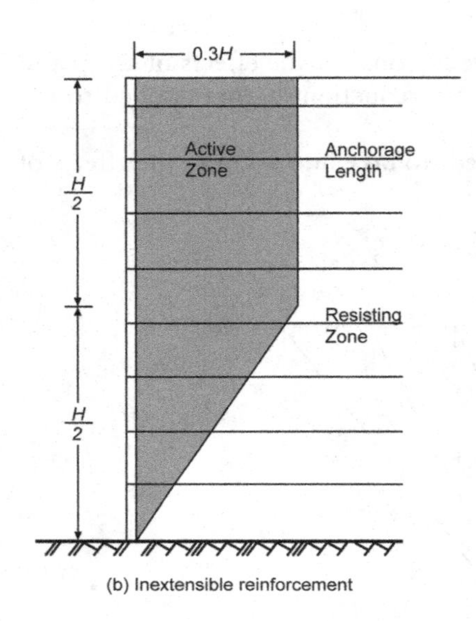

(b) Inextensible reinforcement

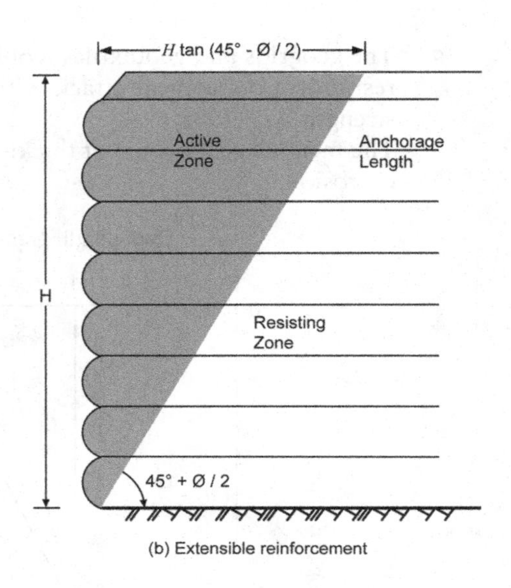

(b) Extensible reinforcement

Figure 20.4 Location of potential failure surfaces.

20.10 Sacrificial thickness for metallic reinforcement

- A sacrificial thickness needs to be applied for corrosion protection with metallic soil reinforcement.

Table 20.10 Sacrificial thickness for reinforcing strips (Schlosser and Bastick, 1991).

Type of steel	*Environment*	*Sacrificial thickness (mm) for minimum service life (yrs)*			
		5 yrs	*30 yrs*	*70 yrs*	*100 yrs*
Black steel	Out of water	0.5	1.5	3.0	4.0
	Fresh water	0.5	2.0	4.0	5.0
	Coastal structure	1.0	3.0	5.0	7.0
Galvanised steel	Out of water	0	0.5	1.0	1.5
	Fresh water	0	1.0	1.5	2.0
	Coastal structure	0	N/A	N/A	N/A

20.11 Reinforced slopes factors of safety

- Different factors of safety are calculated depending on whether the soil reinforcement is considered an additional reducing moment or a reduction to the overturning moments.
- Both are valid limit equilibrium equations.

Table 20.11 Use of the different factors of safety for a reinforced slope (Duncan and Wright, 1995).

Factor of safety using limit equilibrium equation form	*Application to reinforcement design*	*Comment*
$\frac{\text{Soil resisting moment}}{\text{Overturning moment} - \text{reinforcement moment}}$	Allowable force	Preferable
$\frac{\text{Soil resisting moment} + \text{reinforcement moment}}{\text{Overturning moment}}$	Ultimate force	Divide by FS calculated in analysis

20.12 Soil slope facings

- A facing is required on soil slopes depending on the batter.
- A face protection is required to prevent erosion.

Table 20.12 Soil slope stabilisation.

Consideration	*Wall type and facing required*				
Typical slope angle	~90°	70°	45°	Φ°_{cv}	$<< \Phi^{\circ}_{cv}$
Slope	1V:0.01H	1V:0.36H	1V:1H	~1 V:2H to 1V: 1.7H	< 1V:2H
Design	Vertical wall	Battered wall	Reinforced slope	Unreinforced slope	
Type of facing	Active	Active /passive	Passive	No facing	
Wall type	Concrete, embedded Anchors, soil nail (active plate), reinforced soil wall	Gabion, Crib	Geocells, revetments, rock facings Geomesh, Soil nail (passive bent bars), reinforced soil slope		Vegetation

- A soil nail process is a usually a top down process while a reinforced soil wall is a bottom up construction.
- Soil nails have some stiffness that can take up shear forces and bending moments while reinforced earth strips are flexible.

20.13 Wall types for cuttings in rock

- The wall types and facing required is dependent on the stability based on the joint orientations.
- If flattening the slope is not a feasible option at a given site then a facing unit and wall is required.

Table 20.13 Wall type and facings required for cut rock slopes.

Consideration	*Wall type and facing required*				
Rock weathering	*Fresh to slightly*	*Slightly to distinctly*		*Distinctly to extremely*	*Extremely to residual*
Typical cut slope	1V:0.1H (84°)	1V:0.25H (76°)	1V:0.5H (63°)	1V:1.0H (45°)	1V:1.5H) (34°
Maximum slope angle	≤90°	80°	70°	50°	40°
Design if adverse jointing or space limitations	Vertical wall	Battered wall	Reinforced slope →		
Type of facing	Active	Active/passive		Passive	No facing (erosion protection

- Berms for maintenance may be required with a steeper slope.
- Actual slope is governed by the rock strength, joint orientation and rock type.
- Rock trap fences/netting may be required at any slope.

20.14 Drilled and grouted soil nail designs

- Soil nails are either driven or drilled and grouted type. The latter has a larger area and tensile strength, and with a larger spacing.
- An excavated face of 1.0 to 1.5 m is progressively made with soil nails installed with a shotcrete face before excavating further. About 5 kPa cohesion in clayey sand has shown to be sufficient to allow 1m of excavation to proceed.
- For soils without sufficient cohesion the order can be reversed, i.e. shotcrete before nailing.

Table 20.14 Drilled and grouted nails – typical designs (adapted from Phear et al., 2005 and Clouterre, 1991).

Material type	*Typical slope angle*	*Facing type*	*Length*	*Area per nail (m^2)*	*Nails per m^2*
Weak rocks	70 to 90°	Hard	0.6 to 1.0 H	1.5 to 2.5	0.4 to 0.7
Soils	70 to 90°	Hard	0.8 to 1.2 H	0.7 to 2	0.5 to 1.4
Natural soils	45 to 70°	Flexible	0.6 to 1.0 H	1 to 3	0.3 to 1.0
Natural soils and fills	30 to 45°	None	0.8 to 1.2 H	2 to 6	0.1 to 0.5

- Typical strength of a drilled and grouted nail is 100 to 600 kN.
- Table assumes a level ground at the top.
- In high plasticity clays the length may need to be increased to account for creep. An active bar (i.e. bar with a plate) instead of a passive facing (i.e. bent bar) may be required.
- Limitation of soil nails:
  - Some minor movement is acceptable.
  - No water table, or water table can be reduced.

20.15 Driven soil nail designs

- Driven or fired soil nails have a lower tensile capacity than driven or drilled and grouted type. The latter has a larger area and tensile strength, and with a larger spacing.
- Driven nails are usually not applicable in weak rocks or boulder fills.

Table 20.15 Driven nails – typical designs (adapted from Phear et al., 2005 and Clouterre, 1991).

Typical slope angle	*Facing type*	*Length*	*Area per nail (m^2)*	*Nails per m^2*
70 to 90°	Hard	0.5 to 0.7 H	0.4 to 1.0	1 to 2.5
45 to 70°	None	0.5 to 0.7 H	0.7 to 1.2	0.8 to1.4

- ○ Typical strength of a driven nail is 50 to 200 kN.
- ○ Table assumes a level ground at the top.
- ○ Gravel or rock fills would typically have some difficulty. Using a sharpened edge angle iron instead of a bar provides a stiffer inclusion that may work for small enough particle sizes. But limited driveability in such materials limits its use to temporary or surficial stability control

20.16 Sacrificial thickness for metallic reinforcement

- Sacrificial nail thickness or other barriers need to be applied for corrosion protection based on service life.
- For driven nail barriers are not possible.

Table 20.16 Corrosion protection for soil nails (Schlosser et al., 1992).

Environment	*Sacrificial thickness (mm) for minimum service life (yrs)*		
	≤18 months	*1.5 to 30 yrs*	*100 yrs*
A little corrosive	0	2 mm	4 mm
Fairly corrosive	0	4 mm	8 mm
Corrosive	2 mm	8 mm	Plastic barrier
Strongly corrosive	Compulsory plastic barrier + sacrificial thickness above		

20.17 Design of facing

- The design of the facing depends on the uniform pressure acting on the facing and tension in the nails at the facing T_O.
- Spacing (S) = maximum of S_V and S_H.

Table 20.17 Design of facing (Clouterre, 1991).

Spacing (S)	*T_o/T_{max}*	*Comments*
S ≤ 1 m	0.6	Usually driven nails
1 m < S < 3 m	0.5 + (S − 0.5)/5	
S ≥ 3 m	1.0	Grouted Nails

- T_{max} = maximum tension in the nail in service = ultimate nail pull-out force.
- S_V and S_H = Vertical and horizontal spacing, respectively.
- Nails are designed with an overall factor of safety against pull out of 1.5 and 1.3 for permanent and temporary walls, respectively.

20.18 Shotcrete thickness for wall facings

- The shotcrete facing for soil nails depends on the load, and the slope angle.

Table 20.18 Typical shotcrete requirements.

Condition	*Shotcrete thickness and design details*	
Life	Temporary: 75 mm to 150 mm	Permanent: 125 mm to 250 mm
Slope	<70°: 50–150 mm	Near vertical 70° to 90°: 150–275 mm
Typical nail	Bent bars <28 mm	Bent bars >28 mm or plate head
Typical mesh	100 mm to 200 mm opening	75 mm to 100 mm opening size
Typical layers of mesh	Steel mesh on one side to side with soil	Steel mesh on either side Mandatory for thickness >150 mm Additional mesh locally behind plate if significant torque
Embedment below finished level	No requirements	0.2 m in rock 0.4 m in soil or H/20 whichever is higher

20.19 Details of anchored walls and facings

- Where horizontal movement needs to be constrained, prestressing is required.
- Soil nails and anchored walls experience different pressures, with the latter designed for greater loads.
- These two types of walls are designed differently. Table below is for walls with near vertical faces.
- The cost of soil nailing may be 50% of the cost of a tieback wall.
- Greater movement can be expected in a soil wall than the tieback wall.

Table 20.19 Typical details of nails and facings.

Design consideration	*Wall type*	
	Soil nailed wall	*Tieback anchored walls*
Prestressing load	Nominal	Significant
Nuts	Torque to 20 kN Load Vertical System, reducing to 5 kN at 70° slope. In some cases a bent bar may be used instead of plates	Torque to 150 kN to 400 kN typically
Bondage	Along entire length	Over free length
Typical length	0.5 to 1.5 × slope height	Long – to competent strata at depth
Typical inclination	10 to 15° to horizontal	20 to 30° to horizontal
Typical plates	150–250 mm square, 15 mm to 20 mm thick Grade 43 steel	200 mm to 300 mm square, 20 to 25 mm thick Grade 43 steel
Anchorage	24 to 36 mm diameter	Strands or specialist bars with plate
Typical shotcrete face	150 mm to 250 mm	200 mm to 300 mm

- Table is for flat ground above retained wall.
- For large spacing between anchors or nails, and for sloping ground >10° or other surcharges increase the plate sizes shown in table.

20.20 Anchored wall loads

- Anchor loads depend on the wall height, material behind the wall, groundwater conditions and surcharge.
- Table below is for wall anchor inclined at 15° to horizontal and with a factor of safety of 1.5.

Table 20.20 Typical anchor loads (taken from graphs in Ortiago and Sayao, 2004).

Height of wall (m)	*Loading*	*Typical anchor load (kN)*	
		$\phi = 25°$	$\phi = 35°$
3	Horizontal top + 20 kPa surcharge	50	40
	Slope at 30° behind wall + surcharge	120	100
	Groundwater at 50% wall height + surcharge	60	50
	Groundwater at 100% wall height + surcharge	70	70
4	Horizontal top + 20 kPa surcharge	80	70
	Slope at 30° behind wall + surcharge	180	150
	Groundwater at 50% wall height + surcharge	110	90
	Groundwater at 100% wall height + surcharge	130	130
5	Horizontal Top + 20 kPa surcharge	130	110
	Slope at 30° behind wall + surcharge	260	220
	Groundwater at 50% Wall Height + Surcharge	170	150
	Groundwater at 100% Wall Height + Surcharge	200	200
6	Horizontal top + 20 kPa surcharge	190	160
	Slope at 30° behind wall + surcharge	350	300
	Groundwater at 50% wall height + surcharge	240	220
	Groundwater at 100% wall height + surcharge	280	280

- Groundwater condition is for a flat top.
- Table based on:
  - Soil cohesion of 10 kPa.
  - Soil unit weight of 18 kN/m^3.

20.21 Anchor ultimate values for load transfer in soils

- For preliminary design, ultimate load transferred from the bond length to the soil may be estimated for a small diameter, straight shaft gravity-grouted anchor from the soil type and density (or SPT blow count value).
- The maximum allowable anchor design load in soil maybe determined by multiplying the bond length by the ultimate transfer load and dividing by a factor of safety of 2.0.
- Values apply to preliminary design of small diameter straight shaft gravity-grouted ground anchors.

- Full bond uniformity along the anchor length does not occur with progressive debonding occurring. The head of the bonded length will mobilise the bond first and as additional displacements take place the base is then mobilised. Longer anchor lengths are less effective than shorter lengths – although increased load capacity and required for global stability.

Table 20.21 Presumptive ultimate values of load transfer to anchors (Sabatani et al., 1999).

Soil type	*Relative density/consistency (SPT range)*	*Estimated ultimate transfer load (kN/m)*
Sand and Gravel	Loose (4–10)	145
	Medium dense (11–30)	220
	Dense (31–50)	290
Sand	Loose (4–10)	100
	Medium dense (11–30)	145
	Dense (31–50)	190
Sand and silt	Loose (4–10)	70
	Medium dense (11–30)	100
	Dense (31–50)	130
Silt-clay mixture with low plasticity or fine micaceous sand or silt mixtures	Stiff (10–20)	30
	Hard (21–40)	60

- Note: (1) SPT values are corrected for overburden pressure.
- Anchor bond lengths for gravity-grouted, pressure-grouted, and post-grouted soil anchors are typically 4.5 to 12 m since significant increases in capacity for bond lengths greater than approximately 12 m cannot be achieved unless specialized methods are used to transfer load from the top of the anchor bond zone towards the end of the anchor.
- Minimum anchor length of 3 m.

20.22 Rock anchor bond stress

- Rock anchors are usually shaft anchors with no pressure on the grout. Pressure injected may be used in soils to increase capacity.

Table 20.22 Typical values for unit rock-anchor bond stress (Schnabel and Schnabel, 2002).

Rock type	*Bond stress between grout and rock (kPa)*
Sandstone	825–1725
Soft shales	200–825
Slates and hard shales	825–1375
Soft limestone	1000–1500
Hard limestone	1375–2000
Granite and basalt	1725–3000

- A factor of safety of 2.0 to 2.5 is applied to the ultimate value.
- Assumes intact rock. Reduce for fractured or weathered rock.
- Littlejohn (1993) suggests 10% UCS for intact rock, but up to 25% UCS for granular weathered low strength rock

20.23 Anchor bond length

- Estimation of anchor capacity should be based on the simplest commonly installed anchor, i.e., the straight shaft gravity-grouted anchor.

Table 20.23 Typical anchor design (Sabatani et al., 1999).

Design element	*Value*	*Comments*
Design load	260 kN to 1160 kN	Anchor tendons of this capacity can be handled without need for heavy or specialized equipment. Stressing equipment can be handled by one or two workers without the aid of mechanical lifting equipment
Drill hole	<150 mm	Except for hollow stem augered anchors typically ~300 mm
Total anchor length	9 to 18 m	Minimum unbonded length of 3 m for bar tendons and 4.5 m for strand tendons
Ground anchor inclination	10 to 45° below horizontal; 10 to 30° common	Anchor bond zone must be developed behind potential slip surfaces and in layers that can develop necessary design load
First pass assumption	15° inclination; Bond length of 12 m in soil or 7.5 m in rock	For site with no restrictions on right-of-way.

  - These minimum unbonded lengths required to avoid unacceptable load reduction resulting from seating losses during load transfer and prestress losses due to creep in the prestressing steel or the soil.
  - Steep inclinations may be necessary to avoid underground utilities, adjacent foundations, right-of-way constraints, or weak layers. Anchors should be installed as close to horizontal as possible to minimize vertical loads resulting from anchor lock-off loads, however grouting of anchors installed at angles less than 10° is not common unless special grouting techniques used.

Chapter 21

Soil foundations

21.1 Foundation descriptions

- Foundations transfer loads from the structure to the ground in a safe manner.
- The safe foundation must consider both bearing and settlement with regard to the function of the structure.
- The foundation is selected to be economical in term of type, size, materials, time and constructability.
- Anything built by man is a structure and in Civil Engineering this has a wide meaning. Each such structure would have its foundation.
- Structural foundation is the default meaning for "foundation" but other types of foundation structures exists, and pavement and earth structures were described in previous chapters.

Table 21.1 Foundation descriptions.

Application	*Description*	*Examples*
Structure	Space frame of man-made material of steel or concrete, but can be timber. Small volume	Building foundations are the most common, but can be dam, retaining wall, or tanks. The structure is relatively thin compared to its contents or volume of load contained within or by the structure
Pavement structure	Processed natural materials of medium volume	Rigid (Concrete) on sub-base Flexible (Asphalt and granular)
Earth structure	Mainly natural materials (soil and rock) of significant volume	Cut and embankment slopes, reinforced soil structures, soil nails, dams and tunnels. Embankments on weak ground (ground improvement or piled embankments)

21.2 Techniques for foundation treatment

- The soil foundation supports structures such as rigid concrete footings for a building or an embankment for a road. Techniques for fill loading are covered in the table below.

Table 21.2 Dealing with problem foundation grounds with fill placed over.

Improved by	*Specific methods*
Reducing the load	Reducing height of fill Use light weight fill Transfer load (see below)
Replacing the problem materials with more competent materials	Removal of soft or problem materials. Replace with suitable fill/bridging layer Bridging layer may be a reinforced layer Complete replacement applicable only to shallow depths (3 m to 5 m depending on project scale) Partial replacement for deeper deposits
Increasing the shear strength by inducing consolidation/ settlement	Preloading; surcharging; staged loading Use of wick drains with the above Vacuum consolidation For predominantly granular materials: vibro-compaction, impact compaction, dynamic compaction
Reinforcing the embankment or its foundation	Berms or flatter slopes for slope instability Sand drains, stone columns Lime and cement columns Grouting Electro osmosis; thermal techniques (heating, freezing) Geotextiles, geogrids or geocells at the interface between the fill and ground
Transferring the loads to more competent layers	Pile supported structures such as bridges and viaducts Load relief piled embankments with load transfer platforms

- Relative order of cost depends on the site specifics and proposed development. Time and land constraints often govern rather than the direct costs.
- Further discussions on specialist ground treatments are not covered.
- The foundation soil may often require some treatment prior to loading.
- Treatment by compaction was covered previously.

21.3 Types of foundations

- The foundations are classified according to their depth.
- Typically when the embedded length $>5 \times$ bearing surface dimension, then the foundation is considered deep.
- Deep foundations are more expensive but are required where the surface layer is not competent enough to support the loads in terms of bearing strength or acceptable movement.

Table 21.3 Foundation types.

Classification	*Foundation type*	*Typically use*
Shallow	Strip	Edge beams for lightly loaded buildings
	Pad	To support internal columns of buildings
	Raft	To keep movements to a tolerable amount
Deep	Driven piles	Significant depth to competent layer
	Bored piles	Large capacity required. Lateral loading

- Combinations and variations of the above occur, i.e. piles under some edge beams, or pad foundations connected by ground beams.
- Long piles are considered greater than 20 pile diameters, but the separation of long and short is more based on the pile rigidity (EI).

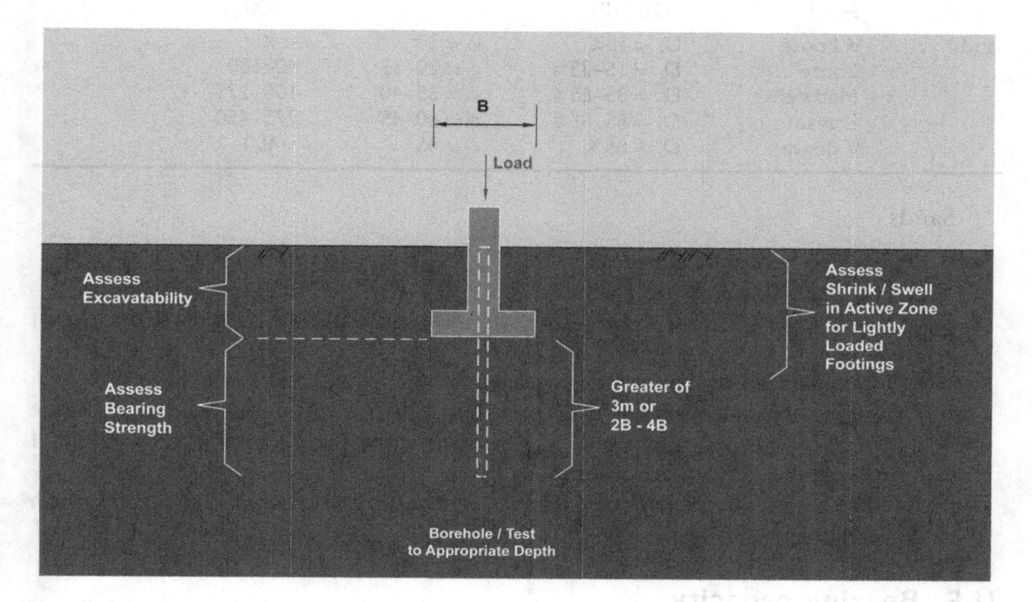

Figure 21.1 Foundation investigation.

21.4 Strength parameters from soil description

- The bearing value is often assessed from the soil description in the borelog. The presumed bearing value is typically given in the geotechnical engineering assessment report based on the site conditions, but often without the benefit of specifics on the loading condition, depth of embedment, foundation geometry, etc. Considerations of these factors can optimise the design and is required for detailed design.
- The use of presumed bearing pressure from the soil description is simple – but not very accurate. Therefore use only for preliminary estimate of foundation size.
- The table is for natural material and assumes an allowable settlement of 25 mm.
- When select material is placed as structural fill and compacted to 98% relative compaction, the bearing value is typically 100 kPa minimum for a competent material underlying the structural fill.

Table 21.4 Preliminary estimate of bearing capacity.

Material	*Description*	*Strength*		*Presumed bearing value (kPa)*
Clay	V. Soft	0–12 kPa		<25
	Soft	12–25 kPa		25–50
	Firm	25–50 kPa		50–100
	Stiff	50–100 kPa		100–200
	V. Stiff	100–200 kPa		200–400
	Hard	>200 kPa		>400
Sands*	V. Loose	$D_r < 15\%$	$\phi < 30°$	<40
	Loose	$D_r = 15–35\%$	$\phi = 30–35°$	40–100
	Med dense	$D_r = 35–65\%$	$\phi = 35–40°$	100–275
	Dense	$D_r = 65–85\%$	$\phi = 40–45°$	275–450
	V. dense	$D_r > 85\%$	$\phi > 45°$	>450

Sands

- *For clayey sands reduce ϕ by 5°.
- *For gravelly sands increase ϕ by 5°.
- *Water level assumed to be greater than B (width of footing) below bottom of footing.
- *For saturated or submerged conditions – half the value in the table.
- Based on a foundation width greater than 1 m and settlement = 25 mm. Divide by 1.2 for strip foundation. The bearing value in sands can be doubled, if settlement = 50 mm is acceptable.
- For $B < 1$ m, the bearing pressure is reduced by a ratio of B (Peck, Hanson and Thornburn, 1974).

21.5 Bearing capacity

- Terzaghi presented the general bearing capacity theory, with the ability of the soil to accept this load dependent on:
  - The soil properties – cohesion (c), angle of friction (ϕ) and unit weight (γ).
  - The footing geometry – embedment (D_f) and width (B).

Table 21.5 Bearing capacity equation.

Consideration	*Cohesion*	*Embedment*	*Unit weight*	*Comments*
Bearing capacity factors	N_c	N_q	N_γ	These factors are non-dimensional and depend on ϕ. See next table
Ultimate bearing capacity (q_{ult})	$c\,N_c+$	$q\,N_q+$	$0.5\,\gamma B\,N_\gamma$	Strip footing
	$1.3\,c\,N_c+$	$q\,N_q+$	$0.4\,\gamma B\,N_\gamma$	Square footing
	$1.3\,c\,N_c+$	$q\,N_q+$	$0.3\,\gamma B\,N_\gamma$	Circular footing

- Surcharge (q) resisting movement = γD_f.
- Modifications of the above relationship occurs for:
  - Water table.
  - Shape, depth and inclination factors.
  - Soil layering.
  - Adjacent to slopes.

21.6 Bearing capacity factors

- The original bearing capacity factors by Terzaghi (1943) have been largely superseded by those of later researchers using different rupture surfaces and experimental data.
- For piles, a modified version of these bearing capacity factors is used.

Table 21.6 Bearing capacity factors (Vesic, 1973 and Hansen, 1970).

Friction angle	*Bearing capacity factors*		*Vesic*	*Hansen*
ϕ	N_c	N_q	$N\gamma$	$N\gamma$
0 (Fully undrained condition)	5.14	1.00	0.00	0.00
1	5.4	1.09	0.07	0.00
2	5.6	1.20	0.15	0.01
3	5.9	1.31	0.24	0.02
4	6.2	1.43	0.34	0.05
5 (Clay undrained condition)	6.5	1.57	0.45	0.07
6	6.8	1.72	0.57	0.11
7	7.2	1.88	0.71	0.16
8	7.5	2.06	0.86	0.22
9	7.9	2.25	1.03	0.30
10 (Clay ~ undrained condition)	8.3	2.47	1.22	0.39
11	8.8	2.71	1.44	0.50
12	9.3	2.97	1.69	0.63
13	9.8	3.26	1.97	0.78
14	10.4	3.59	2.29	0.97
15 (Clay residual strength)	11.0	3.94	2.65	1.18
16	11.6	4.34	3.06	1.43
17	12.3	4.77	3.53	1.73
18	13.1	5.3	4.07	2.08
19	13.9	5.8	4.68	2.48
20 (Soft clays effective strength)	14.8	6.4	5.4	2.95
21	15.8	7.1	6.2	3.50
22	16.9	7.8	7.1	4.13
23	18.0	8.7	8.2	4.88
24	19.3	9.6	9.4	5.75
25 (Very stiff clays)	20.7	10.7	10.9	6.76
26	22.2	11.9	12.5	7.94
27	23.9	13.2	14.5	9.32
28	25.8	14.7	16.7	10.9
29	27.9	16.4	19.3	12.8
30 (Loose sand)	30.1	18.4	22.4	15.1
31	32.7	20.6	26.0	17.7
32	35.5	23.2	30.2	20.8
33	38.6	26.1	35.2	24.4
34	42.2	29.4	41.1	28.8
35 (Medium dense sand)	46.1	33.3	48.0	33.9
36	51	37.8	56	40.0
37	56	42.9	66	47.4
38	61	48.9	78	56
39	68	56	92	67
40 (Dense sand)	75	64	109	80
41	84	94	130	95
42	94	85	155	114
43	105	99	186	137
44	118	115	225	166
45 (Very dense gravel)	134	135	272	201

- The Terzaghi bearing capacity factors are higher than those of Vesic and Hansen.
- The next 2 sections provide simplified versions of the above for the bearing capacity of cohesive and granular soils.

21.7 Bearing capacity of cohesive soils

- For a surface footing the ultimate bearing capacity (q_{ult}) = N_c C_u (strip footing).
- For a fully undrained condition in cohesive soils $\phi = 0°$ and $N_c = 5.14$ (strip) is the minimum value for rough foundations in plane strain (long) conditions. While applicable for concrete poured on to ground, this N_c value can be reduced to 3.14 (π) when a side thrust occurs together with vertical loading as with an embankment loading.
- The bearing capacity increases with the depth of embedment. The change of N_c with the depth of embedment and the type of footing is provided in the table below.
- Often this simple calculation governs the bearing capacity as the undrained condition governs for a clay.

Table 21.7 Variation of bearing capacity coefficient (N_c) with the depth (Skempton, 1951).

Embedment ratio (z/B)	*Bearing capacity coefficient (N_c)*	
	Strip footing	*Circular or square*
0	5.14 ($2+\pi$).	6.28 (2π).
1	6.4	7.7
2	7.0	8.4
3	7.3	8.7
4	7.4	8.9
5	7.5	9.0

- z = Depth from surface to underside of footing.
- B = Width of footing.

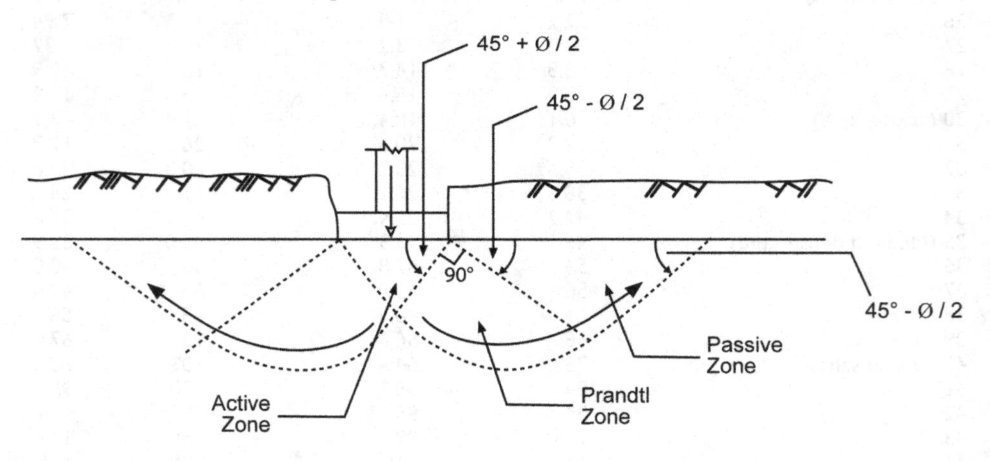

Figure 21.2 General shear failure.

21.8 Bearing capacity of granular soils

- In granular soils, the friction angle is often determined from the SPT N-value. Methods that directly use the N-value to obtain the bearing capacity, therefore can provide a more direct means of obtaining that parameter.
- The table below assumes the foundation is unaffected by water. Where the water is within B or less below the foundation then the quoted values should be halved. This practice is considered conservative as some researchers believe that effect may already be accounted for in the N-value.
- The allowable capacity (FS = 3) is based on settlements no greater than 25 mm. For acceptable settlements of 50 mm say, the capacity can be doubled while for settlements of 12 mm the allowable capacity in the Table should be halved.
- The footing is assumed to be at the surface. There is an increase bearing with embedment depth. This can be up to 1/3 increase, for an embedment = Footing width (B).
- The corrected N-value should be used.

Table 21.8 Allowable bearing capacity of granular soils (adapted from Meyerhof, 1956).

Foundation width B (m)	*Allowable bearing capacity (kPa)*					
	N = 5 *Loose*	*N = 10*	*N = 20* *Medium dense*	*N = 30*	*N = 40* *Dense*	*N = 50* *Very dense*
1	50	100	225	350	475	600
2			200	300	425	525
3				275	375	475
4	25	75	175		350	450
5				250		

- Note the above is based on Meyerhof (1956), which is approximately comparable to the charts in Terzaghi and Peck (1967). Meyerhof (1965) later suggests values ~50% higher, due to the conservatism found.
- For footing widths <6 m, the settlement predictions by Terzaghi and Peck (1967) overestimate by about 2.18 (Sivakugan et al., 1998).

21.9 Settlements in granular soils

- Design of shallow foundations in granular material generally governed by settlements rather than bearing capacity.
- Settlements may be estimated from the SPT N-value in granular soils.
- The settlement estimate is based on the size and type of foundation.

Table 21.9 Settlements in granular soils (Meyerhof, 1965).

Footing Size	*Relationship for settlement*
$B < 1.25$ m	1.9 q/N
$B > 1.25$ m	$2.84\ q/N\ [B/(B + 0.33)]^2$
Large rafts	2.84 q/N

- N = average over a depth = width of footing (B).
- q = applied foundation pressure.

21.10 Upper limits of settlement in sands

- Settlement ratios (settlement/pressure) in sands have been produced by Burland et al. (1977).
- Probable settlement is 50% of upper limit and in most cases the settlement unlikely to exceed 75% of upper limit.
- N-values are not corrected for overburden.
- Significant scatter for the loose sand

Table 21.10 Settlement ratios in sands (from Burland et al., 1977).

Foundation width B (m)	*Settlement/pressure (mm/kPa)*		
	N < 10 *Loose*	*N = 10–30* *Medium dense*	*N > 30* *Dense*
0.5	0.25	0.06	0.03
1	0.3	0.75	0.04
2	0.4	0.1	0.05
4	0.5	0.15	0.07
7	0.6	0.17	0.09
10	0.7	0.19	0.11
20	0.9	0.25	0.15

21.11 Factors of safety for shallow foundations

- Factor of Safety (FS) accounts for uncertainties in loading, ground conditions, extent of site investigation (SI) and consequences of failure. This is the traditional "working stress" design.
- FS = Available Property/Required Property. A nominal (expected, mean or median) value is used.
- Allowable Bearing Capacity = q_{ult}/FS.

Table 21.11 Factors of safety for shallow foundations (Vesic, 1975).

Loading and consequences of failure	*Factor of safety based on extent of SI*		*Typical structure*
	Thorough SI	*Limited SI*	
Maximum design loading likely to occur often. Consequences of failure high	3.0	4.0	Hydraulic structures, silos Railway bridges, warehouses, retaining walls
Maximum design loading likely to occur occasionally. Consequences of failure serious	2.5	3.5	Highway bridges, Light industrial buildings, public buildings
Maximum design loading unlikely to occur	2.0	3.0	Apartments, office buildings

- The industry trend is to use FS = 3.0 irrespective of the above conditions.
- For temporary structures, the FS can be reduced by 75% with a minimum value of 2.0.
- Limit state design uses a partial load factor on the loading and a partial performance factor on the resistance. Design resistance effect ≥ Design action effect.
- Ultimate limit states are related to the strength. Characteristic values are used.
- Serviceability limit states are related to the deformation and durability.
- Shear failure usually governs for narrow footing widths, while settlement governs for large footings (typically 2.0 m or larger).

21.12 Factors of safety for driven pile foundations

- The factors of safety applied to piled foundations is different from that of shallow foundations

Table 21.12 Factors applied to driven piles by AASHTO.

Method	*Approach*	*Partial factor AASHTO (2007) LRFD*	*Factor of safety prior to 2007 AASHTO (1992) – ASD*
Static analysis		0.4	3.5
Dynamic Formula	Gates	0.4	3.5
Wave equation		0.5	2.75
Dynamic test	Min 2% or 2 No.	0.65	2.25
Static test	or 100% dynamic	0.75	2.0
	and >2% Dynamic	0.80	1.9

- LRFD – Load and resistance factor design
- ASD – Allowable stress design

21.13 Pile characteristics

- The ground and load conditions, as well as the operating environment determine a pile type.
- The table provides a summary of some of the considerations in selecting a particular pile type.

Table 21.13 Pile selection considerations.

Pile type		*Typical working load (kN)*	*Cost/ metre*	*Penetration*	*Lateral/ tension capacity*	*Vibration level*
Driven	Precast	250–2000 kN	Low	Low	Low	High
	Prestressed	500–3500 kN	Medium	Medium	Low	High
	Steel H-pile	500–2500 kN	High	High	High	High
	Timber	100–500 kN	Low	Low	Medium	Medium
Cast In situ	Bored auger	Up to 6 MPa on shaft	High	Medium/High	High	Low
	Steel tube	Up to 8 MPa on shaft	Medium	High	Medium	High
	Micro piles	250 to 1000 kN	High	High	Low	Low

- Prestressing concrete piles reduces cracking due to tensile stresses during driving. Prestressing is useful when driving through weak and soft strata. The pile is less likely to be damaged during handling as compared to the precast concrete piles.
- Piles with a high penetration capability would have high driving stresses capability.
- Micro piles (100 mm to 250 mm) are useful in limited access or low headroom conditions
- There are many specialist variations to those summarised in the table.

21.14 Working loads for tubular steel piles

- Steel tube piles are useful where large lateral load apply, e.g. jetties and mooring dolphins.
- They can accommodate large working loads and have large effective lengths.
- The working load depends on the pile size, and grade of steel.

Table 21.14 Maximum working loads for end bearing steel tubular piles (from Weltman and Little, 1977).

Outside diameter	*Typical working load (kN) per pile*		*Approximate maximum effective length (m)*	
	Mild steel	*High yield stress steel*	*Mild steel*	*High yield stress steel*
300 mm	400–800 kN	600–1200 kN	11	9
450 mm	800–1500 kN	1100–2300 kN	16	14
600 mm	1100–2500 kN	1500–3500 kN	21	19
750 mm	1300–3500 kN	1900–5000 kN	27	24
900 mm	1600–5000 kN	2400–7000 kN	32	29

- Loads are based on a maximum stress of 0.3 × minimum yield stress of the steel.
- The effective length is based on axial loading only.
- The loads shown are reduced when the piles project above the soil level.

21.15 Working loads for steel H piles

- Steel tube piles are useful as tension piles.
- They can accommodate large working loads. While H-piles have high driveability, it is prone to deflection if boulders are struck, or at steeply inclined rock head levels.

Table 21.15 Maximum working Loads for end bearing steel H-piles (from Weltman and Little, 1977).

Size	*Typical working load (kN) per pile*		*Approximate maximum effective length (m)*	
	Mild steel	*High yield stress steel*	*Mild steel*	*High yield stress steel*
200 × 200 mm	400–500 kN	600–700 kN	5	4
250 × 250 mm	600–1500 kN	800–2000 kN	7	6
300 × 300 mm	700–2400 kN	1000–3500 kN	8	7

21.16 Load carrying capacity for piles

- The pile loads are distributed between the base and shaft of the pile.
- Piles may be referred to as end bearing or frictional piles. These represent material idealisations since end-bearing would have some minor frictional component, and frictional piles would have some minor end-bearing component. The terms are therefore a convenient terminology to describe the dominant load bearing component of the pile.
- The % shared between these two load carrying element depends on the pile movement and the relative stiffness of the soil layers and pile.

Table 21.16 Pile loads and displacements required to mobilise loads.

Load carrying element	*Symbols*	*Required displacements*
Shaft	Q_s = Ultimate shaft load (Skin friction in sands and adhesion in clays)	0.5% to 2% of pile diameter – typically 5 mm to 10 mm
Base	Q_b = Ultimate base load	5% to 10% of pile diameter for driven piles; greater than 10% for bored piles – typically 25 mm to 50 mm
Total	Ultimate load (Q_{ult}) = $Q_s + Q_b$	Base displacement governs

- Choice of the factor of safety should be made based on the different response of pile and base. Maximum capacity of shaft is reached before the base.
- If the foundation is constructed with drilling fluids and there is uncertainty on the base conditions, then design is based on no or reduced load carrying capacity on the base.
- If the movement required to mobilise the base is unacceptable then no base bearing capacity is used.
- The shaft would carry most of the working load in a pile in uniform clay, while for a pile in a uniform granular material the greater portion of the load would be carried by the base.
- The movement >10% of pile diameter for large bored piles seems illogical and should be applied only for piles <750 mm diameter.

21.17 Pile shaft capacity

- The pile shaft capacity varies from sands and clays.
- Driven piles provide densification of the sands during installation while bored piles loosen the sands.
- The surface of bored piles provides a rougher pile surface/soil interface (δ), but this effect is overridden by the loosening/installation (k_s) factor.
- When construction issues require that permanent steel liners are used, then the extremely low radial stresses for large diameter bored piles results in a low skin friction (Lo and Li, 2003).

Table 21.17 Shaft resistance for uniform soils (values adapted from Poulos, 1980).

Soil type	*Relationship*	*Values*	
		Bored	*Driven*
Clay	Shaft adhesion $C_a = \alpha C_u$	$\alpha = 0.45$ (Non-fissured) $\alpha = 0.3$ (Fissured) $c_a = 100$ kPa maximum	$\alpha = 1.0$ (Soft to firm) $\alpha = 0.75$ (Stiff to very stiff) $\alpha = 0.25$ (Very stiff to hard)
Sands	Skin friction $f_s = k_s \tan \delta\sigma'_v$ k_s = Earth pressure coefficient δ = Angle of friction between pile surface and soil σ'_v = Vertical effective stress	Not recommended (Loose) $k_s \tan \delta = 0.1$ (Medium dense) $k_s \tan \delta = 0.2$ (Dense) $k_s \tan \delta = 0.3$ (Very dense)	$k_s \tan \delta = 0.3$ (Loose) $k_s \tan \delta = 0.5$ (Medium dense) $k_s \tan \delta = 0.8$ (Dense) $k_s \tan \delta = 1.2$ (Very dense)

- Values shown are approximate only for estimation. Use charts for actual values in a detailed analysis.
- In layered soils and driven piles, the shaft capacity varies:
  - The adhesion decreases for soft clays over hard clays – due to smear effects for drag down.
  - The adhesion increases for sands over clays.
  - Table in sands applies for driven displacement piles (e.g. concrete). For low displacement (e.g. steel H piles) the values reduce by 50%.

21.18 Pile frictional values from sand

- For sands, the frictional value after installation of piles is different than before the installation (ϕ_1).
- The in situ frictional value before installation is determined from correlations provided in previous chapters.

Table 21.18 Change of frictional values with pile installation (Poulos, 1980).

Consideration	*Design parameter*	*Value of ϕ after installation*	
		Bored piles	*Driven piles*
Shaft friction	$k_s \tan \delta$	ϕ_1	$\frac{3}{4}\phi_1 + 10$
End bearing	N_q	$\phi_1 - 3$	$(\phi_1 + 40)/2$

21.19 Earth pressure coefficient after pile installation

- The earth pressure coefficient (k_s) after pile installation applies in the analysis of granular soils.

Table 21.19 Earth pressure coefficient (k_s) after various pile installations.

Pile type	*Earth pressure coefficient k_s*
Bored pile	1.0 K_o (typically ~ 0.7)
Driven – low displacement pile	1.4 K_o(typically ~ 1.0)
Driven – high displacement pile	1.8 K_o (typically ~ 1.2)

- When casing is extracted a k_s value of 1.0 is used if wet concrete is placed.
- CFA piles has values less than for bored piles
- δ is generally derived from the critical angle of friction ϕ_{crit}. A factor of 0.7 to 1.0 applies depending on the soil type, pile installation and the roughness of the pile – soil interface.

21.20 End bearing of piles

- The end bearing resistance (q_b) of a pile depends on the cohesion (C_u) for clays and the effective overburden (σ'_v) for sands.
- There is currently an ongoing discussion in the literature on critical depths, i.e. whether the maximum capacity is achieved at a certain depth.
- N_q values from Berezantsev et al. (1961).
- The bearing capacity of bored piles in sands is ½ to 1/3 that of the bearing capacity of a driven pile.

Table 21.20 End bearing of piles.

Soil type	*Relationship*	*Values*	
		Bored	*Driven*
Clay	$q_b = N_c C_u \omega$	$N_c = 9$ $\omega = 1.0$ (Non-fissured) $\omega = 0.75$ (Fissured)	$N_c = 9$ $\omega = 1.0$
Sands	$q_b = N_q \sigma'_v$ $q_b = 10$ MPa maximum	$N_q = 20$ (Loose) $N_q = 30$ (Medium dense) $N_q = 60$ (Dense) $N_q = 100$ (Very dense)	$N_q = 70$ (Loose) $N_q = 90$ (Medium dense) $N_q = 150$ (Dense) $N_q = 200$ (Very dense)

- Assumptions on frictional angles:
  - Loose – 30°.
  - Medium dense – 33°.
  - Dense – 37°.
  - Very dense – 40°.

21.21 Pile shaft resistance in coarse material based on N-value

- Estimates of the pile shaft resistance in granular materials can be determined from the corrected SPT N-value.
- The N-value is the average corrected value along the length of the pile.

Table 21.21 Pile shaft resistance in granular materials (Meyerhof, 1976; Chang and Broms, 1991).

Type of pile	*Displacement*	*Shaft resistance (kPa)*
Driven	High to average, e.g. concrete and including sheet piles	2 N
Driven	Low, e.g. steel H piles	N
Bored	Negligible	0.67 N
Bored	(Residual soils in Singapore)	2 N

21.22 Pile base resistance in coarse material based on N–value

- Estimates of the pile base resistance in granular materials can be determined from the corrected SPT N-value.
- The N-value is the corrected value for 10D below and 4D above the pile point.
- D = Diameter of pile.
- L = Length of pile in the granular layer.

Table 21.22 Pile base resistance in granular materials (Meyerhof, 1976; Chang and Broms, 1991).

Type of pile	*Type of soil*	*Base resistance (kPa)*
Driven	Coarse sand and gravel	40N L/D ≤ 400N
Driven	Fine to medium sand	40N L/D ≤ 300N
Bored	Any granular soil	14N L/D
Bored	(Residual soils in Singapore)	30–45N

21.23 Design parameters for pipe piles in cohesionless siliceous soils

- The following table applies for design and constructing fixed offshore platforms as a guide. Specifically for steel cylindrical pipe pile foundations. Limiting values apply for long piles.
- Shaft friction acts on both inside and outside of tubular piles. Skin friction $f_s = K \tan \delta\sigma'_v$.
- K = 0.8 for both tension and compression for open ended pipe piles driven unplugged.
- K = 1.0 for full displacement (plugged or closed end).
- A pile can be driven in an unplugged condition but act plugged under static loading (refer to later tables on plugging). A plugged pile behaves similarly to a closed ended pile.
- Open ended steel piles are preferred (closed ended) in offshore as open ended has less installation effort and therefore can be driven to greater depths for the same soil condition. Uplift and tension usually govern the design.

Table 21.23 Design parameters for cohesionless siliceous soils (API, 1993).

Soil description	*Density*	*Soil-pile friction angle δ (degrees)*	*Limiting skin friction values (kPa)*	*Bearing capacity factor N_q*	*Limiting unit end bearing values (MPa)*
Sand	Very loose	15	48	8	1.9
Sand-silt	Loose				
Silt	Medium				
Sand	Loose	20	67	12	2.9
Sand-silt	Medium				
Silt	Dense				
Sand	Medium	25	81	20	4.8
Sand-silt	Dense				
Sand	Dense	30	96	40	9.6
Sand-silt	Very dense				
Gravel	Dense	35	115	50	12.0
Sand	Very dense				

- Sand-silt includes those soils with significant fractions of both sand and silt. Strength values generally increase with increasing sand fraction and decrease with increasing silt fractions.

21.24 Pile interactions

- The driving of piles in sands increases the density around the piles depending on the soil displaced (depending on the diameter of pile). Adjacent and later piles are then more difficult to install. Steel H piles are considered low displacement.
- The driving of piles in clays may produce heave.

Table 21.24 Influence of driven piles (after Broms, 1996).

Location	*Influence zone at which density increases*	*Typical pile spacing*
Along shaft	4–6 pile diameters (D)	3D for frictional piles with lengths = 10 m 5D for frictional piles with lengths = 25 m
At base of pile	3–5 pile diameters below pile	2D for end bearing piles

- The spacing can be reduced if pre-drilling is used.
- The above should be considered when driving piles in groups or adjacent to existing piles.
- Pile groups in a granular soil should be driven from the centre outwards to allow for this densification effect.
- Bored piles have 2D or 750 mm minimum spacing, while driven piles are 2.5D spacing in sands.
- Screw piles would be nominally less than for end bearing piles, approximately 1.5D.
- 10 pile diameters is the distance often conservatively used to avoid the effects of pile installation on adjacent services and buildings.
- For cohesionless soils, loose soil is densified to a distance approximately 3D, while on dense soils, dilatancy decreases the friction angle to a distance approximately 3D.
- For cohesive soils, soil is remoulded a distance approximately 2D, during pile driving. This excess pore pressure generated during pile driving lowers shearing resistance, but excess pore pressure dissipates over time with a regain in soil strength (refer set up factors).

21.25 Influence zone for end bearing piles in sands

- The influence zone is material dependent.
- The influence zone by Yang (2006) was from an analytical study.

Table 21.25 Influence zone of end bearing piles (Yang, 2006).

Material type	*Influence zone above pile tip*	*Influence zone below pile tip*	*Side influence from pile side*	*Reference*
Clean sand	1.5–2.5D	3.5 to 5.5D		Yang (2006)
Silty sands (more compressible)	0.5 to 1.5D	1.5 to 3.0D		Yang (2006)
Loose $\phi \sim 30°$	~7D	<1.5D	3.5D	FIP (1986)
Medium $\phi \sim 35°$	~10D	1.5–2.5D	5D	FIP (1986)
Dense $\phi \sim 40°$	~15D	2.5–4.0D	6D	FIP (1986)

- pile diameters D.
- the pile influence zone as adapted from Broms (1981) is illustrated in Figure 21.3.

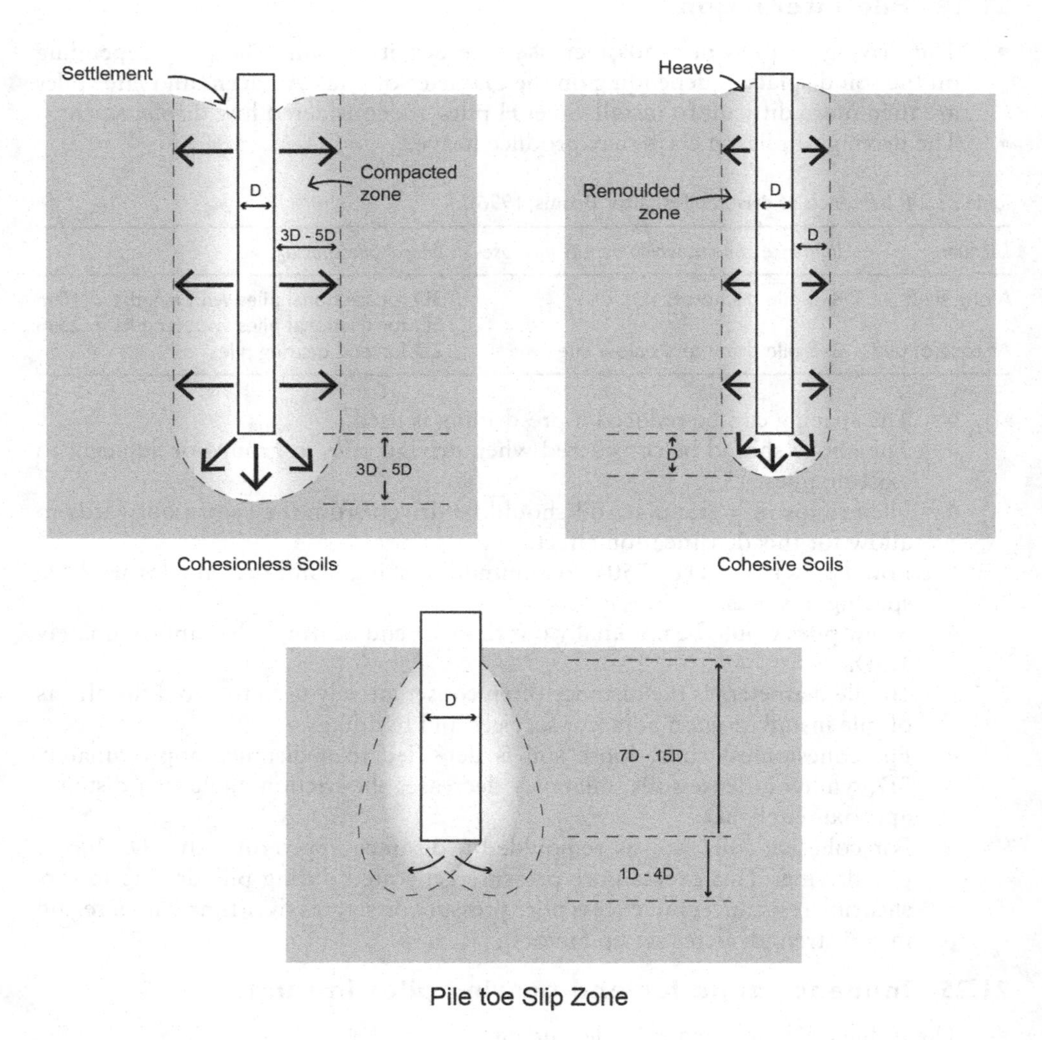

Figure 21.3 Driven pile influence zone.

21.26 Point of fixity

- The point of fixity needs to be calculated to ensure suitable embedment when lateral loads apply. For reinforced concrete piles this point is required to determine the extent of additional reinforcement at the top of the pile.
- The point of fixity is based on the load, pile type, size, and soil condition. The table below is therefore a first approximation only.

Table 21.26 Typical depth to the point of fixity for pile width (D).

Soil condition	*Strength*	*Depth to point of fixity*
Sands	Very loose	≥11D
	Loose	9D
	Medium dense	7D
	Dense	5D
	Very dense	3D
Clay	Soft	≥9D
	Firm	7D
	Stiff	6D
	Very stiff	5D
	Hard	4D

21.27 Uplift on piles

- The uplift capacity is typically taken as 75% of the shaft resistance due to cyclic softening.
- Piles on expansive clay sites experience uplift. The outer sleeve (permanent casing) may be used to resist uplift in the active zone.

Table 21.27 Uplift design.

Depth	*Load*		*Comment*
Surface to depth of desiccation cracking	No shaft capacity resistance	Uplift	Use 1/3 of active zone
Surface to depth of active zone	Swelling pressures (U_s) from swelling pressure tests. Apply U_s to slab on ground + $0.15U_s$ to shaft. Use C_u if no swell test	Uplift	Typically 1.5 m (coastal) to 5.0 m (arid) depending on climate and soil
Below active zone	75% Downward shaft resistance + Dead load	Resistance	Due to cyclic softening

- Air space may be used below the main beam (a suspended floor system) or a void former below the slab may be used to resist slab uplift.

21.28 Plugging of steel piles

- The pile shaft capacity is determined from the perimeter, and its length.
- The pile base capacity is determined from the cross sectional area.
- The pile must be assessed if in plugged or unplugged mode, as this determines the applied area for adhesion and end bearing. Plugged and unplugged piles have the entire cross section or annulus only used, respectively.
- For H-pile sections, the soil is plugged if sufficient embedment occurs. The outer "plugged" perimeter and area is used.
- For open-ended steel pile sections, a soil plug occurs if sufficient embedment and the full plugged cross sectional area are used.
- The plugging should be estimated from the type of soil and its internal friction. The plug forms when the internal side resistance exceeds the end bearing resistance of the pile cross-sectional area.

- The table below is a first estimation guide only and subject to final design calculations as pile plugging can be highly variable.

Table 21.28 Initial estimate guidance pile plugs based on diameter of open pile.

Strength of material	*Likely pile plug*	*Comment*
Very soft clay	25 to 35 pile diameters	10 m to 14 m plug formed for a 400 mm diameter tubular pile (Trenter and Burt, 1981). Under weight of hammer
Soft to stiff clays	*10 to 20 pile diameters*	*Paikowsky and Whitman (1990)*
Very stiff to hard clays	<15 pile diameters	Assumed
	>30 pile diameters	Assumed
Medium dense to dense sands	*20 to 35 pile diameters*	*Paikowsky and Whitman (1990)*
Sands	Internal diameter	Rigid plugging criterion based on field evidence
Loose $D_r < 35\%$	<0.2 m	Jardine et al. (2005)
Medium dense	0.2–0.7 m	
Dense $D_r = 65–85\%$	0.7–1.2 m	
Very dense $D_r > 85\%$	1.2 m	

  - Internal soil plugging for very soft clay showed the internal soil plug moved down with the plug and achieved a final length of 70% of the length of pile for 400 mm diameter pile.
  - For dense sand 40 to 50% of driven length likely.
  - The above is highly variable and caution is required. Other calculations must be performed. Refer to Jardine et al. (2005) for detailed design calculations.

21.29 Time effects on pile capacity

- Pile driving often produces excess pore water pressures, which takes some time to dissipate. Pile capacities often increase with time as a result.

Table 21.29 Soil set up factors (adapted from Rausche et al., 1996).

Predominant soil type along pile shaft	*Range in soil set up factor*	*Recommended soil set up factor*
Clay	1.2–5.5	2.0
Clay-sand	1.0–6.0	1.5
Sand-Silt	1.2–2.0	1.2
Fine sand	1.2–2.0	1.2
Sand	0.8–2.0	1.0
Sand-gravel	1.2–2.0	1.0

  - The time to achieve this increased capacity can vary from a few days in sands to a few weeks in clays.
  - Time dependent changes can be assessed only on a site specific basis, as in some materials, e.g. shales and silts, some relaxation can also occur. This results in a reduction in capacity.

21.30 Piled raft foundations for buildings

- Bearing piles which are driven or embedded into competent strata are less economically attractive as depth to such strata increases. Higher level foundation systems such as piled raft are then considered.
- A piled raft foundation consists of a system of piles interconnected with a rigid raft to provide a uniform settlement profile and load distribution.
- These frictional piles are relatively short and used to enhance the load capacity and control differential settlements.

Table 21.30 Piled raft design (Horikoshi and Randolph, 1998; Randolph, 2003; Reul and Randolph, 2004).

Design element	*Optimal design*
Pile location for uniform load	Distributed over central 16% to 25% of raft area
Pile stiffness	Approximately equal to raft alone
Pile capacity	40% to 70% of total applied load
Non uniform loads	Varying pile lengths
Layer of soft soil below raft	Use short piles extending through soft layer with longer piles in central 25% to 40% of raft area. Pile lengths greater than 70% of width of raft

21.31 Piled embankments for highways and high speed trains

- Piled supported embankments provide a relatively quick method of constructing embankments on soft ground.
- The design consists of determining the pile size (length and width), the pile cap, the load transfer platform (thickness and number of layers and strength of geotextile) for the height of fill and the ground conditions
- There is a minimum fill height where the load may be low, but the support may require closer pile spacing than a higher fill height. This may seem contradictory to the client.
- A minimum fill height allows for arching within the embankment and keeps the settlement throughs between the piles at a reasonably small size.

Table 21.31 Piled raft design dimensions for low embankments (Brandl, 2001).

Design element	*Minimum fill height (H_o) between pile top (surface of piled caps) and surface of railway sleepers/roadway surface*	
Pile cap size = a – s	Typical applications	Movement sensitive systems e.g. High speed trains (v > 160 m/hr)
Pile spacing (a)	$H_o \geq a$	$H_o \geq 1.25a$
Spacing between pile caps (s)	$H_o \geq 1.5s$	$H_o \geq 2.0s$
Fill height	$H_o \geq 1.0$ m	$H_o \geq 1.5$ m

- Load Transfer Platform (LTP) used to transfer the load on to the pile
- Typically LTP thickness = 500mm with at least 2 No. uniaxial geogrids (placed different directions)
- For Geosynthetics used to cap the deep foundations, the allowable strain <3% in long term creep

- For low embankments, there may be dynamic effects of loading on ground:
  - 2–3 m for highways.
  - 4–5 m for high speed trains.

21.32 Dynamic magnification of loads on piled rafts for highways and high speed trains

- The LTP acts as a geosynthetic soil cushion. This reduces the dynamic load on piles for low embankments.
- The table provides this dynamic magnification factor for the loads.

Table 21.32 Dynamic magnification factor for dynamic loads on top of piled railway embankment (Brandl, 2001).

Height of fill	*Dynamic magnification factor* Φ	
	Without geosynthetic cushion	*With geosynthetic cushion on top of pile caps*
$H_o \geq 4.0$ m	1.0	1.0
$H_o \geq 3.0$ m	1.5	1.0
$H_o \geq 2.0$ m	2.5	1.5
$H_o \geq 1.5$ m	3.0	2.0
$H_o \geq 1.0$ m	Not applicable	2.5

21.33 Allowable lateral pile loads

- The allowable lateral pile loads depends on the pile type and deflection.

Table 21.33 Allowable lateral pile loads (USACE, 1993).

Pile type	*Considerations*	*Deflection (mm)*	*Allowable lateral load (kN)*
Timber	No deflection	–	45
Concrete	criteria	–	65
Steel		–	90
Timber	Some deflection	6	40
Concrete	limitations	12	60
		6	50
		12	75
Timber – 300 mm free	Deflection	6	7
Timber – 300 mm fixed	constrained	6	20
Concrete 400 mm – Medium sand		6	30
Concrete 400 mm – Fine sand		6	25
Concrete 400 mm – Clay		6	20

- For group piles the pile in the front row sustain the maximum loading and decreases for each pile row away from the front row and dependent on the pile spacing. This shadowing effect is considered in AASHTO (2007) using p multipliers to progressively reduce each row.
- For a pile group, the outer pile typically carries a greater share of load than the inner piles. The closer spacing has a lower multiplier e.g. 0.7 and 1.0p – multiplier for 3 and 5 pile diameters, respectively. This indicates greater interaction and decreasing efficiency for closer piles.

21.34 Load deflection relationship for concrete piles in sands

- The deflection is limited by the pile sizes and strength of the soil.

Table 21.34 Load deflection for prestressed concrete piles in sands (from graphs in Barker et al., 1991).

Pile size	*Deflection (mm) for friction angle (°) and load (kN)*								
	$\phi = 30°$ (Loose)			*$\phi = 35°$ (Medium dense)*			*$\phi = 40°$ (Very dense)*		
	50 kN	*100 kN*	*150 kN*	*50 kN*	*100 kN*	*150 kN*	*50 kN*	*100 kN*	*150 kN*
250 * 250 mm	10	30	>30	7	22	>30	5	15	30
300 * 300 mm	5	17	30	4	11	20	4	9	15
350 * 350 mm	4	10	18	3	7	13	3	6	9
400 * 400 mm	3	7	12	3	5	8	2	4	7
450 * 450 mm	2	5	8	2	3	6	2	3	4

- Bending moments for the piles range from approximately:
  - 225 kNm to 75 kNm for 150 kN to 50 kN load in loose sands.
  - 200 kNm to 50 kNm for 150 kN to 50 kN load in medium dense sands.
  - 175 kNm to 50 kNm for 150 kN to 50 kN load in very dense sands.
- No significant differences in bending moments for various pile sizes in sands.

21.35 Load deflection relationship for concrete piles in clays

- The deflection of piles in clays is generally less than in sands.

Table 21.35 Load deflection for prestressed concrete piles in clays (from graphs in Barker et al., 1991).

Pile size	*Deflection (mm) for undrained strength (kPa) and Load (kN)*								
	$C_u = 70$ kPa (Stiff)			*$C_u = 140$ kPa (Very stiff)*			*$C_u = 275$ kPa (Hard)*		
	50 kN	*100 kN*	*150 kN*	*50 kN*	*100 kN*	*150 kN*	*50 kN*	*100 kN*	*150 kN*
250 * 250 mm	5	17	>30	3	8	14	1	3	6
300 * 300 mm	3	10	21	2	5	9	<1	2	4
350 * 350 mm	2	7	14	1	4	6	<1	1	3
400 * 400 mm	2	5	10	<1	3	4	<1	<1	2
450 * 450 mm	1	4	7	<1	2	3	<1	<1	2

21.36 Bending moments for PSC piles in stiff clays

- The induced bending moments of PSC clays is dependent on the deflection and pile size.

Table 21.36 Bending moments for prestressed concrete piles in clays (from graphs in Barker et al., 1991).

Pile size	*Bending moment (kNm) for undrained strength (kPa) and load (kN)*								
	$C_u = 70$ kPa (Stiff)			*$C_u = 140$ kPa (Very stiff)*			*$C_u = 275$ kPa (Hard)*		
	50 kN	*100 kN*	*150 kN*	*50 kN*	*100 kN*	*150 kN*	*50 kN*	*100 kN*	*150 kN*
250 * 250 mm	50 kNm	125	225	25	75	150	25	50	100
450 * 450 mm	75 kNm	175	275	75	125	200	50	100	175

Chapter 22

Rock foundations

22.1 Rock bearing capacity based on RQD

- The rock bearing capacity is dependent on the rock strength, defects and its geometry with respect to the footing size.
- The table below is a first approximation based on RQD, which is a function of the defects and the strength to a minor extent.

Table 22.1 Bearing pressures (Peck, Hansen and Thorburn, 1974).

RQD (%)	*Rock description*	*Allowable bearing pressures(MPa) lesser of below values*	
0–25	Very poor	1–3	
25–50	Poor	3–6	UCS
50–75	Fair	6–12	or allowable stress
75–90	Good	12–20	of concrete
>90	Excellent	20–30	

  - This method is commonly used but not considered appropriate for detailed design.

22.2 Rock parameters from SPT data

- The SPT values in rock are usually the extrapolated values, as driving refusal would have occurred before the given values.
- SPT > 50 is useful in soils where very dense is implied, but the value has limited use in rock material.
- At high SPT values one should be obtaining rock cores, however one may need to still rely on SPT in "soft" rocks due to significant core loss or budget limitations (coring may be 5 times cost of solid augering)

Table 22.2 Rock parameters from SPT data.

Strength	*Symbol*	*Point Load Index I_s (50) (MPa)*	*Extrapolated SPT value $(N_o)^*_{60}$*	*Allowable bearing capacity*
Extremely low to low	EL, VL, L	<0.3	60–150	500 kPa to 2.5 MPa
Medium	M	0.3–1.0	100–350	2 to 12 MPa
High	H	1.0–3.0	250–600	
Very to extremely high	VH, EH	>3	>500	>10 MPa

- To obtain N* values, SPT refusal values are required in both seating and test drive (refer Chapter 4). This is an extrapolated value and not driven value as hammer damage may occur unless a cone tip is used. Note that some procedures recommend refusal in the seating drive only – but this is insufficient data.
- Higher values of allowable bearing capacity are likely with more detailed testing from rock core samples.
- The bearing capacity of some non-durable rocks can decrease when its overburden is removed and the rock is exposed and subject to weathering and/or moisture changes.

22.3 Bearing capacity modes of failure

- The mode of failure depends on the joint spacing in relation to the footing size.
- Driven piles therefore have a higher bearing capacity due to its relative size to joint spacing.
- Bored piles (drilled shafts) have a lower bearing capacity than driven piles due to its relative size.

Table 22.3 Failures modes in rock (after Sowers, 1979).

Relation of joint spacing (S) to footing width (B)	*Joints*	*Orientation*	*Failure mode*
S < B	Open	Vertical to sub-	Uniaxial compression
S < B	Closed	vertical 90° to 70°	Shear zone
S > B	Wide		Splitting
S > B Thick rigid layer over weaker layer	N/A	Horizontal to	Flexure
S < B Thin rigid layer over weaker layer	N/A	sub-horizontal	Punching

- A different bearing strength applies for all of the above failure modes, even for a rock with similar rock strength. This is presented in the tables that follow.
- When RQD → 0, one should treat as a soil mass and above concepts do not apply.
- These failure modes form the basis for evaluating the rock bearing capacity.

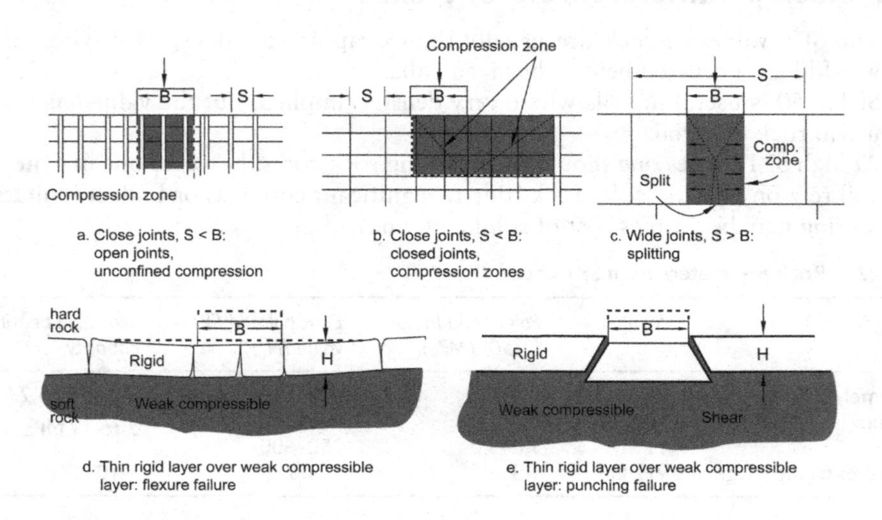

Figure 22.1 Bearing capacity failures modes (Sowers, 1979).

22.4 Compression capacity of rock for uniaxial failure mode

- This is a uniaxial compression failure condition (S < B).
- The table applies for open vertical to sub-vertical joints.

Table 22.4 Ultimate bearing capacity with failure in uniaxial compression.

Failure mode	*Strength range*	*Design ultimate strength*
Uniaxial compression with RQD < 70%	15% to 30% UCS	Use 15% UCS
Uniaxial compression with RQD > 70%	30% to 80% UCS	Use 30% UCS

- Factors of safety to be applied to shallow foundations.
- For deep foundations, piles have the effect of confinement, and the design ultimate strength ~ allowable bearing capacity.
- An alternative approach to this uniaxial failure condition is presented below.

22.5 Ultimate compression capacity of rock for shallow foundations

- This applies for the uniaxial compression failure mode, i.e. open joints with S < B.
- It uses the ultimate bearing capacity $= q_{ult} = 2c\tan(45° + \phi/2)$. This is the Mohr Coulomb Failure criterion for the confining stress $\sigma_3 = 0$.
- The table assumes the cohesion, $c = 10\%\ q_u$ (Chapter 9) for all RQD Values.
- This applies to shallow foundations only, and a factor of safety is required for the allowable case.

Table 22.5 Ultimate bearing capacity (using above equation from Bell, 1992).

Angle of friction	*q_{ult} (kPa) using q_u values 1 MPa–40 MPa*				
	Low	*Medium strength*		*High*	*Very high*
	1 MPa	*5*	*10*	*20*	*40 MPa*
30°	0.2	0.8	1.5	3.1	6.1
40°	0.2	1.1	2.2	4.4	8.7
50°	0.3	1.6	3.1	6.3	13
60°	0.5	2.4	4.8	9.7	19

- The ultimate capacity seems unrealistically low for values of low strength rock, i.e. where $q_u = 1$ MPa. However it is approximately consistent for 15% UCS (RQD = 70%) given in the previous table.
- This suggests that these methods are not applicable for rocks classified as low to extremely low strength (I_s (50) < 0.3 MPa) – "soft" rocks.

22.6 Compression capacity of rock for a shear zone failure mode

- This condition applies for closely spaced joints (S < B).
- A Terzaghi type general bearing capacity theory is used with the following parameters:
  - The soil properties – cohesion (c), angle of friction (ϕ) and unit weight (γ).
  - The footing geometry – embedment (D_f) and width (B).

- However, the shape factors for square and circular footings are different, as well as the bearing capacity factors.
- The bearing capacity factors for rock are derived from wedge failure conditions, while the slip line for soils are based on an active triangular zone, a radial shear zone and a Rankine passive zone.

Table 22.6 Bearing capacity equation.

Consideration	*Cohesion*	*Embedment*	*Unit weight*	*Comments*
Bearing capacity factors	N_c	N_q	$N\gamma$	These factors are non-dimensional and depend on ϕ. See next table
Ultimate bearing capacity (q_{ult})	$1.00\ c\ N_c +$	$\gamma\ D_f\ N_q +$	$0.5\ \gamma\ B\ N\gamma$	Strip footing (L/B = 10)
	$1.05\ c\ N_c +$			Strip footing (L/B = 5)
	$1.12\ c\ N_c +$			Strip footing (L/B = 2)
	$1.25\ c\ N_c +$	$\gamma\ D_f\ N_q +$	$0.8\ \gamma\ B\ N\gamma$	Square footing
	$1.2\ c\ N_c +$	$\gamma\ D_f\ N_q +$	$0.7\ \gamma\ B\ N\gamma$	Circular footing

- Most shallow rock foundations have $D_f \sim 0$ (i.e. at the rock surface) and the embedment term becomes zero irrespective of the N_q value.
- The unit weight term is usually small due to the width (B) term and is usually neglected except in the case of high frictional rock, i.e. $\phi \geq 50°$.

22.7 Rock bearing capacity factors

- These bearing capacity factors have been based on wedge theory. It is different from the bearing capacity factors of soils.

Table 22.7 Bearing capacity factors (from graphs in Pells and Turner, 1980).

Friction angle	*Bearing capacity factors*		
$\phi°$	*N_c*	*N_q*	*N_γ*
0	4	1	0
10	6	2	1
20	8	4	5
30	15	9	15
40	25	20	45
50	50	60	160
60	110	200	1000

22.8 Compression capacity of rock for splitting failure

- A splitting failure condition applies for widely spaced and near vertically oriented joints.
- Joint spacing (S) > footing width (B). The joint extends below the below footing for a depth H.
- The ratio of the joint depth to the footing width (H/B) is used to provide a joint correction factor for the bearing capacity equation.

Table 22.8 Ultimate bearing capacity with failure in splitting (Bishnoi, 1968; Kulhawy and Goodman, 1980).

Foundation type	*Ultimate bearing capacity (q_{ult})*	*Correction factor (J) based on discontinuity spacing (H/B)*									
Circular	$1.0\,J\,c\,N_{cr}$										
Square	$0.85\,J\,c\,N_{cr}$	H/B	0	1	2	3	4	5	6	7	8
Continuous strip	$1.0\,J\,c\,N_{cr}/(2.2+0.18\,L/B)$	J	0.41	0.52	0.67	0.77	0.85	0.91	0.97	1.0	1.0

- J = Joint correction factor.
- N_{cr} = Bearing capacity factor.
- L = Length of footing.
- B = Width of footing.

22.9 Rock bearing capacity factor for discontinuity spacing

- The bearing capacity factor in table 22.7 for the wedge failure does not allow for discontinuity spacing.
- This table is to be used with table 22.8, and applies when the joints are more widely spaced than the foundation width.

Table 22.9 Bearing capacity factors (from graphs in Bishnoi, 1968; Kulhawy and Goodman, 1980).

Friction angle	*Bearing capacity factors (N_{cr}) with discontinuity spacing (S/B)*						
φ°	*Previously tabulated N_c (Table 22.7)*	*0.5*	*1.0*	*2*	*5*	*10*	*20*
0	4	4	4	4	4	4	4
10	6	4	4	4	6	6	6
20	8	4	4	5	9	9	8
30	15	4	4	6	15	15	15
40	25	4	4	8	20	25	25
50	50	4	6	10	25	40	50
60	110	4	8	15	35	50	110

22.10 Compression capacity of rock for flexure and punching failure modes

- This table applies for a rigid layer over weaker layers. The top layer is considered rigid for S > B while the layer is thin for S < B.
- The stress of the underlying layer also needs to be considered.

Table 22.10 Ultimate bearing capacity with failure in flexure or punching.

Failure mode	*Strength range*	*Design ultimate strength*
Flexure	Flexural ~5% to 25% UCS	Use 10% UCS
Punching	Tensile ~50% flexural strength	Use 5% UCS

- Factor of safety needs to be applied and is the same for piles and shallow foundations.

22.11 Factors of safety for design of deep foundations

- The factor of safety depends on:
  - Type and importance of structure.
  - Spatial variability of the soil.
  - Thoroughness of the subsurface program.
  - Type and number of soil tests performed.
  - Availability of on site or nearby full-scale load test results.
  - Anticipated level of construction inspection and quality control.
  - Probability of the design loads actually occurring during the life of the structure.

Table 22.11 Typical factors of safety for design of deep foundations for downward loads (Coduto, 1994).

Classification of structure	*Design life*	*Acceptable probability of failure*	*Design factors of safety, F.S.*			
			Good control	*Normal control*	*Poor control*	*Very poor control*
Monumental	>100 yrs	10^{-5}	2.3	3.0	3.5	4.0
Permanent	25–100 yrs	10^{-4}	2.0	2.5	2.8	3.4
Temporary	<25 yrs	10^{-3}	1.4	2.0	2.3	2.8

- Monumental structures are large bridges or extraordinary buildings.
- Permanent structures are ordinary rail and highway bridges and most large buildings.
- Temporary structures are temporary industrial or mining facilities.

22.12 Control factors

- The control factors referenced in the above table are dependent on the reliability of data derived from subsurface conditions, load tests and construction inspections.

Table 22.12 Typical Factors of safety for design of deep foundations for downward loads (Coduto, 1994).

Factors	*Good control*	*Normal control*	*Poor control*	*Very poor control*
Subsurface conditions	Uniform	Not uniform	Erratic	Very erratic
Subsurface exploration	Thorough	Thorough	Good	Limited
Load tests	Available	Not available	Not available	Not available
Construction inspection	Constant monitoring and testing	Periodic monitoring	Limited	None

- Examples of good and very poor control are:
  - Bored piles constructed with down the hole inspection for clean out and confirmation of founding layers – good control.
  - Bored piles constructed with drilling fluids without the ability for even a down the hole camera inspection – very poor control.

22.13 Ultimate compression capacity of rock for driven piles

- The ultimate bearing capacity $= q_{ult} = 2\, q_u \tan^2(45° + \phi/2)$.
- The design compressive strength $= 0.33–0.8 q_u$ (Chapter 9).
- The table below uses $0.33 q_u$ for RQD < 70% and $0.5 q_u$ for RQD > 70%.

Table 22.13 Ultimate bearing capacity for driven piles (using above equation from Tomlinson, 1996).

Angle of friction	*RQD%*	*q_{ult} (kPa) using q_u values 1 MPa–40 MPa*				
		1 MPa	*5*	*10*	*20*	*40 MPa*
30°	<70	0.4	1.9	3.9	7.8	15
	>70	0.6	2.9	5.9	12	24*
40°	<70	0.8	3.9	7.9	16	Concrete strength governs*
	>70	1.2	6.0	12	24*	
50°	<70	1.6	8.0	16	Concrete strength governs*	
	>70	2.5	12	25*		
60°	<70	3.8	19	Concrete strength governs*		
	>70	5.8	29*			

- Note this ultimate capacity is significantly higher capacity than the previous table for shallow foundations.
- A passive resistance term, $\tan^2(45° + \phi/2)$, enhances the pile capacity.
- The capacities are 1 to 8 times the previous table based on low to high friction angles respectively for RQD < 70% and 3 to 12 times for the RQD > 70%.
- Concrete strength of 40 MPa was the previous upper limit but high strength is now considered 70 MPa in the 21st century.

22.14 Shaft capacity for bored piles

- The shaft capacity increases as the rock quality increases.
- Seidel and Haberfield (1995) provides the comparison between soils and rock capacity.
- The shaft adhesion $= \psi (q_u P_a)^{1/2}$.
- P_a = atmospheric pressure ~100 kPa.
- ψ = adhesion factor based on quality of material.
- q_u = Unconfined compressive strength of intact rock (MPa).

Table 22.14 Shaft capacity for bored piles in rock (adapted from Seidel and Haberfield, 1995).

Adhesion factor ψ	τ = *Ultimate side shear resistance (MPa)*	
	(Seidel and Haberfield, 1995)	*Other researchers*
0.5	$0.1(q_u)^{0.5}$	
1.0 (Lower bound)	$0.225(q_u)^{0.5}$	Lesser of $0.15q_u$ (Carter and Kulhawy, 1987) and $0.2(q_u)^{0.5}$ (Horvath and Keney, 1979) Dyveman & Valsangkar, 1996
2.0 (Mean)	$0.45(q_u)^{0.5}$	
3.0 (Upper bound)	$0.70(q_u)^{0.5}$	

22.15 Shaft resistance roughness

- The shaft resistance is dependent on the shaft roughness.
- The table below was developed for Sydney Sandstones and shales.

Table 22.15 Roughness class (after Pells et al., 1980).

Roughness class	*Grooves*		
	Depth	*Width*	*Spacing*
R1	<1 mm	<2 mm	Straight, smooth sided
R2	1–4 mm	>2 mm	50–200 mm
R3	4–10 mm	>5 mm	
R4	>10 mm	>10 mm	

- Roughness can be changed by the procedures used e.g. reduction factors of (0.9–1.0) for polymer slurry and (0.7–0.9) for bentonite slurry as compared to without any drilling fluid (Seidel and Collingwood, 2001) and roughness heights of 3–15 mm for UCS = 1–10 MPa, but 1–5 mm for UCS < 1 MPa or UCS > 10 MPa.
- Above R4 condition is used in Rowe and Armitage (1984) for a rough joint.

22.16 Shaft resistance based on roughness class

- The shaft resistance for Sydney sandstones and shales can be assessed by applying the various formulae based on the roughness class.
- τ = Ultimate side shear resistance (MPa).
- q_u = Unconfined compressive strength of intact rock (MPa).

Table 22.16 Shaft resistance (Pells et al., 1980).

Roughness class	τ = *Ultimate side shear resistance (MPa)*
R1 R2	$0.45(q_u)^{0.5}$
R3	Intermediate
R4	$0.6(q_u)^{0.5}$

22.17 Design shaft resistance in rock

- Different researchers have derived various formulae. In some instance for specific types of rock. This should be considered when using below table.
- The summary table below combines the concepts and rules provided above by the various authors.
- The formula has to be suitably factored for a mix of conditions, e.g. low quality rock with no slurry and grooving of side used.

Table 22.17 Shaft capacity for bored piles in rock (modified from above concepts).

Typical material properties	*Construction condition*	*τ = Ultimate side shear resistance (MPa)*
Soil, RQD << 25%		$0.1(q_u)^{0.5}$
Low quality rock RQD <25%, clay seams defects <60 mm	Slurry used, straight, smooth sides	$0.2(q_u)^{0.5}$
Medium quality rock RQD = 25%–75% defects 60–200 mm		$0.45(q_u)^{0.5}$
High quality rock RQD >75% defects >200 mm	Artificially roughened by grooving	$0.70(q_u)^{0.5}$

22.18 End bearing capacity of rock socketed piles

- End bearing capacity of rock socketed piles depends on the displacement. Some relationships shown.

Table 22.18 End bearing capacity of bored piles in rock.

End bearing capacity q_{max} (kPa)	*Comment*	*Reference*
$2.7q_u$	For rock to a depth > 1D below base of socket being intact or tightly jointed.	Rowe and Armitage (1987)
$2q_u$	For embedment into rock < 1D	
$4.83q_u^{0.51}$	$r^2 = 0.81$ from database of 39 shaft load tests on rocks of relatively low strength	Zhang and Einstein (1998)
$1.1q_u$ or $2.2q_u$ Intact $1.0q_u$ or $2.2q_u$ Slightly fractured $0.7q_u$ or $1.3q_u$ $0.4q_u$ or $0.7q_u$ Fractured 1.2 MPa or 2.5 MPa Crumbly	Maximum pressure for normalised displacement (δ/D) of 1% to 2%, respectively No q_u	CIRIA (1999)

22.19 Load settlement of piles

- Some movement is necessary before the full load capacity can be achieved. The full shaft capacity is usually mobilized at approximately 10 mm.
- Due to the large difference in movement required to mobilise the shaft and base, some designs use either the shaft capacity or the base capacity but not both.

- Reese and O'Neil (1989) use the procedure of movement >10 mm, then the load is carried entirely by base while displacement <10 mm then the load is carried by shaft. Therefore calculation of the settlement is required to determine the load bearing element of the pile.
- Often 50% to 90% of the load is required by the shaft capacity for bored piles. Pile driven to refusal would have its base capacity as the dominant load transfer.
- The base resistance should be ignored where boreholes do not extend beyond below foundation or in limestone areas where solution cavities are possible.

Table 22.19 Pile displacements.

Load carrying element	*Displacement required*	
	Typical	*Material specific, e.g. bored piers in clay/mudstones*
Shaft	0.5% to 2% shaft diameter 5–10 mm	1% to 2% of shaft diameter 10 mm maximum for piles with diameters >600 mm
Base	5% to 10% shaft diameter	10% to 20% of base diameter

  - Factor of safety to consider the above relative movements.

22.20 Pile refusal

- Piles are often driven to refusal in rock
- The structural capacity of the pile then governs.
- There is often uncertainty on the pile founding level.
- The table can be used as guide, where all the criteria are satisfied, and suitably factored when not all of the factors are satisfied.

Table 22.20 Estimate of driven pile refusal in rock.

Rock property				*Likely pile penetration into rock (m)*
*SPT value, N**	*I_s (50)MPa*	*RQD (%)*	*Defect spacing (mm)*	
> 400	>1.0 0.3–1.0	>75% 50–75%	>600	<<B <B
200–400	0.1–0.3	25–50%	200–600	B–3B 2B–4B
100–200			60–200	3B–5B 5B–7B
<100	<0.1	<25%	<60	>5B

  - As the structural capacity and driving energy determines the pile refusal levels, the table should be factored downwards for timber piles and upwards for steel piles.
  - For example a 450 mm prestressed concrete pile is expected to have arrived at refusal (2–3 mm set) within 3m of an N* ~ 100 sandstone, but an H pile requires N* > 200 to achieve that set.

- For shales that N* value is likely to be ~160 (Refer Table 1.11).
- A refusal criterion is not the same as a driving criteria and adequate pile capacity can be obtained without pile refusal being attained.

22.21 Limiting penetration rates

- The pile refusal during construction may be judged by the penetration rates.
- This varies according to the pile type.

Table 22.21 Penetration rate to assess pile refusal.

Pile Type	*Maximum blow count (mm/blow)*
Concrete	2–3 mm
Timber	6–8 mm
Steel – H	1–2 mm
Steel – pipe	1–2 mm
Sheet piles	2–3 mm

- Piles in granular material with SPT refusal of N > 50 (very dense) is expected to have a pile set of 5 mm to 20 mm, and 10 mm to 40 mm for dense sands and gravels.

22.22 Pile installation

- Pile refusal production rates depend on type of soil or rock, pile type and size as well as equipment used.

Table 22.22 Pile production rates.

Material type	*Typical production rates*	*Comments*
Soils	2 m/minute for displacement piles SPT < 30	Augers and drilling buckets for hard soils and up to medium strength rocks Core barrels in boulders, gravels or high strength rock
XW rock	2 m/minute for driven PSC piles up to "soft" rock	Lower rate for spliced piles
HW rock	0.4 m/hr for open ended tubular steel piles	
MW rock	0.2 m/hr for open ended tubular steel piles	Bored piles would have higher rates. Conventional drilling tools to UCS = 50 MPa, with air roller core barrels for UCS = 50 to 100 MPa
SW/FR rock	0.2 m–1.0 m/hr for bored piles UCS > 100 MPa	Specialised cluster drilling required

- Single down the hole hammers has a high productivity at a higher cost for piles up to 750 mm diameter.
- Presence of water may have significant cost implications for bored piles, especially in granular material.

Chapter 23

Movements

23.1 Types of movements

- Some movements typically occur in practice, i.e. stress and strain are interrelated. If the load is applied and soil resistance occurs, then some nominal movement is often required to mobilise the full carrying capacity of the soil or material.
- The large factors of safety in the working stress design, typically captures the acceptable movement, i.e. deformations are assumed kept to an acceptable level. Limit equilibrium and conditions can then be applied in the analysis. However, many design problems (e.g. retaining walls) should also consider deformation within the zone of influence.
- In the limit state design, movements need to be explicitly checked against allowable movements for the serviceability design case.

Table 23.1 Types of movement.

Design application	*Parameter*	*Typical movement*
Shallow foundations	Allowable bearing capacity	25 mm for building
Deep foundations	Shaft friction	10 mm for shaft friction to be mobilised
Retaining walls	Active and passive earth pressure Coefficient	0.1% H for K_a to be mobilised in dense sands 1% H for K_p to be mobilised in dense sands
Reinforced soil walls	Frictional and dilatancy to transfer load to soil reinforcement	25 to 50 mm for geogrids 50 to 100 mm for geotextiles
Pavements	Rut depth based on strain criterion related to number of repetitions	20 mm rut depths in paved major roads 100 mm rut depths in mine haul roads
Embankment	Self-weight settlement	0.5 to 1.0% Height of embankment
Drainage	Total settlement	Varies with cross-fall 100 mm to 500 mm differential to achieve required gradients. 0.5 m/s minimum self-cleaning velocity. Maximum gradient of 5% to avoid lining of drain or erosion effects
Pipes	Gradient	1.0% to 2.5% as too flat or steep and solids left behind leading to blocking of pipe

23.2 Foundation movements

- The immediate settlement is calculated using elastic theory.
- Consolidation settlements occur with time as water is expelled from the soil.
- Creep settlement (also called secondary compression) occurs as a change of structure occurs.
- Effects of compaction discussed in chapter 12.
- Compaction induced settlement is due to removal of air while consolidation is removal of water.

Table 23.2 Types of movements.

Principal soil types	*Type of movements*			
	Immediate	*Consolidation*	*Creep*	*Swell*
Rock	Yes	No	No	Some
Gravels	Yes	No	No	No
Sands	Yes	No	No	No
Silts	Yes	Minor	No	Minor
Clays	Yes	Yes	Yes	Yes
Organic	Yes	Minor	Yes + Degradation	Minor

- ○ Immediate and consolidation settlements are dependent on the applied load and the foundation size.
- ○ Self-weight settlement can also occur for fill constructed of the above materials. The settlement will depend on the material type, level of compaction and height of the fill.

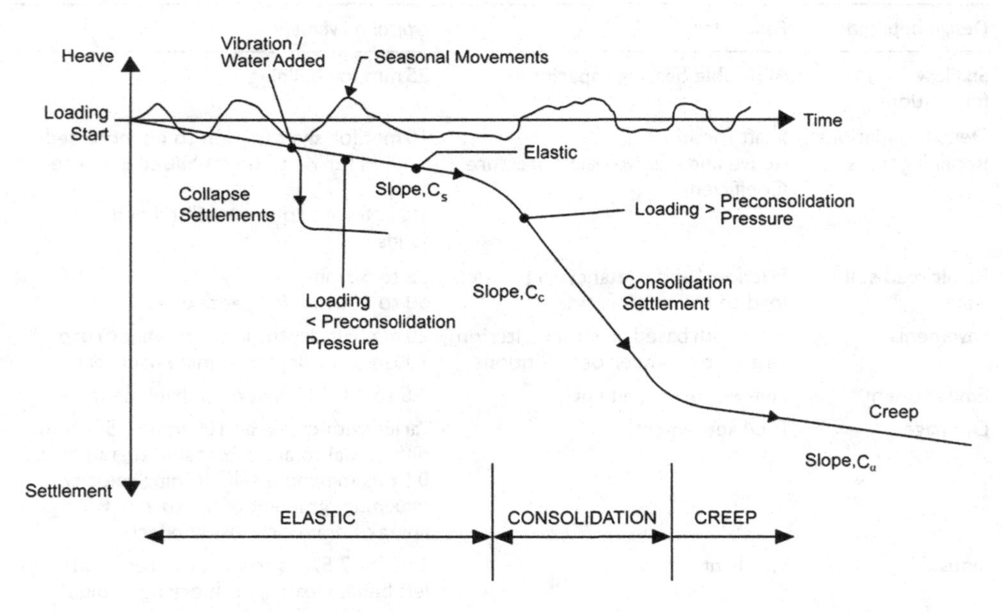

Figure 23.1 Foundation movements.

23.3 Immediate to total settlements

- The settlement estimates are usually based on the settlement parameters from the oedometer test.
- This is mainly for consolidation settlements, but may also be applied to elastic settlements for overconsolidated soils.
- For stiff elastic soils, a factor of safety of 2.5 is assumed.
- Secondary settlement is neglected in this table. Saturated soil is assumed.

Table 23.3 Immediate, consolidation and total settlement ratio estimates (after Burland et al., 1978).

Type of soil	*Immediate settlement, (Undrained)* ρ_u	*Consolidation settlement* ρ_c	*Total settlement* $\rho_T = \rho_u + \rho_c$	*Ratio* ρ_u/ρ_T
Soft yielding	$0.1\,\rho_{oed}$	ρ_{oed}	$1.1\,\rho_{oed}$	<10–15%
Stiff elastic	$0.6\rho_{oed}$	$0.4\rho_{oed}$	ρ_{oed}	33–67%

- $\rho_u/\rho_T \rightarrow 70\%$ for deep layers of overconsolidated clays.
- $\rho_u/\rho_T \rightarrow 25\%$ for decreasing thickness of layer and increasing non-homogeneity and anisotropy.

23.4 Consolidation settlements

- One-dimensional settlements $= \rho_{od} = \rho_{oed}$ from the odeometer test (refer chapter 11).
- Consolidation settlement $(\rho_c) = \mu\,\rho_{oed}$.
- μ = settlement coefficient based on Skempton's pore pressure coefficient and the loading geometry.
- The table shows a simplified version of this consideration.

Table 23.4 Correction factors based on Skempton and Bjerrum (Tomlinson, 1995).

Type of clay	*Description*	*Correction factor*
Very sensitive	Soft alluvial, estuarine and marine	1.0–1.2
Normally consolidated		0.7–1.0
Overconsolidated	London Clay, Weald, Oxford and Lias	0.5–0.7
Heavily overconsolidated	Glacial Till, Keuper Marl	0.2–0.5

23.5 Typical self-weight settlements

- The self-weight settlements occur for all placed fills – even if well compacted.
- The self-weight settlement of general fills is assumed to occur over 10 years, although refuse fills take over 30 years to stabilise.
- Depth of fill – H.
    - Additional compactive effort (say 95% to 100% of standard compactive effort) does not reduce the above self-weight settlement, although a higher modulus or strength may be achieved.

Table 23.5 Typical potential self-weight settlements (Goodger and Leach, 1990).

Compaction	*Material*	*Self-weight settlement*
Well compacted	Well graded sand and gravel	0.5% H
	Shale, chalk and rock fills: clay	0.5% H
	Well controlled domestic refuse placed in layers	10% H
	Mixed refuse	30% H
Medium compacted	Rock fill	1.0% H
Lightly compacted	Clay and chalk	1.5% H
	Clay placed in deep layers	1.0–2.0% H
Compacted by scrapers	Opencast backfill	0.6–0.8% H
Nominally compacted	Opencast backfill	1.2% H
Uncompacted	Sand	3.5% H
	Clay fill (Pumped)	12.0% H
Poorly compacted	Chalk	1.0% H

- ○ The self-weight is generally a mechanical process except for refuse where biological and chemical decomposition dominate the settlement process.

23.6 Limiting movements for structures

- The maximum allowable movement depends on the type of structure.

Table 23.6 Typical limiting settlements for structures.

Type of structure	*Maximum allowable vertical movement*	*Reference*
Isolated foundations on clays	65 mm	Skempton and
Isolated foundations on sands	40 mm	Macdonald (1955)
Rafts clays	65 to 100 mm	
Rafts on sands	40 to 65 mm	
Buildings with brick walls		Wahls, 1981
• L/H ≥ 2.5	75 mm	
• L/H ≤ 1.5	100 mm	
Buildings with brick walls, reinforced with reinforced concrete or reinforced brick	150 mm	
Framed structures	100 mm	
Solid reinforced concrete foundations of smokestacks, silos, towers	300 mm	
Bridges	50 mm	Bozozuk, 1978
At base of embankments on soft ground		
• Rail	100 mm	
• Road	200 mm	

- ○ Movement at the base of an embankment is not equivalent to movement at the running surface, which can be 10% or less of that movement.
- ○ Irrespective of the magnitude of the movements, often the angular distortion may dictate the acceptable movements. Cracks may become visible at values significantly below these values shown. These cracks may be aesthetic and can

affect the market value of the property although the function of the building may not be compromised.

23.7 Limiting angular distortion

- The angular distortion is the ratio of the differential settlement to the length.

Table 23.7 Limiting angular distortion (Wahls, 1981).

Category of potential damage	δ/L
Machinery sensitive to movement	1/750
Danger to frames with diagonals	1/600
Safe limit for no cracking of buildings	1/500
First cracking of panel walls Difficulties with overhead cranes	1/300
Tilting of high rigid building becomes visible	1/250
Considerable cracking of panel and brick walls	1/150
Danger of structural damage to general buildings	
Safe limit for flexible brick walls L/H > 4	

23.8 Relationship of damage to angular distortion and horizontal strain

- The damage is usually a combination of different stains.
- The relationship between horizontal strains, ε_h ($\times 10^{-3}$) and angular distortion ($\times 10^{-3}$) is shown in Boscardin and Cording (1989) for different types of construction and severity.

Table 23.8 Distortion factors (after Boscardin and Cording 1989).

Distortion factor	*Type of construction*	*Upper limit of*	
		Angular distortion ($\times 10^{-3}$)	*Horizontal strains*, ε_h ($\times 10^{-3}$)
Negligible	All	<1.6	0
Slight		<3.2	0
Moderate to severe		<6.6	0
Severe to very severe		≥6.6	0
Negligible	All	0	<0.7
Slight		0	<1.5
Moderate to severe		0	<3.0
Severe to very severe		0	≥3.0
Moderate to severe	Deep mines	0	3
		2	2.7
Moderate to severe	Shallow mines and tunnels, Braced cuts	2	2.7
		4.5	1.5
Moderate to severe	Building settlement	6.1	0.4
		6.6	0.0

23.9 Movements at soil nail walls

- The wall movements are required for the active and passive state to apply. The type of soil and its wall movement governs the displacement. This was tabled in Chapter 19.
- The displacement of the wall facing depends on the type of soil and the wall geometry.
- At the top of a wall, the horizontal displacement $(\delta_h) = \delta_v(L/H)$.

Table 23.9 Displacements of soil nail wall (Clouterre, 1991).

Movement	*Soil type*		
	Intermediate soils (rock)	*Sand*	*Clay*
Vertical displacement (δ_v)	H/1000	2H/1000	4H/1000
Distance from wall to zero movement	$0.8H(1 - \tan\eta)$	$0.8H(1 - \tan\eta)$	$0.8H(1 - \tan\eta)$

- High plasticity clays may produce greater movements.
- Batter angle of facing $= \eta$.

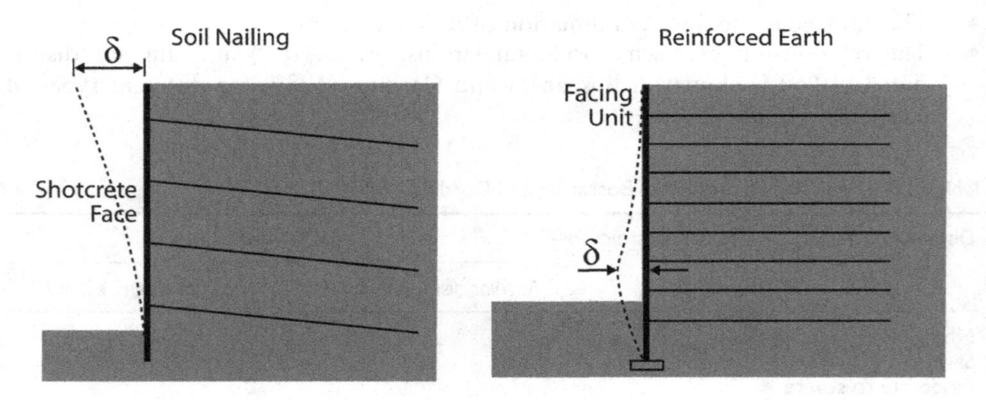

Figure 23.2 Comparison of movement between soil nailing and reinforced soil walls.

23.10 Tolerable strains for reinforced slopes and embankments

- The reinforcing elements must be stiff enough to mobilise reinforcement forces without excessive strains.
- The allowable long term reinforcement tension load $= T_{lim} \leq E_{secant} \times \varepsilon_{tol}$.
- Secant modulus of reinforcement $= E_{secant}$.
- Tolerable strain $= \varepsilon_{tol}$.

Table 23.10 Tolerable strains for reinforced slopes and embankments (Duncan and Wright, 2005).

Reinforced application	*Considerations*	*Tolerable strains, ε_{tol} (%)*
Reinforced soil walls		10
Reinforced slopes	Embankments on firm foundation	10
Reinforced embankments	On non-sensitive clay, moderate crest deformation tolerable	10
	On non-sensitive clay, moderate crest deformation not tolerable	5–6
	On highly sensitive clays	2–3

- Steel reinforcement is inextensible for all practical purposes, and reinforcement stiffness is not a governing criteria.

23.11 Movements in inclinometers

- The loading from the embankment results in a lateral movement.
- Table provides relative movement when failure likely.

Table 23.11 Relative movements below embankment.

Measurement	*Symbols/relationship*
Horizontal movement	δ_H
Vertical movement	δ_v
Inclinometer at side of embankment on soft clay	$\delta_H/\delta_v \sim 0.3$ (typical)
	~ 0.6 (initial trigger level)

- Horizontal ruptures occur on or before 200 mm at a rigid interface, e.g. concentrated movement just above inclinometer in rock.
- In any observational method, the short term condition applies. However the stability analysis used for design should not be applied directly as factors of safety and conservatism would have been used in the analysis. Small factors of safety with realistic parameters should be applied to assess observational trigger limits.

23.12 Acceptable movement in highway bridges

- The movement criteria for bridges stated below do not consider the type or size of bridge.

Table 23.12 Movement criteria for bridges (Barker et al., 1992, Moulton et al., 1978, Bozozuk, 1978).

Movement criteria	*Acceptable movement (mm)*	
	Vertical	*Horizontal*
Not harmful	<50	<25
Ride quality affected	60	
Harmful but tolerable	100–50	50–25
Usually intolerable	>100	>50

23.13 Acceptable angular distortion for highway bridges

- Angular distortion (A) = δ/S:
  - δ – Differential settlement between foundations.
  - S – Span length.

Table 23.13 Angular distortion criteria for bridges (Barker et al., 1992, Moulton et al., 1978).

Value of angular distortion	*Continuous span*	*Single span*
0.000 to 0.001	100%	100%
0.001 to 0.003	97%	100%
0.003 to 0.005	92%	100%
0.005 to 0.008	85%	95%

  - $A \leq 0.004$ is acceptable for continuous span bridges.
  - $A \leq 0.008$ is acceptable for single span bridges.

23.14 Serviceability and ultimate piles design

- Piles are based on an ultimate or serviceability load.
- Likely settlement shown below – see table 21.16.

Table 23.14 Typical displacements based on design load applied.

Criteria	*Typical pile displacement at loads*
Serviceability	1% diameter
Ultimate	10% diameter

23.15 Tolerable displacement for slopes and walls

- The literature is generally vague on tolerable movements.

Table 23.15 Movements just before a slide (data from Skempton and Hutchinson, 1969).

Type of system	*Total movement (cm)*
Small to large walls	20–40
Medium to large landslides	40–130

23.16 Observed settlements behind excavations

- The settlements behind a wall depend on the type of soil, and distance from the excavation face.
- The table applies to soldier piles or braced sheet piles with cross bracing or tie backs.

Table 23.16 Observed settlements behind excavations for various soils (Peck, 1969, O'Rouke et al., 1976).

Type of soil	*Settlement/Maximum depth of excavation (%)*	*Distance from excavation/Maximum depth of excavation (%)*
Medium to dense sands with interbedded stiff clays with average to good workmanship	0.3 0.1 0.0	0 1.2 2.0
Sand and soft to hard clay with average workmanship	1 0.5 0.0	0 0.7 2.5
Very soft to soft clay to a limited depth with construction difficulties or significant depth below bottom of excavation but $N_b < N_{cb}$	2 1 0.5 0.0	0 1.2 2.3 4.0
Very soft to soft clay to a significant depth below bottom of excavation and $N_b > N_{cb}$	Above soft clay scenario is lower bound envelope, i.e. greater settlements expected	

- Data was obtained from excavations using standard soldier piles or sheet piles braced with cross-bracing or tie backs.

23.17 Settlements adjacent to open cuts for various support systems

- These are empirically derived values for horizontal movements at the crest of an excavation.
- This may be conservative for residual soils, and with recent advances in construction procedures.

Table 23.17 Horizontal movements for varying support systems (Peck, 1969).

Type of wall		*Horizontal movement as % of excavation height*
Externally stabilised	Cantilever retaining walls	0.5%
	Propped retaining walls	0.2–0.5%
	Tied back walls	0.05 to 0.15%
Internally stabilised	Soil nails	0.1–0.3%

23.18 Tolerable displacement in seismic slope stability analysis

- When seismic factors of safety <1.15 then this initial screening should be replaced by a displacement analysis.

Table 23.18 Tolerable displacement (after Duncan and Wright, 2005).

Slope type	*Tolerable displacement*
Typical slopes and dams	1.0 m
Landfill covers	0.30 m
Landfill base	0.15 m

23.19 Seismic performance criteria

- A performance based design (PBD) may also be adopted for seismic design to supplement the factor of safety approach.
- Nikolaou (2011) provides a proposed PBD framework for seismic design.
- Practically no damage was expected for the OBE, no severe damage for the MDE and no catastrophic failure expected for the MCE.
- In terms of downtime:- interruption expected for the OBE; some interruption for the MDE and; long term interruptions were expected for the MCE.

Table 23.19 Seismic performance criteria (Nikolaou, 2011).

Geocomponent	*OBE – expected event within design life*	*MDE – primary design event*	*MCE – largest event considered*
Foundations	No permanent deformation	Extensive cracking <6 mm	Extensive damage observable offset
	Negligible settlement	Localised damage	Replacement
Critical slope	F.S ≥ 1.2	F.S ~1.0	Shallow failures
	δ = 0 mm	δ ≤ 25 mm	δ ≤ 100 mm
Ordinary slope	δ ≤ 25 mm	δ ≤ 100 mm	Shallow failures
			δ ≤ 900 mm

- Operational Basis Earthquake (OBE)
- Maximum Design Earthquake (MDE)
- Maximum Credible Earthquake (MCE)
- Nikolaou (2011) used return periods of 500, 2,500 and 5,000 years for the case study examined.

23.20 Rock displacement

- A probability of failure of less than 0.5% could be accepted for unmonitored permanent urban slopes with free access (Skipp, 1992).

Table 23.20 Permanent rock displacement for rock slope analysis (Skipp, 1992).

Failure category	*Annual probability*	*Permanent displacement (%)*
Catastrophic	0.0001	3
Major	0.0005	1.5
Moderate	0,001	0.3
Minor	0.005	0.15

23.21 Allowable rut depths

- The allowable rut depth depends on the type of road.
- The allowable rut depth is a serviceability criterion and does not correspond to actual failure of a base course or subgrade material.

Table 23.21 Typical allowable rut depths (QMRD, 1981; AASHTO, 1993).

Type of road	*Paving*	*Allowable rut depth*
Haul type	Unpaved	100 mm
Access	Unpaved	75 mm
Low volume	Unpaved	30 mm to 70 mm
	Paved	20 mm to 50 mm
Major roads	Paved	10 mm to 30 mm

23.22 Levels of rutting for various road functions

- The rutting criteria are based on the design speed of the road to ensure the safety of road users.

Table 23.22 Indicative investigation levels of rutting (Austroads, 2004).

Road Function	*Speed*	*Percentage or road length with rut depth exceeding 20 mm*
Freeways and other high class facilities		10%
Highways and main roads	100 km/h	10%
Highways and main roads	≤80 km/h	20%
Other Local Roads (sealed)	60 km/h	30%

 - Rut measured with a 1.2 metre straight edge.

23.23 Free surface movements for light buildings

- Australian Standards (AS2870) is based on a free surface movement (y_s) calculated from the shrink – swell index test (I_{ss}), the depth of active and cracked zone and the soil suction.
- The free surface movement is used to classify the site reactivity i.e. its reaction to moisture changes.
- This applies for residential buildings and lightly loaded foundations.

Table 23.23 Free surface movements for light buildings (AS 2870, 2011).

Class	*Site classification*	*Surface movement (y_s, mm)*
A	Competent rock	
S	Slight	<20
M	Moderate	20 to 40
H1	High	40 to 60
H2		60 to 75
E	Extreme	>75
P	Problem	

 - Competent rock excludes extremely weathered rocks, mudstones, and clay shales.

- Problem sites includes
  - controlled fill other than sand >0.4 m
  - uncontrolled fill in sand >0.8 m
  - uncontrolled fill in other than sand >0.4 m

23.24 Free surface movements for road pavements

- The free surface movement can be used to classify the road subgrade movement potential.
- Calculations should include the depth of pavement based on the strength criteria design. Should pavements be excessive, a non-reactive subgrade layer (capping layer) is required below the pavement to reduce the reactive movement to an acceptable value.

Table 23.24 Free surface movements for road subgrades (Look, 1992).

Road performance	*Surface movement (y_s, mm)*	
	Flexible pavements	*Rigid pavements*
Acceptable	≤10	≤5
Marginal	10 to 20	5 to 15
Unacceptable	≥20	≥15

- Higher movements would be acceptable at the base of the embankment e.g. 100 mm for a high embankment on soft ground. That movement does not necessarily translate to the surface area. This should be checked based on the embankment height.

23.25 Allowable strains for roadways

- The allowable rutting is based on the number of cycles applied to the pavement layers.
- The design is based on ensuring each layer has not exceeded its allowable strain.
- Limiting strains shown separately.

Table 23.25a Typical allowable strains for pavement layers (Austroads, 2004).

Material	*Allowable strains*
Asphalt	1000 microstrain
Base at 0 to 10,000 cycles	2500 microstrain
Sub Base at 0 to 10,000 cycles	2000 microstrain
Base at 10,000 to 20,000 cycles	3500 microstrain
Sub Base at 10,000 to 20,000 cycles	4000 microstrain
Base at 0 to 20,000 to 30,000 cycles	5000 microstrain
Sub Base at 0 to 20,000 to 30,000 cycles	7000 microstrain

Table 23.25b Typical limiting strains (Austroads, 2007).

Material	*Typical limiting strains*
Asphalt	200 microstrain
Lean mix concrete	450 microstrain
Subgrade	500 to 3000 microstrain ≤1500 microstrain (major roads)

23.26 Limiting strains for mine haul roads

- Mine haul roads have a significantly different limiting strain criteria.
- Below has been developed for semi-permanent and permanent haul roads. Temporary roads would have a different criteria.

Table 23.26 Limiting strains for mine haul roads (Tannant and Regensburg, 2001).

Material	*Typical limiting strain*	*Load repetitions*
Top of surface	2000 microstrain 1500 microstrain ≥2000 microstrain	800,000 loaded trucks 2.5×10^6 permanent haul road 500,000 semi-permanent
Top of subgrade	1500 microstrain and generally <2500 microstrain	For permanent mine haul roads

- May also be used for heavily loaded areas at container ports.
- 800,000 load repetitions typical for a permanent haul road with 400,000 loaded trucks over its lifespan.

23.27 Tolerable deflection for roads

- Typical values to determine whether maintenance is required for flexible pavements and based on deflection testing is provided in table.

Table 23.27 Elastic deflection (85th percentile) test results (Lay, 1990).

Type of Road	*Traffic*	*Deflection (mm)*
Residential	30 commercial vehicles/day	2.0
Major road	10^5 ESA	1.5
	10^6 ESA	0.9
	10^7 ESA	0.65

- ESA – equivalent standard axles in one direction.
- Subtract 0.1 mm for AC surface layers over 50 mm thickness.

23.28 Tolerable deflection for roads based on CBR

- The CBR values influence the deflection. If the CBR is known the corresponding deflections required for overlay design is determined.
- Tolerable deflection for the normal design standard (20 mm rut) for both Benkelman Beam and FWD with 40 kN loading is shown.

Table 23.28 Tolerable deflection (QTMR, 2012).

Design traffic (ESA)	*Deflection (mm)*				
CBR %	*3*	*5*	*7*	*10*	*20*
10^5 ESA	1.2	1.0	0.9	0.8	0.6
10^6 ESA	1.1	0.9	0.8	0.7	0.55
10^7 ESA	0.9	0.8	0.7	0.65	0.55

  - The outer wheel paths are susceptible to environmental (moisture) changes. Deflections factored if tested at end of dry period (factor of 1) or end of wet period (factor of 1.1 to 1.2 generally). The higher value applies to high rainfall environments. For silty subgrades a factor of 1.2 to 1.6 applies for low to high rainfall, respectively.

23.29 Tolerable deflection for proof rolling

- Proof rolling is used to identify and correct problem areas prior to the placement and compaction of stronger, more expensive materials. The overlying materials will not perform adequately if the underlying subgrade does not provide the required support.
- Proof rolling is not appropriate in non-cohesive soils and most appropriate where there are 20% of fines present in the material. Any material with 5% or less fines would not be appropriate for proof load testing.
- The deflection observed is dependent on equipment and ground profile. Underlying material affects deflection. No visible limit applies to a high CBR material

Table 23.29 Deflection criteria for proof rolling.

Equipment	*Mass*	*Load intensity kPa*	*Tolerable Deflection (mm) for CBR %*			
			3%	*5%*	*10%*	*15%*
Water truck	8t rear axle	550	40	25	10	No visible deflection
Static roller 12t	12t	350	25	15	5	

  - No visible limit is less than or equal to 2 mm.
  - A full water truck typically has a gross weight of 12t with 2/3 of load (8t) on the rear axle.
  - A typical roller is 6t/m width of wheel.

23.30 Peak particle velocity

- The level of vibration produced depends mainly on the construction method used and the seismic characteristics of the underlying strata. Key parameters are:
  - Peak particle velocity (mm/s), which is dependent on:
    - Source and type of the vibration (Energy Level),
    - Distance from the source, and
    - Ground conditions.
- The acceptable level of vibration is dependent on the type of structure at risk and its distance from the source or the sensitivity of the local population to vibrations.
- Acceptable distances for a piling operation with a 48 kJ drop hammer would then vary from 25 m to 5 m for an architectural significant building to a buried service (see following table).

Table 23.30 Peak particle velocity (Eurocode, 1993).

Type of property	*Peak particle velocity (mm/s) for vibration type*	
	Continuous	*Transient*
Ruins, buildings of architectural merit	2	4
Residential	5	10
Light commercial	10	20
Heavy Industrial	15	30
Buried Services	25	40

23.31 Vibration from typical construction operations

- Indicative vibration levels with distance from various vibration sources are provided by New, 1986.

Table 23.31 Vibration level with distance from construction sources (New, 1986).

Construction Activity	*Distance with peak particle velocity (mm/s)*				
	2 mm/s	*5 mm/s*	*10 mm/s*	*20 mm/s*	*40 mm/s*
Plate compactor	–	–	6 m	2 m	–
Heavy lorry on poor road surface	2 m	–	–	–	–
2.4 m dia tunnelling machine in soil	3 m	2 m	–	–	–
2.4 m dia tunnelling machine in rock	10 m	6 m	4 m	–	–
48 kJ drop hammer on clay	40 m	20 m	15 m	7 m	5 m
Explosive demolition of 14 storey block	>100 m	60 m	50 m	40 m	25 m

23.32 Perception levels of vibration

- The nuisance consideration is much lower than any of the code vibration levels.
- This depends on the type of ground and frequency level.

Table 23.32 Perception levels of peak particle velocity (Sarsby, 2000).

Perception Level	*Peak particle velocity (mm/s)*			
	5–50 Hz	*Silt (5–10 Hz)*		*Clay (20–40 Hz)*
Imperceptible	<1	<0.3		<0.4
		0.3–1.0	just perceptible	0.4–1.0
Perceptible	1–5	1–3	clearly perceptible	1–3
Distinctly perceptible	5–20	3–10	annoying	3–5
Strongly perceptible	20–50	10–40	unpleasant	5–10
Severe	≥50	≥40	painful	>10

Chapter 24

Appendix – loading

24.1 Characteristic values of bulk solids

- The physical properties of bulk solids are often required in design calculations.

Table 24.1 Characteristic values of bulk solids (AS 3774 – 1996).

Type of bulk solid	*Unit weight (kN/m³)*	*Effective angle of internal friction (°)*
Alumina	10.0–12.0	25–40
Barley	7.0–8.5	26–33
Cement	13,0–16.0	40–50
Coal (black)	8.5–11.0	40–60
Coal (brown)	7.0–9.0	45–65
Flour (wheat)	6.5–7.5	23–30
Fly ash	8.0–11.5	30–35
Iron ore, pellets	19.0–22.0	35–45
Hydrated lime	6.0–8.0	35–45
Limestone powder	11.0–13.0	40–60
Maize	7.0–8.5	28–33
Soya beans	7.0–8.0	25–32
Sugar	8.0–10.0	33–38
Wheat	7.5–9.0	26–32

24.2 Surcharge pressures

- Uniform surcharge loads are applied in foundation and slope stability analysis.
- These would be factored fir number of lanes/tracks.

Table 24.2 Surcharge loads.

Loading source	*Equivalent uniformly distributed pressure*
Railways	40 kPa
Major roads and highways	20 kPa (permanent); 10 kPa (temporary)
Minor roads and ramps	10 kPa
Footpaths	5 kPa
Buildings	10 kPa per storey

24.3 Live load on sloping backfill

- Minimum live load depends on the class off the structure and the backslope.

Table 24.3 Live load for backfill slope (Vertical: Horizontal), (AS4678, 2002).

Classification	*Steeper than 1V: 4H*	*1V: 4H or flatter*
A – High consequence	2.5 kPa	5 kPa
B – Medium consequence H ≥ 1.5 m	2.5 kPa	5 kPa
C – Low consequence	1.5 kPa	2.5 kPa

24.4 Construction loads

- Wheel vehicles provide the greatest ground pressure.
- Tracked vehicles may be heavier, but provide a reduced pressure due to track width. This is useful in trafficking low strength areas.

Table 24.4 Typical wheel loads from construction traffic.

Equipment	*Size*	*Approximate mass (tonnes)*		*Tyre inflation pressure (kPa)*
		Fully laden	*Per wheel*	
Scrapers	Small	25	6	200–400
	Large	110	28	500–600
Dump trucks	Small	25	4	350–700
	Large	80	20	600–800

24.5 Ground bearing pressure of construction equipment

- The table above is simplified below with some additional equipment shown.

Table 24.5 Ground bearing pressure.

Type of equipment	*Typical bearing pressure (kPa)*
Bulldozer	
– Small	50
– Large	100
Wheeled tractor	180
Loaded Scraper	
– Small	150
– Medium	200
– Large	300
Heavy compaction equipment	Contact pressure
– Smooth drum roller	300–400
– Pneumatic with 4 to 6 tyres	400–700
– Sheepsfoot roller	1400–7000
– Grid roller	1400–8000

24.6 Vertical stress changes

- Soil stresses decrease with increased distance from the loading.
- The shape, type of the foundation, and the layering of the underlying material affects the stress distribution.
- The table below is for a uniform elastic material under a uniformly loaded flexible footing. These Boussinesq solutions are for a uniform pressure in an isotropic homogeneous semi-infinite material.

Table 24.6 Vertical stress changes (originally from Janbu, Bjerium and Kjaernsli, 1956, but here from graphs in Simons and Menzies, 1977).

Depth below base of footing (z) in terms of width (B)	*Footing shape in terms of length (L)*	*Change in stress Δp in terms of applied stress q*
z/B = 0.5	Square (L = B)	Δp/q = 0.70
	L = 2B	Δp/q = 0.82
	L = 5B	Δp/q = 0.82
	L = 10B	Δp/q = 0.82
	L = ∞	Δp/q = 0.82
z/B = 1.0	Square (L = B)	Δp/q = 0.33
	L = 2B	Δp/q = 0.49
	L = 5B	Δp/q = 0.56
	L = 10B	Δp/q = 0.56
	L = ∞	Δp/q = 0.56
z/B = 2.0	Square (L = B)	Δp/q = 0.12
	L = 2B	Δp/q = 0.20
	L = 5B	Δp/q = 0.28
	L = 10B	Δp/q = 0.30
	L = ∞	Δp/q = 0.30
z/B = 3.0	Square (L = B)	Δp/q = 0.06
	L = 2B	Δp/q = 0.11
	L = 5B	Δp/q = 0.17
	L = 10B	Δp/q = 0.20
	L = ∞	Δp/q = 0.22
z/B = 5.0	Square (L = B)	Δp/q = 0.02
	L = 2B	Δp/q = 0.04
	L = 5B	Δp/q = 0.08
	L = 10B	Δp/q = 0.11
	L = ∞	Δp/q = 0.14

- There is a 10% change in normal stress at approximately 2B (square foundation). Hence the guideline for the required depth of investigation (Refer Chapter 1).
- For a strip footing the 10% change in stress occurs at approximately 6B.
- For layered systems and / or non-uniform loading, the above stress distribution does not apply. Poulos and Davis (1974) is the standard reference for these alternative solutions.

Chapter 25

Appendix – conversions

25.1 Length, area and volume

- Conversion factors to metric units for space related parameters.

Table 25.1 Conversions for length, area and volume.

Convert	*Imperial to SI (metric) units*		*Multiply by*
Length	inch	mm	25.4
	inch	cm	2.54
	foot	m (metres)	0.3048
	yard	m	0.914
	mile	km	1.609
Area	in^2	mm^2	645.16
	in^2	cm^2	6.452
	ft^2	m^2	0.0929
	Acre	m^2	4047
	Acre	Ha (Hectares)	0.4047
Volume	in^3	cm^3	16.39
	ft^3	m^3	0.02832
	gallon = 0.1605 ft^3	litres = 1000 cm^3	4.546
	US gal	cm^3	3785
Section modulus	in^3	mm^3	0.16837×10^5
	in^3	m^3	0.16837×10^{-4}
Moment of inertia	in^4	mm^4	0.4162×10^6
	in^4	m^4	0.4162×10^{-6}

25.2 Mass, density, force and pressure

- Conversion factors to metric units for mass related parameters.

Table 25.2 Conversions for mass, density, force and pressure.

Convert	*Imperial to SI (metric) units*		*Multiply by*
Mass	lb = 16 oz	kg	0.4536
	ton = 2240 lb	kg	1016
		1000 kg = 1 ton (metric)	

(Continued)

Table 25.2 (Continued)

Convert	*Imperial to SI (metric) units*		*Multiply by*
Density	lb/in^3	g/cm^3	27.68
	lb/ft^3	kg/m^3	16.02
Unit weight	lb/in^3	kN/m^3	271.43
	lb/ft^3	kN/m^3	0.1572
Force	lbf	N	4.448
	tonf	kN	9.964
	kip	kN	4.448
Pressure, stress	lbf/in^2	kN/m^2 (kPa)	6.895
	lbf/ft^2	N/m^2	47.88
	$kips/ft^2$	kN/m^2	47.88
	tsf	kN/m^2	95.76
		100 kPa = 1 bar	

25.3 Permeability and consolidation

- Conversion factors to metric units for flow related parameters.

Table 25.3 Conversions for mass, density, force and pressure.

Convert	*Imperial to SI (metric) units*		*Multiply by*
Velocity	ft/s	m/s	0.305
	mile/hr	km/h	1.609
	knot = 1.689 ft/s	m/s = 1.853 km/h	0.514
Hydraulic conductivity	in/sec	cm/sec	2.54
	ft/sec	m/sec	0.305
	ft/min	m/min	0.305
Flow rate	ft^3/sec (1 cu sec)	m^3/sec	0.02832
	gal/min	m^3/sec	7.577×10^{-5}
Coefficient of consolidation	in^2/sec	cm^2/sec	6.452
	in^2/sec	$m^2/year$	20,346
	ft^2/sec	cm^2/sec	929.03

Chapter 26

References

- References have been tabulated as essential background (General) to understanding the background of the data tables provided.
- References are then specific to whether investigations and assessment, or analysis and design. The references in the latter may not be repeated if already in investigations and assessments.

26.1 General – most used

- Barnes G.E. (2000). "Soil Mechanics – Principles and Practice" 2nd Edition Macmillan Press.
- Barker R.M., Duncan J.M., Rojiani K.N., Ooi P.S.K., Tan C.K. and Kim S.G. (1991), "Manuals for the design of bridge foundations" National Cooperative Highway Research Program Report No. 343, Transportation Research Board, Washington.
- Bowles J.E. (1996), "Foundation Analysis and Design" 5th Edition McGraw-Hill.
- Carter M. (1983), "Geotechnical Engineering Handbook" Pentech Press.
- Das B.M. (1999), "Principles of Foundation Engineering" 4th Edition, Brooks/Cole Publishing Company.
- Hausmann M.R. (1990), "Engineering Principles of Ground Modification" McGraw-Hill Publishers.
- Mayne P., Christopher B. and Defomg J. (2001), "Manual on subsurface Investigations" National Highway Institute, Publication No. FHWA NH1-01-031, Federal Highway Administration, Washington, DC.
- Sowers G.F. (1979), "Introductory Soil Mechanics and Foundations" 4th Edition, Macmillan Publishing Co Inc., New York.
- Tomlinson M.J. (1995), "Foundation Design and Construction" 6th Edition, Longman.
- Waltham A.C. (1994), "Foundations of Engineering Geology" Blackie Academic & Professional.

26.2 Geotechnical investigations and assessment

- AASHTO (1993), "Guide for the Design of Pavement Structures" American Association of State and Highway Officials.
- Adams, B., Look B. and Gallage, C. (2010), Assessment of pile driving refusal using the standard penetration test, Proceedings of the 11th International

Association of Engineering Geologists Congress, Auckland, New Zealand, September, CD-ROM, pp. 2641–2648, Taylor and Francis publishers.

- AGS Anon, (2000), "New focus on landslide risk management" The earthmover & Civil Contractor, December feature article, pp. 53–54.
- ASCE (1993), "Bearing Capacity of Soils" Technical Engineering and Design Guides. US Corps of Engineers No. 7.
- Australian Standard (AS 1726 – 1993), "Geotechnical Site Investigations" Standards Australia.
- Australian Standard (AS 1289 – various years), "Methods of testing soils for engineering purposes", Standards Australia.
- Austroads (1992), "Section 3: Foundations" Bridge Design Code.
- Austroads (2004), "Pavement Design – A guide to the structural design of road pavements" Australian Road Research Board.
- Austroads (2004), "Pavement Rehabilitation Manual" Australian Road Research Board.
- Barton N. (1983), "Application of Q-System and index test to estimate Shear Strength and Deformability of Rock Masses", Proceedings International Symposium of Engineering Geology and Underground Construction, Portugal.
- Beaumont D. and Thomas J. (2007), "Driving tubular steel piles into weak rock – West Australian experience", 10th Australia New Zealand conference in Geomechanics, Brisbane, Australia, pp. 430–435.
- Bell F.G. (1992), "Description and Classification of rock masses" Chapter 3 in Engineering in Rock Masses edited by FG Bell, Butterworth Heinemann.
- Bell F.G. (1992), "Properties and Behaviour of rocks and rock masses" Chapter 1 in Engineering in Rock Masses edited by FG Bell, Butterworth Heinemann.
- Berkman, D.A. (2001), "Field Geologists' Manual" 4th Edition, Monograph No. 9, The Australasian Institute of Mining and Metallurgy, Victoria.
- Bienawski Z.T. (1984), "Rock Mechanics Design in mining and tunnelling" A.A. Balkema, Rotterdam.
- Bienawski Z.T. (1993), "Classification of rock masses for engineering: The RMR system and future trends, In Comprehensive Rock Engineering, (J.A. Hudson Ed.), Volume 3, pp. 553–574, Pergamon Press.
- Bishop A.W. and Bjerrum L. (1960), "The relevance of the triaxial test to the solution of stability problems" ASCE Conference on Shear Strength of Cohesive Soils, Boulder, pp. 43–501.
- Bjerrum, L. (1972), "Embankments on soft ground" Proceedings of ASCE Speciality Conference on Performance of Earth and Earth Supported Structure, Purdue University, pp. 1–54.
- Braun H.M.M. and Kruijne R., (1994), "Soil Conditions" Drainage Principles and Applications (Ed. H.P. Ritzema), International Institute for Land Reclamation and Management Publication, The Netherlands, 2nd Edition.
- Brown E.T., Richards L.R. and Barr R.V. (1977), "Shear strength characteristics of the Delabole Slates", Proceedings of the Conference in Rock Engineering, Newcastle University, Volume 1, pp. 33–51.
- British Standards 8002 (1994), "Code of Practice for Earth Retaining Structures" British Standards Institution.
- Carter M. and Bentley S.P. (1991), "Correlation of Soil Properties" Pentech Press.

- Cedergren H.R. (1989), "Seepage, Drainage and flow nets" 3rd Edition, John Wiley & sons.
- Chandler, R.J. (1988), "The in-situ measurement of the undrained shear strength of clays using the field vane" Vane Shear Strength testing in soils: Field and Laboratory studies, ASTM STP 10154, American Society of Testing and Methods.
- Chandler R.J. and Skempton A.W. (1974), "The design of permanent slopes in stiff fissured clays", Geotechnique 24, pp. 457–464.
- Clayton C.R.I. (1995). "The Standard Penetration Test (SPT): Methods and Use" CIRIA Report 143.
- Concrete Institute (1999), "Industrial Floors and Pavements Guidelines".
- Croney D. and Croney P. (1991), "The design and performance of road pavements" 2nd Edition McGraw Hill.
- Deere D.U., Hendron A.J., Patton F.D. and Cording E.J. (1967), "Design of Surface and Near Surface Construction in Rock", Proceedings 8th U.S. Symposium in Rock Mechanics on Failure and Breakage of Rock (Ed. C. Fairhurst), New York, pp. 237–302.
- Deere D.U. and Miller R.P. (1966), "Engineering Classification and Index Properties for Intact Rock" Technical Report AFWL-TR-65-115, Air Force Weapons Laboratory, New Mexico.
- Dietrich W.E., Bellugi D. and Real de Asua (2001), "Validation of the shallow landslide model, SHALSTAB for forest management, Land use and watersheds: – influence of hydrology and geomorphology in urban and forest areas, Editors M.S. Wignosta and S.J. Burges, pp. 195–127.
- Duncan J.M. and Wright S.G. (2005), "Soil Strength and slope Stability", Wiley Publishers.
- Farrar D.M. and Daley P. (1975), "The operation of earth moving plant on wet fill", Transport and Road Research Laboratory Report 688.
- Fugro Ltd. (1996), "Cone Penetration Test: Simplified Description of the Use and Design Methods for CPTs in Ground Engineering".
- Gay D.A. and Lytton R.L. (1992), "El Paso moisture barrier study", Texas A & M University Report for the Texas Department of Transport.
- Geotechnical Engineering Office (1987), "Geoguide 2: Guide to Site Investigation" Hong Kong.
- Geotechnical Engineering Office (1988), "Geoguide 3 – Guide to rock and soil descriptions", Hong Kong Government.
- Gordon J.E. (1979), "Structures or why things don't fall down" Penguin Books.
- Harr M.E. (1996), "Reliability Based Design in Civil Engineering" Dover publications, New York.
- Heymann G. (1988), "The stiffness of soils and weak rocks at very small strains" PhD Thesis, Department of Civil Engineering, University of Surrey.
- Hilf J.W. (1991), "Compacted Fill" Foundation Engineering Handbook, 2nd edition, Edited by Hsai-Yang Fang, Chapman and Hall, pp. 249–316.
- Hillel D. (1972), "Soil and Water: Physical Principles and Processes" Academic Press.
- Hoek E. and Bray J.W. (1981), "Rock Slope Engineering" 3rd Edition, Institution of Mining and Metallurgy, London.

- Holtz R.D. and Kovacs W.D. (1981), "An introduction to geotechnical engineering" Prentice Hall.
- Hunt R.E. (2005), "Geotechnical Engineering Investigation Handbook" 2nd edition, Taylor and Francis.
- Jaimolkowski M., Lancellotta R., Pasqualini E., Marchetti S. and Nova R. (1979), "Design Parameters for soft clays" General Report, Proceedings 7th European Conference on Sol Mechanics and Foundation Engineering, No. 5, pp. 27–57.
- Kay, J.N. (1993), "Probabilistic Design of Foundations and Earth Structures" Probabilistic Methods in Geotechnical Engineering, Li & Lo (eds), Balkema, pp. 49–62.
- Kulhawy F.H. (1992), "On the evaluation of static soil properties", Proceedings of a Speciality ASCE Conference on Stability and Performance of Slopes and Embankments – II, Volume 1, Berkley, California, pp. 95–115.
- Kulhawy F.H. and Goodman R.E. (1987), "Foundations in Rock" Ground Engineering Reference Book (Ed. F.G. Bell), Butterworths, London, pp. 55/1-13.
- Kulhawy F.H. and Mayne P.W. (1990), "Manual on estimating soil properties for foundation design" Report EL – 6800, Electric Power Research Institute, California.
- Lacey D., Look B. and Williams D. (2013), "Assessment of the zone of influence and relationship between PLT and LFWD tests" 9th ANZ Young Geotechnical Professionals Conference, Melbourne, Australia.
- Ladd C.C., Foote R., Ishihara K., Schlosser F. and Poulos H.G. (1977), "Stress – deformation and strength characteristics", State of the art report, Proceedings of the 9th International Conference on Soil Mechanics and Foundation Engineering, Volume 2, pp. 421–494.
- Larisch M. (2013), "Bored piles in rock and other techniques by piling contractors", Presentation notes.
- Lee I.K., White, W. and Ingles O.G. (1983), "Geotechnical Engineering" Pitman Publishers.
- Look B.G. (1997), "The Standard Penetration Test Procedure in ~~Soil~~ Rock" Australian Geomechanics Journal. No. 32, December pp. 66–68.
- Look B.G. (2004), "Effect of Variability and Disturbance in the measurement of Undrained Shear Strength" 9th Australia New Zealand Conference in Geomechanics, Auckland, New Zealand, Vol. 1, pp. 302–308.
- Look B.G. and Griffiths S.G. (2004), "Characterization of rock strengths in south east Queensland" 9th Australia New Zealand conference in Geomechanics, Auckland, New Zealand, pp. 187 to 194.
- Look B.G. and Griffiths S.G. (2004), "Rock strength properties in south east Queensland" 9th Australia New Zealand conference in Geomechanics, Auckland, New Zealand, pp. 196 to 203.
- Look B.G. (2005), "Equilibrium Moisture Content of volumetrically active clay earthworks in Queensland" Australian Geomechanics Journal, Vol. 40, No. 3, pp. 55–66.
- Look B.G. (2009), "Spatial and statistical distribution models using the CBR tests" Australian Geomechanics Journal, Vol. 44, No. 1, pp. 37–48.

- Look B.G. and Wijeyakulasuriya V. (2009), "The statistical modelling for reliability assessment of rock strength" 17th International Conference on Soil Mechanics and Foundation Engineering, Alexandria, Egypt, Vol. 1, pp. 60–63.
- Marchetti S. (1980), "In situ tests by the flat Dilatometer" Journal of Geotechnical Engineering, ASCE, Vol. 106, GT3, pp. 299–321.
- Marchetti S. (1997), "The Flat Dilatometer: Design Applications" Proceedings of the 3rd International Geotechnical Engineering Conference, Cairo, pp. 421–448.
- Marsh A.H. (1999), "Divided Loyalties" January, Ground Engineering, p. 9.
- Mayne P.W, Coop M.R., Springman S.M., Huang A. and Zornberg J.G. (2009), "Geomaterial Behaviour and Testing – State of the art lecture" Proceedings of the 17th International Conference on Soil Mechanics and Geotechnical Engineering, Alexandria, Egypt, Ed. M. Hanza, M. Shahien and Y. El-Mossallamy, Volume 4, pp. 2777–2872.
- Meigh A.C. (1987), "Cone Penetration Testing: Methods and Interpretation" CIRIA Ground Engineering Report: In-situ Testing.
- Mellish D., Lacey, D., Look B. and Gallage C. (2014), "Testing, Spartial and Temporal varibility in a residual soil profile and its effect on the design strength", Unpublished to date.
- MTRD Report No. 2-2 (1994), "Precision of soil compaction tests: Maximum dry density and optimum moisture content" Department of Road Transport, South Australia.
- NAVFAC (1986), "Soil Mechanics" Design Manual 7.01, Naval Facilities Engineering Command, Virginia.
- NAVFAC (1986), "Foundations and Earth Structures Design Manual" Design Manual DM 7.02, Naval Facilities Engineering Command, Virginia.
- Patton F.D. (1966), "Multiple Modes of Shear Failure in Rock" Proceedings 1st Congress of the International Society for Rock Mechanics, Lisbon, Vol. 1, pp. 509–513.
- Peck R.B., Hanson W.E. and Thornburn T.H. (1974), "Foundation Engineering" 2nd Edition, John Wiley and Sons, New York.
- Pezo, R.F. and Hudson W.R., "Prediction Models of Resilient Modulus for Nongranular Materials," Geotechnical Testing Journal, Vol. 17, No. 3, 1994, pp. 349–355.
- Pettifer G.S. and Fookes P.G. (1994), A revision of the graphical method for assessing the excavatability of rock. Quarterly Journal of Engineering Geology, 27: pp. 145–164.
- Phoon K. and Kulhawy, F.H. (1999), "Characterization of geotechnical variability" Canadian Geotechnical Journal, Volume 36, pp. 612–624.
- Phoon K.-K. and Kulhawy F.K. (2008), "Serviceability limit state reliability-based design", in Reliability based design in geotechnical engineering, Ed. Kok-Kwang Phoon, Taylor and Francis Publishers, pp. 344–384.
- Powell W.D., Potter J.F., Mayhew H.C. and Nunn M.E. (1984), "The Structural Design of Bituminous Roads" Transportation Road Research Laboratory Report RL 1132, TRL, UK.
- Priddle J., Lacey D., Look B. and Gallage C. (2013), "Residual soil properties of south east Queensland", Australian Geomechanics Journal, Vol. 48, No. 1, pp. 67–76.

- Queensland Main Roads (1989), "Controlling moisture in pavements" Technical note 7.
- Riddler N.A. (1994), "Groundwater Investigations" Drainage Principles and Applications (Ed. H.P. Ritzema), International Institute for Land Reclamation and Management Publication, The Netherlands, 2nd Edition.
- Robertson P.K. and Campanella R.G. (1983), "Interpretation of Cone Penetration Tests", Canadian Geotechnical Journal, Vol. 31, No. 3, pp. 335–432.
- Robertson P.K., Campanella R.G., Gillespie D. and Greig J. (1986), "Use of piezometer cone data: Use of in situ tests in Geotechnical Engineering", ASCE Geotechnical Special Publication No. 6, pp. 1263–1280.
- Rowe, P.W. (1972), "The Relevance of Soil Fabric to Site Investigation Practice" Geotechnique, Vol. 22, No. 2, pp. 195–300.
- Sabatani P.J., Bachus R.C., Mayne P.W., Schnedier J.A. and Zettler T.E. (2002), "Evaluation of Soil and Rock properties Geotechnical Engineering Circular No. 5, Report No. HHWA-IF-02-034", Federal Highway Administration, Washington.
- Santamarina C., Altschaeffl A.G. and Chameau, J.L. (1992), "Reliability of Slopes: Incorporating Qualitative Information" Transportation research Record 1343, Rockfall, Prediction, control and landslide case histories, pp. 1–5.
- Schmertmann J.H. (1978), "Guidelines for cone penetration testing – Performance and Design", U.S. Department of Transportation, Federal Highways Administration, Washington, DC.
- Serafim J.L. and Pereira J.P. (1983), "Considerations of the Geomechanics Classification of Bieniawski" Proceedings of the International Symposium of Engineering Geology and Underground Construction, Lisbon, pp. 1133–1144.
- Skempton A.W. (1957), "Discussion on the design and planning of Hong Kong Airport" Proceedings of the Institution of Civil Engineers, Vol. 7.
- Skempton A.C. (1986), "Standard Penetration Test Procedures and the effects in sands of overburden pressure, relative density, particles size aging and over consolidation" Geotechnique 36, No. 3, pp. 425–447.
- Skipp B.O. (1992), "Seismic Movements and rock masses" Chapter 14 in Engineering in Rock Masses edited by FG Bell, Butterworth Heinemann.
- Spears (1980), "Towards a classification of shales" Journal of The Geological Society, No. 137, pp. 125–130.
- Somerville S.H. (1986), "Control of Groundwater for temporary works" CIRIA Report 113.
- Strohm W.E. Jr, Bragg G.H. Jr and Zeigler T.H. (1978), "Design and construction of Compacted Clay Shale Embankments" Technical Guidelines Report FHWA-RD-78-141, Vol. 4.
- Stroud M.A. and Butler F.G. (1975), "The standard penetration test and the engineering properties of glacial materials" Proceedings of the symposium on Engineering Properties of glacial materials, Midlands, U.K.
- Stroud M.A. (1974), "The Standard Penetration Test: Introduction Part 2", Penetration Testing in the U.K., Thomas Telford, London, pp. 29–50.
- Takimatsu K. and Seed H.B. (1984), "Evaluation of settlements in sands due to earthquake shaking" Journal of Geotechnical Engineering, Vol. 113, pp. 861–878.

- US Army Corps of Engineers (1997), "Introduction to probability and reliability methods for use in geotechnical engineering", Technical letter No. 1110-2-547, Department of the Army, Washington, D.C.
- US Department of Transportation (2001), "Load and resistance factor design (LRFD) for highway substructures", NHI Course No. 132068, Publication FHWA HI-98-032, Federal Highway Administration.
- Walsh P., Fityus S. and Kleeman P. (1998), "A note on the Depth of Design Suction Changes in South Western Australia and South Eastern Australia" Australian Geomechanics Journal, No. 3, Part 3, pp. 37–40.
- Vaughan P.R., Chandler R.J., Apted J.P., Maguire W.M. and Sandroni S.S. (1993), "Sampling Disturbance – with particular reference to its effect on stiff clays" In Predictive Soil Mechanics, Ed. G.T. Houlsby and A.N. Schofield, Thomas Telford Publishers London, pp. 685–708.
- Vaughan P.R. (1994), "Assumption, Prediction and reality in Geotechnical Engineering" Geotechnique Vol. 44, No. 4, pp. 573–609.
- Walkinshaw J.L. and Santi P.M. (1996), "Shales and Other Degradable Materials" Transportation Research Board, Landslides: Investigations and Mitigation, Special Report 247, National Academy Press, Washington, pp. 555–576.
- Webster S.L., Brown R.W. and Porter J.R. (1994), "Force Projection Site Evaluation Using the Electric Cone Penetrometer (ECP) and the Dynamic Cone Penetrometer (DCP), Technical Report GL-94-17, U.S. Air Force, FL.

26.3 Geotechnical analysis and design

- AASHTO – AGC – ARTBA (1990), "Guide Specifications and Test Procedures for Geotextiles" Task Force 25 Report, Subcommittee on New Highway Materials, American Association of State Highway and Transportation Officials, Washington, D.C.
- AASHTO (2007), "LRFD Bridge Design Specifications", American Association of State Highway and Transportation Officials, Washington, D.C.
- AASHTO (2006), "Highway Design Manual", American Association of State Highway and Transportation Officials, Washington, D.C.
- Angell, D.J. (1988), "Technical Basis for the Pavement Design Manual" Queensland Main Roads, Pavements Branch, Report RP165.
- API (1993), "Recommended Practice for Planning, Designing and Constructing Fixed Offshore Platforms – Load and Resistance Factor Design" American Petroleum Institute API RP2A, Washington, D.C.
- Australian Geomechanics Society (AGS) Landslide working Group (2007c), "Practice note guidelines for landslide risk management", Australian Geomechanics Journal, Volume 42, No. 1, pp. 63–114.
- Australian Standard (AS 2870 – 1988), "Residential Slabs and Footings Part 1: Construction" Standards Australia.
- Australian Standard (AS 2870 – 1990), "Residential Slabs and Footings Part 2: Guide to design by engineering principles" Standards Australia.
- Australian Standard (AS 2870 – 2011), "Residential Slabs and Footings": Guide to design by engineering principles" Standards Australia.

- Australian Standard (AS3774 – 1996), "Loads on bulk solids containers" Standards Australia.
- Australian Standard (AS4678 – 2002), "Earth Retaining Structures" Standards Australia.
- Austroads (2004), "Pavement Rehabilitation – a guide to the design of rehabilitation treatments for road pavements" Standards Australia.
- Austroads (1990), "Guide to Geotextiles" Technical Report.
- Austroads (2009), "Guide to Pavement Technology Part 4G: Geotextiles and Geogrids", Austroads Publication.
- Australian National Committee of Irrigation and Drainage (2001), "Open Channel Seepage and Control" Volume 2.1: Literature Review of earthen channel seepage.
- Barton N., Lien R. and Lunde J. (1974), "Engineering Classification of rock masses for design of tunnel support" Rock Mechanics, No. 6, pp. 189–239, Norwegian Geotechnical Institute Publishers.
- Barton N. (2006), "Lecture Series on Rock Mechanics and Tunnelling" School of Engineering, Griffith University, Gold Coast Campus.
- Bell F.G. (1992), "Open excavation in rock masses" Chapter 21 in Engineering in Rock Masses edited by FG Bell, Butterworth Heinemann.
- Bell F.G. (1998), "Environmental Geology – Principles and Practice" Blackwell Science.
- Berg R., Christopher B.R. and Perkins S. (2000), "Geosynthetic Reinforcement of the Aggregate Base/Subbase Courses of Pavement Structures" Geosynthetic Materials Association White Paper.
- Boscardin M.G. and Cording E.J. (1989), "Building response to excavation induced settlement", Journal of Geotechnical engineering, ASCE, Vol. 124, No. 5, pp. 463–465.
- Bishnoi B.L. (1968), "Bearing Capacity of a closely Jointed Rock" PhD Dissertation, Georgia Institute of Technology, Atlanta.
- Bozozuk M. (1978), "Bridge Foundations Move" Transportation Research Record 678, Transportation Research Board, Washington, pp. 17–21.
- Brandtl H. (2001), "Low Embankments on soft soil for highways and high – speed trains" Geotechnics for Roads, Rail tracks and Earth structures, Ed. A. Gomes Correia and H. Brandtl, Balkema Publishers.
- British Standards BS6031 (1991), "Code of Practice for Earthworks" British Standards Institution.
- Broms B.B. (1981), "Precast Piling Practice" Thomas Telford Ltd, London.
- Brooker E.W. and Ireland H.O. (1965), "Earth Pressures at rest related to stress history" Canadian Geotechnical Journal, Volume 2, No. 1, pp. 1–15.
- Burland J.B., Broms B.B. and De Mello V.F.B. (1977), "Behaviour of Foundations and Structures" Proceedings 9th International Conference on Soil Mechanics and Foundation Engineering, Tokyo, pp. 495–546.
- Cai M., Kaiser P.K., Uno H., Tasaka Y. and Minami M. (2004), "Estimation of rock mass deformation modulus and strength of jointed hard rock masses using the GSI system", International Journal of Rock Mechanics and Mining Sciences, Volume 41, pp. 3–19.

- Carter J.P. and Kulhawy F.H. (1988), "Analysis and design of drilled shaft foundations socketed into rock" Final Report Project 1493–4, EPRI EL–5918, Cornell University, Ithaca, New York.
- Caterpillar (2012), "Caterpillar Performance Handbook", Edition 42, Caterpillar Inc. USA.
- Cedergren H.R. (1989), "Seepage, Drainage and flow nets" 3rd Edition, John Wiley & sons.
- Chang M.F. and Broms B.B. (1991), "Design of Bored Piles in Residual Soils based on Field Performance data", Canadian Geotechnical Journal, Vol. 28, pp. 200–209.
- CIRIA Report 181 (1999), "Piled Foundations in weak rock" Report No. 181.
- Chowdury R., Flentje P. and Bhattacharya G. (2010), "Geotechnical Slope Analysis", CRC press/Balkema Publishers.
- Clouterre (1991), "Soil Nailing Recommendations – For Designing calculating, constructing and inspection Earth Support Systems using soil nailing" French National Research, English Translation from the Federal Highway Administration, FHWA–SA–93–026, Washington, USA.
- Coduto D.P. (1994), "Foundation Design, Principles and Practices" Prentice Hall.
- Cooke, R.U. and Doornkamp J.C. (1990), "Geomorphology in Environmental Management" 2nd Edition, Oxford University Press, Oxford.
- Cruden D.M. and Varnes D.J. (1996), "Landslides Types and Processes" Landslides Investigations and Mitigations, Ed. Turner and Schuster, Special Report 247, Transportation Research Board, pp. 36–75.
- Department of Transport (1991), "Specification for highway works", HMSO London, UK.
- Delpak, R., Omer, J.R. and Robinson R.B. (2000), "Load/settlement prediction for large diameter bored piles in Mercia Mudstone" Geotechnical Engineering, Proceedings Institution of Engineers, 143, pp. 201–224.
- Duncan, J.M. and Wright S.G. (2005), "Soil Strength and slope Stability" Wiley Publishers.
- Dykeman, P. and Valsangkar A.J. (1996), "Model Studies of Socketed Caissons in Soft Rock" Canadian Geotechnical Journal Volume 33, pp. 347 –75.
- Duncan, J.M. and Wright S.G. (2005), "Soil Strength and slope Stability" Wiley Publishers.
- Ervin, M.C. (1993), "Specification and control of earthworks", Conference ON Engineered Fills held at Newcastle upon Tyne, Edited by BG Clarke, CFP Jones and AIB Moffat, Thomas Telford Publications.
- Eurocode 3 (1993). "Design of steel structures, part 5, piling". ENV 1993-5.
- Federation Internationale de la Precontrainte (1986), "Precast concrete piles", Thomas Telfoird, Londoin.
- Fell R., MacGregor P., Stapledon D. and Bell G. (2005), "Geotechnical Engineering of Dams", AA Balkema Publishers.
- FHWA (1989), "Geotextile Design Examples", Geoservices Inc. report to the Federal Highway Administration, Contract No. DTFH-86-R-00102, Washington, D.C.
- Fleming W.G., Weltman A.J., Randolph M.F. and Elson W.K. (1992), "Piling Engineering" 2nd Edition, Blackie Academic and Professional Service.

- Fookes P.F., Lee E.M. and Milligan G. (2005), "Geomorphology for Engineers" Whittles Publishing and CRC Press.
- Forrester K. (2001), "Subsurface drainage for slope stabilization" American Society of Civil Engineers Publishers.
- Forssblad, L. (1981), "Vibratory soil and rock fill compaction" Robert Olsson Tryckeri Publishers, Sweden.
- Gannon J.A., Masterton G.G.T., Wallace W.A. and Muir Wod D. (1999), "Piled Foundations in weak rock" CIRIA Report No. 181.
- German Society for Geotechnics (2003), "Recommendations on Excavations", Ernst and Sohn Publishers.
- Giroud, J.P. and Noirway L. (1981), "Design of geotextile reinforced unpaved roads" Journal Of Geotechnical Engineering, ASCE Journal, Volume 107, No. GT9, pp. 1233–1254.
- Giroud, J.P., Badu-Tweneboah, K. and Soderman, K.L. (1994). "Evaluation of Landfill Liners", Fifth International Conference on Geotextiles, Geomembranes and Related Products, Vol. 3, pp. 981–986.
- Goodger H.K. and Leach B.A. (1990), "Building on derelict land" Construction Industry Research and Information Association, London.
- Hammitt G.M. (1970), "Thickness requirement for unsurfaced roads and air-fields, bare base support, Project 3782-65." Technical Rep. S-70-5, U.S. Army Engineer Waterways Experiment Station, CE, Vicksburg, Miss.
- Hannigan, P.J., Goble G.G., Thendean G., Likins G.E. and Rausche F. (1998), "Design and Construction of piled Foundations – Volumes 1 and 2", Federal Highway Administration Report No. FHWA–HI-97-013, Washington, DC.
- Hansen J.B. (1970), "A revised and extended formula for bearing capacity" Danish Geotechnical Institute Bulletin, No. 28.
- Hoek E., Kaiser P.K. and Bawden W.F. (1997), "Support of Underground Excavations in Hard Rock" A.A. Balkema, Rotterdam.
- Hoek E. and Brown E.T. (1997), "Practical estimates of rock mass strength", International Journal of Rock Mechanics and Mining Sciences, Volume 34, No. 8, pp. 1165–1186.
- Horner, P.C. (1988), "Earthworks" ICE Construction Guides, Thomas Telford London, 2nd Edition.
- Holtz, R., Barry P. and Berg R. (1995), "Geosynthetic design and construction guidelines" National Technical Information.
- Holtz R.D., Christopher B.R. and Berg R.R. (1995), "Geosynthetic design and construction guidelines" Federal Highway Administration, Virginia.
- Holtz R., Barry P. and Berg R. (1995), "Geosynthetic design and construction guidelines" National Technical Information Service.
- Horikoshi, K. and Randolph, M.F. (1998), "A contribution to optimum design of piled raft", Geotechnique Vol. 48, No. 3, pp. 301–317.
- Horner, P.C. (1988), "Earthworks" ICE Works Construction Guides, 2nd edition, Thomas Telford, London.
- Hutchinson D.J. and Dideerichs M.S. (1996), "Cablebolting in Underground Mines" Bi-Tech Publishers, Richmond, Canada.
- Jardine R., Chow F., Overy R. and Standing J. (2005), "ICP Design methods for driven piles in ands and clays" Thomas Telford Publishing.

- Jardine R., Fourie A., Maswose J. and Burland J.B. (1985), "Field and Laboratory measurements of soil stiffness" Proceedings of the 11th International Conference on Soil Mechanics and Foundation Engineering, San Francisco, Volume 2, pp. 511–514.
- Jardine R.J., Potts D.M., Fourie D.M. and Burland J.B. (1986), "Studies of the influence of non-linear stress strain characteristics in soil structure interaction" Geotechnique, Volume 36, No. 3, pp. 377–396.
- Ingles O.G. (1987), "Soil Stabilization" Ground Engineer's Reference Book, Ed. F.G. Bell, Butterworths Heinemann Publishers.
- Kaiser P.K., Diederichs M.S., Martin C.D., Sharp J. and Steiner W. (2000), "Underground works in hard rock Tunnelling and Mining", GeoEng 2000 Melbourne, Vol. 1, Technomic Publishing Co., pp. 841–926.
- Koerner R. (1998), "Designing with Geosynthetics" (4th Edition) Prentice Hall Publishers.
- Kimmerling R. (2002), "Shallow Foundations" Geotechnical Engineering Circular No. 6, Federal Highway Administration Report No. FHWA–SA–02–054, Washington.
- Kulhawy F.H. and Goodman R.E. (1980), "Design of Foundations on discontinuous rock", Proceedings of the International Conference on Structural Foundations on Rock, International Society for Rock Mechanics, Vol. 1, Sydney, pp. 209–220.
- Kulhawy F.H. and Carter J.P. (1980), "Settlement and Bearing Capacity of Foundations on Rock Masses" Engineering in Rock Masses (Edited by F.G. Bell), Butterworth Heinemann, pp. 231–245.
- Kutzner C. (1997), "Earth and Rock fill Dams – Principles of Design and Construction, A.A. Balkema Publishers.
- Lacey D., Look B. and Williams D. (2013), "Relative modulus improvement due to geosynthetic intrusions within a uniform gravel, determined using a light falling weight deflectometer", Unpublished to date.
- Lawson C.R. (1994), "Subgrade Stabilisation with geosynthetics" Ground Modification Seminar, University of Technology, Sydney.
- Lay M.G. (1990), "Handbook of Road Technology, Volume 1: Planning and Pavements" Gordon and Breach Science Publishers, 2nd Edition.
- Lee K.L. and Singh A. (1971), "Relative Density ad Relative Compaction of Soils" Bulletin No. 272, Highway Research Board, National Academy of Science, Washington.
- Littlejohn S. (1993), "Overivew of rock anchorages", In Comprehensive Rock Engineering, Vol. 4, Excavation, support and monitoring, Edited by John Hudson, Pergamon Press.
- Lo S.C.R. and Li K,S, (2003), "Influence of a Permanent Liner on the Skin Friction of large diameter Bored Piles in Hong Kong Granitic Saprolites", Canadian Geotechnical Journal, Volume 40, pp. 703–805.
- Look B.G., Wijeyakulasuriya V.C. and Reeves I.N. (1992), "A method of risk assessment for roadway embankments utilising expansive materials" 6th Australia – New Zealand Conference on Geomechanics, New Zealand, February, pp. 96–105.
- Look B.G., Reeves I.N. and Williams D.J. (1994), "Development of a specification for expansive clay road embankments" 17th Australian Road Research Board Conference, August, Part 2, pp. 249–264.

- Look B.G. (1996), "The Effect of Volumetrically Active Clay Embankments on Roadway Performance", PhD Thesis, The University of Queensland.
- Look B.G. (2004), "Rock Strength at the coring interface" Australian Geomechanics Journal, Vol. 39, No. 2, pp. 105–110.
- Look B. and Thorley C. (2011), "Landslide failure perspectives in practice in south east Queensland", Australian Geomechanics Journal, Vol. 46, No. 2, pp. 109–124.
- Look B. and Campbell C. (2013), "Variability in selecting design values in Queensland – a calibration of professional opinions" Australian Geomechanics Journal, Vol. 48, No. 1, pp. 51–66.
- Mayne P.W. and Kulhawy F.H. (1982), "Ko – OCR Relationships in soil" ASCE Journal of the Geotechnical Engineering Division, Vol. 108, GT6, pp. 851–872.
- MacGregor F., Fell R., Mostyn G.R., Hocking G. and McNally G. (1994), "The estimation of rock rippability" Quarterly Journal of Engineering Geology 27, pp. 123–144.
- McConnell K. (1998), "Revetment systems against wave attack – a design manual" Thomas Telford Publishing.
- Meyerhof G.G. (1956), "Penetration Test and bearing capacity of cohesion-less soils" Journal of the Soil Mechanics and Foundations Division, ASCE, Vol. 19, No. SM2, pp. 1–19.
- Meyerhof G.G. (1965), "Shallow Foundations" Journal of the Soil Mechanics and Foundations Division, ASCE, Vol. 91, No. SM2, pp. 21–31.
- Meyerhof G.G. (1976), "Bearing Capacity and settlement of pile foundations", Journal of the Soil Mechanics and Foundations Division, ASCE, Vol. 102, No. GT3, pp. 197–228.
- Moulton L.K., Gangarao, H.V.S. and Halvorsen G.T. (1985), "Tolerable movement criteria for highway bridges" Report No. FHRA/RD–85/107, Federal Highway Administration, Washington.
- Mulholland P.J., Schofield, G.S. and Armstrong P. (1986), "Structural design criteria for residential street pavements: interim report based on Stage 1 of ARRB project 392", Australian Road Research Board.
- NAVFAC (1986), "Foundations and Earth Structures" Design Manual 7.02.
- Nelson K.D. (1985), "Design and Construction of small earth dams" Inkata Press, Melbourne.
- New B.M. (1986), "Ground vibration caused by civil engineering works", TRRL research report 53. Transport and Road Research Laboratory, Department of Transportation, UK.
- Nikolaou S. (2011), "Performance based design – concepts in geotechnical earthquake engineering" Geo-Strata, American Society of Civil engineers, pp. 42–47.
- O'Rouke T.D. (1975), "The ground movements relate to braced excavation and their influence on adjacent buildings", US Department of Transport DOT–TST76, T-23.
- Ortiago J.A.R. and Sayao A.S.F.J. (2004), "Handbook of Slope Stabilisation" Springer Publishers.
- Oweis I. and Khera (1998), "Geotechnology of Waste Management", 2nd edition, Cengage Learning Publishers.

- Paikowsky S.G. and Whitman R.V. (1990), "The effects of plugging on pile performance and design" Canadian Geotechnical Journal, Volume 27, No. 4, pp. 429–440.
- Peck R.B. (1969), "Deep excavations and tunnelling in soft ground", Proceedings of the 7th International Conference I Soil Mechanics and Foundation engineering, Mexico, State of the Art Volume, pp. 225–290.
- Phear A., Dew C., Ozsoy B., Wharmby N.J., Judge J. and Barley A.D. (2005), "Soil Nailing – best practice Guidance" CIRIA Publication C637, London.
- Poulos H.G. and Davis E.H. (1980), "Pile Foundation Analysis and Design" John Wiley and Sons, New York.
- Poulos H.G. and Davis E.H. (1974), "Elastic Solutions for Soil and Rock Mechanics" John Wiley and Sons, New York.
- Queensland Main Roads (1990), "Pavement Design Manual" 2nd Edition with amendments to 2005.
- Queensland Main Roads (1999), "Specifications MRS 11.27 – Geotextiles", Queensland Government.
- Queensland Transport and Main Roads (2012), "Pavement Rehabilitation Manual", Queensland Government.
- Quies J. (2002), "A Dam for a dam" Civil Engineering (February).
- Randolph M.F. (2003), "Science and empiricism in pile foundation design", Geotechnique 53, No. 10, pp. 847–875.
- Rausche F., Thendean G., Abou-matar H., Likins G. and Goble G. (1996), "Determination of pile driveability and capacity from penetration tests" Federal Highway Administration Report, No. DTFH61-91-C-00047, Washington.
- Reese L.C. and O'Neil M.W. (1989), "Drilled shafts: Construction and Design, FHWA Publication No. HI-88-042.
- Reul, O. and Randolph, M. (2004). "Design strategies for piled rafts subjected to non uniform vertical loading." Journal of Geotechnical and Geoenvironmental Engineering, Vol. 130, No. 1, pp. 1–13.
- Richards L.R., Whittle R.A. and Ley G.M.M. (1978), "Appraisal of stability conditions in rock slopes", Foundation Engineering in Difficult Ground (Ed. Bell F.G.), Butterworth, London, pp. 192–228.
- Richards, L. (1992), "Slope stability and rock fall problems in rock masses" Chapter 11 in Engineering in Rock Masses edited by F.G. Bell, Butterworth Heinemann.
- Richardson G.R. and Middlebrooks P. (1991), "A simplified Design Method for silt fences" Geosynthetics Conference, St Paul, MN, pp. 879–885.
- Ritchie A.M. (1963), "Evaluation of rock fall and its control" Highway Research Board record, No. 17, Washington, pp. 13–28.
- Rowe R.K. and Armitage H.H. (1987), "A design method for drilled piers in soft rock", Canadian Geotechnical Journal, Vol. 24, No. 1, pp. 126–142.
- Sabatini P.J., Pass D.G. and Bachus R.C. (1999), "Geotechnical engineering circular no. 4 – Ground Anchors and Anchored Systems" Report FHWA-IF-99-015, Federal Highway Administration, Washington, D.C.
- Sarsby R.W. (2000), "Environmental Geotechnics", Thomas Telford Publishers.
- Schnabel H. and Schnabel H.W. (2002), "Tiebacks in Foundation Engineering and Construction" 2nd edition, AA Balkema Publishers.

- Schor H.J. and Gray D.H. (2007), "Land forming" John Wiley and Sons Publishers.
- Schlosser F. and Bastick M. (1991), "Reinforced Earth" Foundation Engineering Handbook, 2nd Edition (Ed. Hsai-Yang Fang), Chapman and Hall Publishers, pp. 778–795.
- Seidel J.P. and Haberfield C.M. (1995), "The axial capacity of pile sockets in rocks and hard soils" Ground Eng., 28(2), pp. 33–38.
- Seidel J.P. and Collingwood B. (2001), "A new socket roughness factor for prediction of rock socket shaft resistance" Canadian Geotechnical Journal, 38, pp. 138–153.
- Selby, M.J. "Hill Slope Materials and Processes" 2nd Edition, Oxford University Press, Oxford.
- Singer M.J. and Munns, D.N. (1999), "Soils: An Introduction" 4th Edition, Prentice – Hall.
- Simons N.E. and Menzies B.K. (1977), "A short Course in Foundation Engineering" Butterworth & Co. Publishers.
- Singh B. and Varshney R.S. (1995), "Engineering for Embankment dams" AA Balkema Publishers, Rotterdam.
- Sivakugan, N., Eckersley, J. D. and Li, H. (1998). "Estimating settlements in granular soils by neural networks," IEAust Civil/Structural Transactions. CE40, 49–52.
- Skempton A.W. (1951), "The bearing capacity of clays" Building Research Congress.
- Skempton A.W. and Bjerrum L. (1957), "A contribution to the settlement analysis of foundations on clay" Geotechnique, No. 7, pp. 168–178.
- Skempton A.W. and Hutchinson J.N. (1969), "Stability of natural slopes and embankment foundations" Proceedings of the 7th International Conference of Soil Mechanics and Foundation Engineering, Mexico, State of the Art Volume, pp. 291–340.
- Skempton A.W. and Macdonald (1956), "The allowable settlement of buildings" Proceedings of the I.C.E., Vol. 5, No. 3, Pt. 3, pp. 737–784.
- Skipp B.O. (1992), "Seismic Movements and rock masses" Chapter 14 in Engineering in Rock Masses edited by FG Bell, Butterworth Heinemann.
- Smith R. (2001), "Excavations in the Atlantic Piedmont" Foundations and Ground Improvement, ASCE Geotechnical Special Publication No. 113, Virginia.
- Soeters R. and van Westen C.J. (1996), "Slope instability: recognition, analysis and zonation" Landslides Investigations and Mitigations, Ed. Turner and Schuster, Special Report 247, Transportation Research Board, National Academy Press, Washington, pp. 129–177.
- Steward J., Williamson R. and Mahoney J. (1977), "Guidelines for use of Fabrics in construction and maintenance of low volume roads" USDA Forest Service Portland Oregon and FHWA Report # TS–78-205.
- Tannant D.D. and Regensburg B. (2001), "Guidelines for mine haul road design".
- The Institution of Civil Engineers (1995), "Dredging: ICE design and practice guide" Thomas Telford Publishing.
- Thompson C.D. and Thompson D.E. (1985), "Real and Apparent relaxation of driven piles" American Society of Civil Engineers, Journal of Geotechnical Engineering, Vol. 11, No. 2, pp. 225–237.

- Tingle, J.S. and Webster S.L. (2003), "Review of Corps of Engineers design of Geosynthetic Reinforced Unpaved Roads", Annual General Meeting CD-ROM TRB, Washington, D.C.
- Transportation Association of Canada (2004), "Guide to Bridge Hydraulics" Thomas Telford Publishing, London, 2nd Edition.
- Transportation Research Board (1996), "Landslides: Investigations and Mitigation" Special Report 247, National Academy Press, Washington.
- Trenter N.A. and Burt N.J. (1981), "Steel pipe piles in silty clay soils at Belawan, Indonesia" Tenth International Conference on Soil Mechanics and Foundation Engineering, Volume 2, pp. 873–880.
- Tynan A.E. (1973), "Ground Vibrations: Damaging effects to Buildings" Australian Road Research Board.
- US Army Corps of Engineers (1993), "Bearing Capacity of Soils" Technical Engineering and Design Guides, No. 7, ASCE Press.
- United States Department of the Interior (1965), "Design of Small Dams" Bureau of Reclamation.
- Van Santvoort G. (1995), "Geosynthetics n Civil Engineering" Centre for Civil engineering Research and Codes Report 151, Balkema Publishers.
- Vesic A.S. (1973), "Analysis of ultimate loads of shallow foundations" Journal of Soil Mechanics and Foundation Division", American Society of Civil Engineers, Vol. 99, No. SM1, pp. 45–73.
- Vesic A.S. (1975), "Bearing Capacity of Shallow Foundations" Chapter 3, Foundation Engineering Handbook, 1st Edition, Editors H.F. Winterkorn and H.Y. Fang, Van Nostrand Reinhold Company Publishers.
- Waters T., Robertson N. and Carter (1983), "Evaluation of Geotextiles" Main Roads Department, Queensland, Internal report R1324.
- Weltman A.J. and Little J.A. (1977), "A review of bearing pile types" DOE and CIRIA Piling Development Group Report PG1, Construction Industry Research and Information Association (CIRIA), London.
- Whiteside P.G.D. (1986), "Discussion on rock fall protection measures" Proceedings Conference of Rock engineering and Excavation in an Urban Environment, Institution of Mining and Metallurgy, Hong Kong, pp. 490–492.
- Woolorton F.L.D. (1947), "Relation between the plasticity index, and the percentage of fines in granular soil stabilization" Proceedings of the 27th HRB Annual Meeting, pp. 479–490. Highway Research Records, Washington, D.C.
- Wyllie D.C. and Norrish N.I. (1996), "Rock Strength Properties and their Measurements" Landslides Investigations and Mitigations, Ed. Turner and Schuster, Special Report 247, Transportation Research Board, National Academy Press, Washington, pp. 372–390.
- Wyllie D.C. (1999), "Foundations on Rock" Routledge, New York.
- Yang J. (2006), "Influence Zone for End Bearing of Piles in Sands", Journal of Geotechnical and Geoenviromental Engineering", American Society of Civil Engineers, Vol. 132, No. 9, pp. 1229–1237.
- Zhang L. and Einstein H.H. (1998), "End Bearing Capacity of Drilled Shafts in Rock" Journal of Geotechnical and Geoenviromental Engineering, ASCE, Vol. 124, No. 7, pp. 574–584.

Index

Printed in the United States
by Baker & Taylor Publisher Services

Printed in the United States
by Baker & Taylor Publisher Services